Energy, Environment, Natural Resources and Business Competitiveness

Life starts and ends as a question.

Without an answer.

The only evidence is that energy can sustain life.

Energy, Environment, Natural Resources and Business Competitiveness

The Fragility of Interdependence

DIMITRIS N. CHORAFAS

GOWER

Gower Applied Business Research
Our programme provides leaders, practitioners, scholars and researchers with thought provoking, cutting edge books that combine conceptual insights, interdisciplinary rigour and practical relevance in key areas of business and management.

Published by
Gower Publishing Limited
Wey Court East
Union Road
Farnham
Surrey, GU9 7PT
England

Gower Publishing Company
Suite 420
101 Cherry Street
Burlington,
VT 05401-4405
USA

www.gowerpublishing.com

British Library Cataloguing in Publication Data
Chorafas, Dimitris N.
Energy, environment, natural resources and business competitiveness : the fragility of interdependence.
1. Energy policy--European Union countries. 2. Energy industries--European Union countries. 3. Energy industries--Environmental aspects--European Union countries. 4. Renewable energy sources--European Union countries. 5. Agriculture and energy--European Union countries. 6. Overpopulation--Environmental aspects.
7. International organization.
I. Title
333.7'9'094-dc22

ISBN: 978-0-566-09234-3 (hbk)
ISBN: 978-0-566-09235-0 (ebk)

Library of Congress Cataloging-in-Publication Data
Chorafas, Dimitris N.
Energy, environment, natural resources and business competitiveness : the fragility of interdependence / Dimitris N. Chorafas.
p. cm.
Includes index.
ISBN 978-0-566-09234-3 (hbk) 1. Energy development. 2. Power resources--Economic aspects. 3. Power resources--Environmental aspects. 4. Energy policy. I. Title.
HD9502.A2C476 2010
333.79--dc22

2010043786

MIX
Paper from responsible sources
FSC® C018575
www.fsc.org

Printed and bound in Great Britain by the MPG Books Group, UK

Contents

List of Figures

List of Tables

Preface

Philosophers say that every generation leaves both assets and liabilities to the next. Alert people comment that the way we are going we will leave our children and grandchildren with an almost unresolvable challenge of energy supplies, plenty of waste of all sorts polluting the air they breathe and the water they drink, as well as the appalling problem of a huge and uncontrollable explosion in world population that defies rational thinking.

These themes go beyond the narrow definition of energy, but they correlate among themselves and have the broad topic of energy at their core. This book is aimed at those who have a professional, academic or general interest in all or any of seven critical issues:

1. energy sources, their nature, contribution, pollution and other consequences,
2. environmental problems associated with power production and usage,
3. the financing and controlling of energy-related projects and processes,
4. the future direction of agriculture, whose produce is now used as an energy source,
5. complex social and technical issues resulting from lack of family planning – and, therefore, of demands for energy,
6. the impact of both energy and an exploding population on the pollution of air, land and water,
7. truth and hype about the most talked about environmental after-effects of the aforementioned subjects.

Taking the risk of displeasing some readers, this book harbours no taboos and condones no hype. Its contribution is the holistic approach of its contents, including some hitherto unpublished personal opinions from knowledgeable sources – some of whom asked not to be acknowledged, because we no longer live in a time of free expression.

- People at large no longer think of themselves as free to speak their mind without the risk of censorship, condemnation or reprisals, and
- those who still express their opinion and welcome criticism want the latter to be constructive, not a nervous response to something not liked or, even worse, lacking the background and information for a balanced judgement.

Expressing one's factual opinion is important, both to individuals and society because only *free* men and women can do so and moreover, quite literally, one is what one writes. Whether this writing brings good news or bad news is immaterial, and the same is true whether one writes it as a comedy or as a tragedy. In fact, in most human affairs, comedy and tragedy merge into one. Nearly 25 centuries ago, Socrates put it this way: 'Comedy and tragedy are the same thing, and they should be written by the same authors.'

The decision to turn a satirical eye on the climate-change circus, to which we are all overexposed, was not taken lightly. Though this seems to be an issue with dramatic consequences for planet Earth – as well as for *Homo sapiens* and its future on this planet – the current mass response to poorly documented environmental issues, as well as the associated state policies, are also a subject of ridicule. Al Gore's Hollywood extravaganza and the UN's *Fiascopenhague* are two examples.

Underlying all this nonsense is too much wishful thinking and a lot of basic misconceptions. Since the end of World War II, a mere two-thirds of a century ago, humankind has been living with the expectation that growing global demand from a rapidly increasing population seeking higher standards of living will result in greater and greater wealth for everybody. But there are limits to growth, and these were been surpassed long ago.

The ongoing rapid depletion of the earth's resources, and most particularly of fossil fuels, has so far had only a minor effect on people's awareness of future shortages. Although environmental pollution has become a widespread social theme, the scientific documentation of pollution incidents is (at best) wanting, and what has happened so far in terms of environmental conservation is of little practical value. Not only ordinary people, but also politicians and intellectuals are ignoring the fact that the world now faces major *unconventional challenges* which range:

- from runaway birth rates and their disastrous effect on planet Earth,
- to the excesses of consumerism, associated with ever-increasing demands for energy, food and luxuries, such as vacations at the antipodes of the globe.

Most of the case studies included in this book come from the European Union and North America, but there are also examples from other parts of the world. To meet the requirements of its audience, the text examines the positive and negative aspects of current trends and policies, in each one of its themes – and it does so in the light of documented ongoing experiences. Attention is paid to the costs and after-effects of what is happening now, so that valid conclusions can be drawn for the future.

Precisely for this reason, it has been a deliberate choice to consider a wider time horizon than the first decade of this century. Several examples have been taken from the mid-1980s and early 1990s when things started going the wrong way in terms of energy – including spoilage, air pollution, soil poisoning, scarcity of clean water and mountains of man-made waste. Coupled with a wild population explosion (more than half the people living in Iran and Turkey are less than 30 years old, and in Nigeria younger than 15), these abuses may well signal the end of an age.

Carbon dioxide emissions and the greenhouse effect caused by the burning of hydrocarbon fuels have changed the way in which a growing number of people look at the current state of energy supplies, but the study, development and use of alternative energy sources has been proceeding with neither a vision nor a longer-range plan. In addition, experience from the last ten years shows that, to be really effective, new energy solutions must be:

- cunning,
- properly calculated,
- thoroughly tested, and

- sustainable in the long run.

So far, none of these prerequisites has been met in an able manner. We are simply bumping from one alternative energy approach to the next, only to discover that each has a downside which makes questionable the wisdom of employing it. Worse still, there are no energy solutions radical enough and able to fulfil the world's rapidly growing energy consumption requirements.

In fact, the widespread hype about different energy models pays no attention to the notion that coal, oil and gas are not just 'commodities', but the fuel on which almost every country pins its hopes for further development and a higher standard of living. Far from being alien to these near-sighted judgements which conveniently forget the need for conserving energy, the European Union is at the heart of the storm because:

- coal, of which it has plenty of supplies, is characterized by an unacceptable level of carbon dioxide emissions, and
- the European Union's own oil and gas resources are limited, overexploited and on their way to extinction – apart from unconventional gas discoveries.

While biofuels are currently promoted as alternatives to fossil fuels (said by many to be the number one reason for what is – rightly or wrongly – perceived as global warming), as far as Europe is concerned, the advent of biofuels should be seen as a strategic inflection point. For the first time in over 60 years there is a chance of attaining three goals:

- solving a large part of the European energy scarcity problem,
- reducing the effect of fossil fuels on the environment, and
- promoting European agriculture by directing part of it to aims that are attainable, profitable and lasting – as well as sources of employment in the European Union.

The reason why this has not been properly appreciated so far lies in the fact that there is plenty of *misinformation* about energy sources and their usage, as well as about objectives that might bring about an agricultural renaissance. Misinformation has been widespread because of the wrong policy of telling people what they want to hear. As politicians and conquerors have found over the centuries, it is easier to tell a big lie than a small one.

Carbon dioxide emissions are the visible tip of the iceberg of a long list of excesses characterizing the years following World War II. 'The epoch of the masses is the epoch of the colossal,' wrote José Ortega y Gasset. 'We are living ... under the brutal empire of the masses.'[1] In this empire, which has become global, little attention has been paid to a couple of basic truths:

- We cannot negotiate with nature, and
- we cannot ignore it, for in the end it will win.

This is the message that this book brings to the reader, and it explains that the salient reason for the stress on the earth's resources, and its environment, is the ongoing

1 José Ortega y Gasset, *The Revolt of the Masses*, W.W. Norton, New York, 1932.

wild reproduction of *Homo sapiens*, coupled with an unprecedented spoilage. The two constitute a furious demonstration of the empire of the masses and they are disquieting, even alarming, because:

- civilization is measured by the quality of its constituents, not by big numbers and the upheaval they produce;
- education makes the difference in culture, but functional illiteracy is gaining ground these days, while educational discipline has taken leave of absence;
- the post–World War II 'green' and 'blue' revolutions promised way beyond what they could deliver and can now be ranked among the big failures of the last few decades; and
- improvements in well-being call for a lot of energy production and for all sorts of consumption, and there is no country which can claim to have a realistic energy policy consistent with environmental values.

While the real reasons for a deteriorating environment on planet Earth are still obscure, because scientific evidence is ambiguous, the effort of confronting environmental degradation depends too heavily on loud demonstrations and useless 'congresses' than on doing basic homework. To make matters worse, in the middle of all that noise, the world's salient problem is not even confronted, let alone solved. *Universal birth control* is the answer – but that's a big taboo.

In 500 BC, when human civilization started to flourish, the estimated total world population was only about 100 million. Today, no more than 25 centuries down the line, which in chronological time is a triviality, there are nearly 7 billion people – a 7,000 per cent increase – and the next milestone is said to be 9 billion.

Niels Bohr, the nuclear physicist, once remarked that we are all actors as well as spectators in the unfolding of the natural order. We cannot separate ourselves from the world around us, no matter how hard we try. Worse still, we find it very difficult to separate ourselves from past habits that make up the status quo, examples being:

- consumerism,
- overuse of all sorts of resources, and
- disregard for family planning and associated widespread pollution.

No how the effort is measured, it will take exceptional people and an unprecedented amount of willpower to turn things around. Advance global planning and constant global training are other basic necessities. The reader should, however, be aware that the moment one leaves the beaten path the enemies of progress unite.

This book challenges current thinking and its effect on policies and practices. Readers accustomed to mainstream books and articles which blame fossil fuels for a deteriorating global environment might like to read a contrary opinion: while fossil and other fuels may play a role, it is the lack of global family planning that has created the current stress. If these matters are left to their own devices, things will become worse, not better.

I am indebted to a long list of knowledgeable people and organizations for their contribution to the research which made this book feasible. I also thank several senior executives and experts for their constructive criticism during the preparation of the manuscript, most particularly to Dr Heinrich Steinmann, Dr Nelson Mohler and Martin West.

Let me take this opportunity to thank Martin West for proposing this research theme, supervising it and furthering it through the publication channels; Kevin Selmes for seeing this book all the way to publication; and to Linda Cayford for the editing work. To Eva-Maria Binder goes the credit for compiling the research results, typing the text and making the camera-ready artwork and index.

Dr Dimitris N. Chorafas
Valmer and Entlebuch
January 2011

PART I

The Pillars of a Realistic Energy Policy

CHAPTER 1 Energy Means Power

1. Nothing Hurts as Much as the Truth

As an old adage has it, 'Nothing hurts as much as the truth'. The truth is that when we say 'we want to save planet Earth for future generations', we are all liars. When the so-called Greens lament about coal-fired plants and nuclear energy, but in wintertime run around their homes in a T-shirt instead of lowering the central heating and putting on an additional pullover, hypocrisy has a ball.

Human nature is such that everyone wants to have his cake and eat it too. Talk about climate change makes good public relations, particularly when accompanied by grand promises, empty words and pizzazz (see also Hollywood's extravaganza in Chapter 14). The real question, often forgotten in arguments about energy and environmental pollution, is to what degree we are prepared to *forego our comfort* in order to obtain environmental protection outcomes that are:

- satisfactory,
- measurable, and
- commendable.

This is, of course, a tall order because rapid population growth, urbanization and rising consumption are putting immense pressure on the earth's natural resources and its environment (see section 2). With raw materials becoming scarcer because of overconsumption and with global warming a threatening force, politicians say that they are ready to tackle the problem – but they don't really mean it because their talk is not followed by the right decisions and actions.

Many people look at technology as being the solution, but technological advances have unpredictable outcomes. As far as the preservation of energy and the conservation of the environment are concerned, computers and communications are part of the problem. Very few people are aware of how much energy technology consumes, even though this is a real and current issue.

Processing large amounts of data requires a great deal of power. 'In two to three years we will saturate the electric cables running into the building,' suggests Alex Szalay of Johns Hopkins University.[1] In 2006 the US National Security Agency came close to exceeding its power supply, which would have blown out its electrical infrastructure – cryptographic machinery, networks of computers, huge databases and the wizards mining them included.

Information technology companies are aware of the challenge. This is why firms like Microsoft and Google have had to put some of their enormous data centres next to

1 'Handling the Cornucopia', *The Economist*, 25 February 2010.

hydroelectric plants, with the aim of ensuring easy access to enough energy at a reasonable price. But if alternative energy is used to satisfy technology's power hunger, there will be so much less of it to substitute for coal-fired plants.

Moreover, people are careless in their use of energy. Alan Meier, a scientist at Lawrence Berkeley National Laboratory did a study that showed 10 per cent of electricity in buildings is wasted in appliances and installations that were turned off, but remained on standby power.[2] This and plenty of other examples demonstrate that nobody *really* cares about energy conservation (see also section 7).

As far as pollution is concerned, people are just as lousy with their habits. A case in point is vacations. They travel thousands of miles polluting the atmosphere to enjoy a few days in the South Pacific or attend 'earth-saving' global conferences that are more 'love affairs' than anything else. Look at *Fiascopenhague* (Chapter 14). Ecologically speaking, jet travel has been a disaster.

Not that governments are more careful. As the twentieth century came to a close, energy consumption in the United States stood at nearly 2.3 billion tons of oil equivalent. At 0.9 billion, China's consumption at that time was a fraction of it; but it caught up. Ten years later, in 2009:

- overall US consumption has been slightly lower than in 1999, at about 2.2 billion,
- but China surpassed it by a nod having experienced a 2,444 per cent rise in tons of oil-equivalent power consumption in just a decade.[3]

Confronted with these vertiginous numbers, the 'solutions' proposed by the Greens and other environmentalists are half-baked at best. 'Innovations' like alternative energy, electric vehicles, treatment of dirty water and the like have short legs. It's not that these ideas are wrong, it's just that they don't even scratch the surface of *the earth's sustainability problem* which is much deeper and can be expressed in four bullet points:

- overpopulation,
- galloping urbanization,
- ever-greater demands for energy, and
- widespread pollution all over the globe.

These four themes correlate among themselves and with other ills. Overpopulation and urbanization have contributed, and continue to contribute, a great deal to global pollution. The same is true of uncontrolled energy consumption and of chemical-based intensive agriculture whose advent has been propelled by the need to feed billions of people who were irresponsibly brought into the world in the first place.

The rapid degradation of the earth's environment would have been better controlled *if* environmental protection vigilantes had taken sufficient care to eliminate the roots of the problem rather than cutting some minor branches or doing something merely for show – like the initiative of capping and trading CO_2 (Chapter 10). By allowing himself

2 *UCLA Magazine*, April 2010.

3 Still, in terms of tons of oil-equivalent consumption *per person*, Canada comes first followed by the United States, South Korea and Russia (statistics from *The Economist*, 12 June 2010).

to destabilize the delicate balance of natural forces and resources, man has acted against himself. He has created conditions which *definitely favour*:

- famine,
- illiteracy,
- illness, and
- unprecedented pollution.

The equilibrium achieved through millennia of natural evolution has been broken. This did not come by chance, but as a result of dangerous transgressions in the domain of demographic instability which (surprisingly) no international conference on environmental protection from Kyoto 1997 to Copenhagen 2009 has ever tackled.

People able to look into the future and with the courage to openly express their opinions say that the largest and fastest-growing developing countries should be the first to exercise birth control (Chapter 13) and that the West should help them do so. These countries are (in alphabetical order): Bangladesh, Brazil, China,[4] Colombia, Egypt, India, Iran, Indonesia, Nigeria, Mexico, Pakistan, Thailand and Turkey.[5]

It does not take a genius to understand the important link between rapid population growth and the increasing scarcity of food, energy and minerals. The negative effects of excessive population growth also show up, in a big way, in the horde of difficulties encountered in economic development, including falling educational standards, widespread, poorly controllable illnesses (like AIDS) and all sorts of breakdowns in social order.

2. The Correlation between Global Population and Energy Demand

Since the dawn of mankind, energy has meant power. This has been one of the reasons why civilization started near or alongside waterways. The energy supply problems faced by one community were shared by every other because energy is the sort of tool which leverages the man's reach. 'Man is a tool-using animal,' wrote the nineteenth-century Scottish historian Thomas Carlyle. 'Without tools he is nothing, with tools he is all.'

In old times, the vast majority of man's tools were powered by muscle. Certainly, well before the steam engine, human societies used energy from natural sources such as wood, coal, water and wind. But even if wood and coal were in abundance, the energy they released could not be used much beyond producing heat. In the physical sense of transferred motion, useful work could be done only by:

- wind, and
- moving water.

4 After successfully controlling the population explosion under Mao, China has lost the will to continue his wise demographic policies.

5 As outlined in a meeting focusing on the current challenges facing planet Earth.

Wind and moving water, however, are erratic and not always found where they are needed (keep that in mind when we talk about alternative energies in Chapter 5). Therefore, for millennia muscle power remained the main energy source in both transport and production, and was also one the most basic reasons for the practice of slavery. 'The Northern States are jealous of our slaves and our prosperity; we owe them nothing,' said to Ralph Waldo Emerson to Achille Murat, son of the ex-King of Naples, nephew of Napoleon and landlord of Liponia in the US South.[6]

The steam engine changed all that, but did not relieve the need for the further development of energy sources.[7] A simple but ingenious device – James Watt's controller (developed 1782–1784) – harnessed the power of steam, opened new horizons and made it feasible to convert the heat generated by burning coal and wood into an energy form that could do productive work.

The law of unanticipated consequences, however, was at work. Quite unexpectedly, the domestication of energy stimulated the need for more and more of it. Also, by raising *Homo sapien*'s standard of living it boosted their numbers,[8] – further contributing to the exponential growth of energy demands.

- Initially, this was a virtuous cycle of energy production and larger communities working towards improved well-being,
- but as the billions of *Homo sapiens* multiplied, it turned into a vicious cycle, leading to greater and greater pollution of the earth's land, air and water supplies.

Today, many politicians, scientists and ordinary people (particularly the young and the Greens) seem convinced that the earth's climate is changing for the worse because of hydrocarbons. But they fail to see that the real underlying cause is human activity. This is what creates excessive emissions of carbon dioxide and other greenhouse gases.

- *If* we do not correlate the explosion in human population with skyrocketing energy consumption,
- *then* we attack the symptoms rather than the core of the disease.

In addition, such one-track thinking blocks the opportunity to look at alternative hypotheses. While emissions are the devil incarnate, according to the majority opinion, a minority remain sceptical. To their mind, dependable data suggest that the earth's mean temperature is changing because of natural reasons – the more potent being solar radiation – and that the earth may be cooling rather than heating up.

On the basis of longer-term evidence, a growing number of scientists believe that the argument that 'present-day temperatures are higher than they have been for thousands of years' is in fact too shaky to be meaningful. Nor is it the first time that the upper limit of the earth's average temperature range has been broken.

What this means in practical terms is that the scientific community is deeply divided in regard to real origins of global warming. Argument and counterargument can better

---

6 Robert D. Richardson, Jr, *Emerson. The Mind on Fire*, University of California Press, Berkeley, 1995.

7 In ancient Greece, in the third-century BC Hero devised a steam engine, but it was only used as a toy and for amusement purposes.

8 *Homo sapiens* means knowledgeable man. This is, of course, an overstatement – but it is also an established term.

be appreciated if we understand that in their background lies the nature of science itself. While, in the longer run, science is objective, in the short term scientists, and politicians above all,[9] are not necessarily so.

- For politicians, the difference is made by public opinion, and
- for scientists, the doubt arises from missing or biased time series.

The message brought by the bulleted statement above is that arguments will persist for decades. 'What is truth?' That was Pontius Pilate's answer to Jesus's assertion that: 'Everyone that is of the truth heareth my voice.'[10] This sounds suspiciously like the modern argument over climate change.

Even if challenged, however, existing scientific evidence that climate change is man-made has to be taken seriously because the stakes are so high that delays might be lethal. But at the same time the measures taken must be holistic and global – not partial and half-baked, as is currently the case – including the measures adopted by the various UN environmental conferences.

Even if the theory that present-day climate change is largely due to solar radiation ultimately prevails, the filling of the atmosphere with CO_2, methane, ethanol, CFCs and other gases makes a bad situation worse. It also shows that the man-made contribution to the destruction of nature is indeed a serious threat. The greater is the number (in billions) of the earth's *Homo sapiens*, the greater will be the disaster.

3. Current Problems Go Well Beyond Pollution

Awareness about excessive pollution is relatively recent, but its real causes have been identified in the wrong way, making the 'medicine' discussed during the various international 'tourist and climate' events out of focus and ineffectual. 'Environmental summits' come a dozen a dime these days, but while each costs hundreds of millions of pounds, dollars and euros, they produce nothing really concrete. Whatever is decided is vague and non-enforceable.

Summits or no summits, practically everyone knows that, short of taking draconian measures, there is no possibility of the large majority of countries caring, or even being able, to stabilize their populations at less than their present sizes. The corollary to this is that *if* a country is not in charge of its reproduction rates, *then* it is unable to:

- create resources able to support education, food and energy for a fast-growing population, and
- put its pollution under lock and key, whether this concerns the land, drinking water or the air.

Wastage has become a policy. Some studies point out that a baby born in the United States costs 30 times more in terms of resources than a baby born in Nepal or Bangladesh.

9 Who manipulate the scientists through well-financed projects and are subsequently manipulated by them (see section 5).

10 John 18:38

Numbers, however, give a tricky message because what they suggest is turned on its head by the galloping needs presented by the population explosion for food, energy, learning, health care and the higher standard of living promoted by globalization.[11]

According to reliable estimates, between now and 2050 the world's population will rise by 30 per cent or more to about 9 billion, but demand for agricultural produce will rise by double that rate and demand for meat will increase by 100 per cent. These demands for 'more and better food' will not be met without farmers clearing large stretches of new land, including wetlands and vital forest areas, and without a huge amount of pollution.

It is indeed difficult to understand why governments and politicians don't raise the alarm that the status quo is accompanied by important downsides, such as growing illiteracy worldwide, waves of illegal immigration and global and massive drugs traffic, as well as the internationalization of crime and terrorism.

Illiteracy rates around the world are growing by leaps and bounds, and this statement also applies to the developed countries.[12] In large stretches of the developing world matters are even worse than they are in the West, as not only do states lack the money and human resources to educate the massive new generation, but many pupils alsocome to classes suffering from poor nutrition, anaemia and an endemic lack of parental care. When they grow up, these children join the huge wave of massive immigration to the developed world, much of which is illegal and seen as linked to other undesirable phenomena like drugs, people and arms trafficking. This is largely a consequence of the immigrants' home countries' policies that have led to urban neglect, as well as their failure to control population growth (Chapter 13).

In addition, leaving aside the health and social hazards of the global massive drugs trade, plenty of the money made out of it is used to finance acts of global terrorism. According to some estimates, money from the international opium trade (originating in Afghanistan) finances the Taliban to the tune of $1.1 billion per year. Drug trafficking is an integral component in the fundamentalists' and fanatical imams' effort to achieve their great ambition of exporting 'their' revolution.

- The actors of terrorism are not just the poor and the oppressed.
- They are also the educated people, including medical doctors and scientists as terrorist events in Britain document.

Neither are the poorer nations helping themselves, as they should, by cutting down their military spending, avoiding waste and corruption, improving their economic management and showing respect for the environment in which they live – or for the privacy of citizens in other parts of the globe whose homeland they invade as illegal immigrants. Heads of state do not seem to have asked themselves the question: 'Can my country's surplus of hungry mouths be stopped at national borders?'[13]

---

11 D.N. Chorafas, *Globalization's Limits. Conflicting National Interests in Trade and Finance*, Gower, Farnham, 2009.

12 D.N. Chorafas, *Education and Employment in the European Union: The Social Cost of Business*, Gower, Farnham, 2011.

13 Quite to the contrary, there are indications that in several sub-Saharan nations young children are taught how to infiltrate the West as illegal immigrants, how to escape being caught by the authorities, and, when caught, evade questions aimed to identify their country of origin.

As this text will document in chapter after chapter, vital resources – including food, education and energy – are not available in sufficient amounts to confront the population explosion and come out ahead. Ultimately, the quality of life and security of the people is more vital than runaway reproduction. And that means sufficient food, learning, employment and respect for social and ecological values. I call this concept *lasting development*. Affordable and sustainable energy supplies are an integral part of it.

An unexpected consequence of globalization has been the boom in demand for 'more' food and energy. But not every major mistake is due to globalization. During the last three decades we have created a world in which all aspects of production – from raw materials and manufacturing to food and energy, as well as distribution, financing, branding and usage of manufactured goods – have become commodities that can be accessed from anywhere by anyone with the financial means to do so.

4. Decisions on Energy and on Biocides are Political

Any decision to curb emissions, reduce energy wastage, as well as monitor and punish anyone who steps over the control line is going to involve making one of the major social choices. At a much higher level of priority and of importance, the decision to curb the curve of exponential reproduction of *Homo sapiens* is also political, but it is more often a decision for the developing countries than the industrializing ones. Such a decision is much more urgent to developing countries than to industrialized ones.

Yet, not only has such urgency not been brought in perspective, but also over a dozen years – from Kyoto 1997 to Copenhagen 2009 – developing countries have been given special dispensation to continue polluting the earth because they insist that 'they need to do so for their development'. This happens even though everybody knows that:

- they have to control their rapidly rising emissions, and
- they should pay much more attention to renewable energy sources than they have done so far.

China, for instance, is building coal-fired power stations at a blazing rate. Even though it also has a wind-generation capacity, coal-based electricity production and the emissions from an unprecedented rapid increase in automotive traffic are clogging its air, while rapidly growing and rampant industrial activity pollutes its water resources and its landscape. In short, China is well on its way to becoming the biggest polluter on earth. Neither is there any evidence of any governmental plan to cut down pollution.

Britain, by contrast, has declared through the mouth of Tony Blair, its former CEO, that the term 'gas emissions' will become a bad memory of the past. Looking at the bare statistics, British greenhouse-gas emissions have fallen by 15 per cent since 1990, which seems to be inside its target under the Kyoto Protocol. The bad news is that most of this decline has been the result not of a big, steady and well-rounded effort but a one-off switch from coal-fired power stations to gas. By 2008 that decline had almost halted.

- Emissions are now falling by less than 1 per cent per year, and
- the government has admitted that it will fail to meet a self-imposed target of reduction in CO_2 even though the 2007–2011 recession has cut economic activity.

The cases of Britain and China make interesting reading when contrasted to one another. It is awfully wrong that at the various 'environmental' conferences and summits, developing countries have been given *carte blanche* to increase their levels of pollution at great cost to the environment. As for verbal 'commitments' made in these conferences to limit the level of pollution 10, 20 or 30 years down the line – they aren't even worth a bucket of warm piss, as a former US vice president, John Nance Garner, answered in the 1930s when asked about the importance of the vice presidency.

It is a testament to the shallowness of the commitments made by all of humankind that nothing significant has been done, despite widely aired good intentions. More than two decades ago a *Newsweek* article phrased its warning in the following way:

> *Nature does not care if the globe is populated by bald eagles or Homo sapiens or six-eyed telepathic thugs. What nature cares about is that the ecosystem lives. Should man sour the environmental conditions now slanted in our favor, creatures will rise up in our stead that thrive on murky greenhouse air, or dine on compounds human metabolisms find toxic.*'[14]

The core of the matter is that a long list of issues connected to pollution boil down to one word: *chemicals*. However, this is a word that politicians don't want to hear as being connected to anything bad. More than 30 years ago Rachel Carson made the point: 'Can anyone believe it is possible to lay down a barrage of poisons on the surface of the earth without making it unfit for all life? They should not be called "insecticides," but "biocides".'[15]

Both developed and less developed nations are using these biocides in massive volumes. There is a horde of them: herbicides, pesticides, insecticides and more. None is a smart poison selecting only its intended target species. Instead, they:

- *poison all life* with which they come in contact, and
- this includes man, their developer and persistent user.

This chemical war waged by man against nature and against himself is never-ending, and all life is caught in its violent crossfire. Peaceful coexistence during the Iron Curtain years has shown that the old fears of extinction of humankind by nuclear fallout were exaggerated. The central problem of our age is not a sudden burst, but the steady, day-after-day contamination of humanity's cradle with substances of incredible potential for harm. CO_2 is only one part of the problem. A bigger share is taken by biocides which:

- accumulate in the tissues of plants and animals, and
- penetrate the germ cells to shatter or alter the very material of heredity on which the future depends.

Therefore, one of the greatest scientific challenges humanity now faces is to understand the fundamental nature of the system that supports life on planet Earth in *all* its aspects. This system, the *biosphere*, is a complex interplay of inextricably linked biological,

14 *Newsweek*, 24 July 1989. As stated in the Preface, in order to provide a better perspective, it has been a deliberate decision to span the historical references and statistics over a quarter of a century, starting in the late 1980s.

15 Rachel, Carson, *Silent Spring*, New York: Houghton Mifflin, 1962.

chemical and physical processes, integrated over space and time, and morphing under various forms of energy.

While ordinary people may not necessarily understand the fundamentals of biocides, they cannot fail to notice the direct result of the increasingly complex effects of human habitation and its global impact. Hydrogen, oxygen, carbon, nitrogen, phosphorous and sulphur are the six elements which determine life on planet Earth, and each of these elements follows a closed loop, or cycle, by:

- first increasing molecular energy states, as the elements are incorporated into living cells, and
- then decreasing energy levels as the various organisms decompose and turn into minerals.

These cycles are the expression of life – the metabolic system of the planet. Their various patterns are the consequence of biological, chemical and physical processes that operate across a wide spectrum of timescales, but escape the naked eye as well as the mind of non-specialists and of political activists who are dedicating their energy to the wrong problem. The contamination of our world is not just a matter of CO_2 pollution but of much broader damage to the biosphere created simultaneously:

- by billions of men and women, and
- by their activities, ranging from food production and energy consumption to widespread pollution.

Worse still, the progressive build-up of dangerous chemicals in human bodies adds up to cumulative poisoning. Under normal conditions, supply and removal throughout the entire biosphere are largely the consequence of an everyday undisturbed biological activity. Even if the supplier and the remover are different types of organism, they are part of the same system. This is the basis of *interdependence* which contributes to the *preservation* of species and of life, but finds itself under present and real danger because of the activities of billions of *Homo sapiens*.

5. 'Climategates' Hurt the Cause of Environmentalists

'The more extraordinary the event, the greater the need of it being supported by strong proofs,' said mathematician and astronomer Pierre-Simon Laplace in the late eighteenth century. These days, as the preceding sections have indicated, the extraordinary event is the threefold issue of an expanding human population, energy and shortages, and the need for environmental protection. This fits squarely with Laplace's dictum, but in contrast to what is required by a rigorous scientific discipline:

- the strong proofs of cause and effect have not been established, and
- as far as existing studies are concerned, many voices are being raised against the prevailing bias and misguidance.

One of the most damaging incidents to the cause of those asking for a Green Bill has been the so-called 'Climategate' incident. Scientists at the Climatic Research Unit (CRU) of the University of East Anglia in Britain have been accused of burying evidence that did not support their claims about global warming.[16]

Bias of that type, which manipulates or misuses scientific data, is not unheard of. One of the most ludicrous cases took place in 1953, and it involved America's big tobacco companies, worried about new scientific research linking their products to cancer. More precisely, they worried that it might hurt sales, Big Tobacco made sure that the public was fed with plenty of disinformation, raising doubts in the public mind about the validity of health-linked research.[17]

Since then, techniques employed to bias the facts, including scientific findings, have been based on various sorts of organized disinformation, displaying an enduring but disgraceful willingness to stick to weak, undocumented or outright discredited arguments that play well. Examples include:

- the nuclear winter,
- missile defence,
- acid rain, and
- the ozone layer.

In plenty of debates, which sometimes border on pseudoscience, a relatively small group of people – some of them eminent – becomes involved in various sorts of disinformation. Typically, though not always, these people work through organizations often created specifically to handle a 'hot' issue that seems to present an opportunity for profitable and longer–term exploitation.

Money, of course, plays a key role. It is no secret that politicians – but not only politicians – influence the thinking and findings of some scientists through generous project sponsorships. As the well-known American journalist Walter Lippman said in his time, in a democracy there is no greater power than public opinion.

- Many questionable 'truths' and postulates are fabricated in the name of public opinion, and
- projects, even scientific projects, are led by an inner circle of a tunnel-visioned selected few.

When this happens, the outcome of such projects is wrong ten times out of ten. And it is definitely influencing the findings of studies on environmental issues (today's 'name of the game'[18]) towards what is 'expected', and these are brought to life not only by a constellation of pressure groups but also by whole parties, such as the Greens.[19] As for the environmental projects' deliverables, for years almost nobody seems to have bothered to systematically examine the differences between cases

16 *Financial Times*, 18 February 2010.

17 Even though the tobacco companies' own in-house scientists were confirming what the public-health researchers had found out about health damage. 'Double Danger for Big Tobacco', *Business Week*, 13 September 2004.

18 In a way not too dissimilar to the role played by nuclear weapons systems during the long decades of the Cold War.

19 Which in some countries collect a hefty 10 per cent of the popular vote.

- where the science is clear, or nearly so, and
- where there is, or was, a fair amount of hype and of disinformation.

This is now changing. In 2010, in Britain, there has been an investigation into the scientific probity of the aforementioned Climatic Research Unit (CRU) at the University of East Anglia. The inquiry panel looked at 11 CRU publications from the past couple of decades, spent time talking to the researchers, and examined other CRU documentation. Based on its findings, the inquiry panel concluded that if there was any malpractice at CRU they probably would have found it; they did not. Instead, they found 'dedicated if slightly disorganised researchers ill-prepared for public attention'.[20] The problem is that they received a great deal of public attention in regard to their environmental findings which have now been put in question. The inquiry panel also criticized the CRU researchers for:

- unwillingness to pass on data to their critics,
- failing to specify which weather stations they were using,
- keeping quiet about IPCC[21] discussions that should have been made public, and
- deleting files rather than having them surrendered under Britain's Freedom of Information Act.

The fact that the inquiry panel found CRU's procedures on freedom of information poor is evidence that those working on this project were not the sort of staff able to provide strong proofs for an extraordinary event, as advised by Pierre-Simon Laplace. An article in *The Economist* put this in the following way:

> *Informed by a warts-and-all account of peer review ... the committee took the researchers' harsh behaviour towards critics and leniency towards allies not as unduly partial or aggressive, but as part of the 'rough and tumble of interaction in an area of science that has become heavily contested'.*[22]

To put it mildly, other investigations, too, have revealed serious weaknesses in environmental research and its findings. The same issue of *The Economist* makes reference to the Dutch Environmental Assessment Agency which found errors in chapters of reports by the IPCC (see Chapter 14).

The Dutch Environmental Assessment Agency also raised questions about the IPCC researchers concentrating on bad or worst-case possibilities, rather than on a range of outcomes. The irony is that the IPCC is financed by taxpayers' money and is supposed to be objective and impartial. Those who pay for it need to have the most critical information available to them in an unbiased form, in order to understand more clearly what is going on.

These audits clearly indicate that from the IPCC to the CRU, passing from a number of other policy-making centres and sponsored environmental projects, various save-the-planet entities have allowed themselves to become part of the problem. *Environmental*

20 'Climate Science and its Discontents: A Place in the Sun', *The Economist*, 14 April 2010.

21 Intergovernmental Panel of Climatic Change.

22 'The Controversies in Climate Science: Science Behind Closed Doors', *The Economist*, 8 July 2010.

alarmism has taken over where Cold War hysteria left off. Bias, if not outright disinformation, brilliantly describes the mindset of project researchers who need to produce shocking news to justify their budget.

6. Scientists against Environmental Alarmism

Serge Galam, physicist and researcher at the French Conseil National de Recherche Scientifique (CNRS), along with plenty of other physicists and meteorologists, maintains that we are not currently in a position to say whether the earth is indeed going through a period of global warming or cooling. It is this uncertainty that underlies the fact that strong statements being made about global warming are *not* scientifically validated.[23]

In his book *Les dérangements du temps*, Emmanuel Garnier, a climate specialist also at the CNRS, demonstrates that focusing only on environmental variables (like the movement of glaciers) leads to the wrong conclusions and that this is accentuated by using short timeframes. The evidence provided by past centuries and millennia should also, most definitely, be taken into account.[24] One example is the symbol so often used as proof of environmental warning: the Alpine glacier of Grand Aletsch which has retreated by 3.4 km in 50 years. Garnier, however, points out that during Christ's time it was even smaller than today. In addition, from AD 950 to 1200 there has been a warm-climate period, followed by small glacial intervals in the fourteenth, mid-sixteenth and eighteenth centuries – most likely associated with solar events.

Garnier's thesis in *Les dérangements du temps* is that there have consistently been steady changes from significantly warm to cold weather and vice versa. The year 1608 saw a glacial *blitzkrieg*, but 1665 to 1680 were exceptionally hot years. In addition, each century experienced storms like those which are taken today as undisputable evidence of CO_2's malfeasance. For example, Garnier points out the following facts:

- The long heatwave of the ninth and tenth centuries promoted the Vikings' discovery of North America.
- Between 1500 and 2009 there were 22 hurricanes of force 10 to 12.
- In the decade of 1560 an exceptional drought lasted 350 days.
- The floods of 1784 were catastrophic.[25]
- There have been also been notable cool periods during the above-mentioned timeframe.

The earth's global average temperature has in fact been decreasing over the past 8–10 years. In an editorial contradicting the contents of Al Gore's Hollywood movie *An Inconvenient Truth* (see Chapter 14, section 3), the *Executive Intelligence Review* (*EIR*) noted that the earth is cooling, not warming – and cooling perhaps enough to bring on another Little Ice Age. According to Lyndon LaRouche, the *EIR*'s editor, 'the cooling shown by

23 Serge Galam *Les scientifiques ont perdue le Nord. Reflexions sur le rechauffement climatique*, Plon, Paris, 2008. Galam supported his thesis brilliantly in an interview he gave to *24 Hours* on 22 February 2010.

24 Emmanuel Garnier, *Les dérangements du temps. 500 ans de chaud et de froid en Europe*, Plon, Paris, 2010.

25 Ibid.

satellite temperature monitoring negates the entire globally averaged temperature increase of 0.6°C for the past 150 years, as hyped by many'.[26]

The facts provided by LaRouche, Garnier and Galam provide counterevidence to global warming and, as such, they should not be taken lightly. Scientists espousing this line of thought that negates environmental alarmism maintain that the recent inactivity of the sun is consistent with forecasts made by Russia's Pulkovo Observatory in St Petersburg.

Khabibullo Abdusamatov, head of the Observatory's space research lab, warned that because of unprecedented low solar radiation over the past 30 years, global temperatures would drop slightly rather than rise. And they would continue to decrease, even if industrial emissions of carbon dioxide were to reach record levels. According to Abdusamatov, by 2041 solar activity will reach its minimum in a 200-year cycle. Thereafter, approximately in the 2055–2060 timeframe, the earth will undergo a period of global cooling, similar to the one experienced in the late seventeenth century, which will last for half a century. There is no evidence that environmental alarmists have factored such projections into their deliberations.[27]

Thorough scientific research should not have failed to take into consideration opinions that support the growing conviction among climate specialists, geologists, chemists and other scientists about this alternative cause of climate change. It is widely known that over hundreds of thousands of years climatic changes have determined a golden horde of shifts between:

- ice ages,
- global warming ages,
- interglacial ages, and
- inter-global warming periods.

Indeed, it would not just be silly, but also highly unscientific, to write off the fact that, as many scientists believe, it is *primarily* solar cycles and only *secondarily* human activity that are responsible for changes in temperature.

Even if this theory is contested by those who think that man-made CO_2 and other greenhouse gases (GHGs) bear the responsibility, until scientific evidence is unshakeable it is wrong to opt wholeheartedly for either of these opposing hypotheses. Science simply does not work in that way. Its remit is:

- the methodological search of facts and counterfacts,
- the examination and test under conditions of disbelief, and
- the building up of tentative theories until a nasty new fact relegates them to oblivion.

However, when politics mixes with the scientific method, the latter turns on its head – and those involved in projects tend to forget that everything they do (or decide) has consequences. Some of these may be unexpected, but may have a high impact. For instance, a drastic and immediate reduction of CO_2, as conceived and implemented by Tony Blair, poses enormous economic problems for Britain's industry, and its effects are

26 *EIR*, 17 July 2009. EIR is published by Lyndon LaRouche, an American Democratic Party member.

27 Ibid.

particularly felt when it is unilateral.[28] Politicians, particularly European politicians, have a problem in deciding on their priorities. Is the priority employment and dynamic economic activity? Or is it the blind follow-up of unproven theories? By trying to have it both ways, they fall into their own trap.

Through failing to understand that reality is never as tidy as theory, in the second half of May 2010 the European Commission came up with the suggestion that the European Union should consider unilaterally moving to a 30 per cent cut in CO_2 emissions. Unsurprisingly, the French and German governments swiftly declared their opposition to that plan – and for good reasons.

Nor is the sponsored projects community blameless. When the IPCC[29] says that global warming is 'without ambiguity' the result of human activity, discarding solar activity and other possible causes, it provides clear evidence that it is a pressure group rather than a true scientific outfit.[30]

Serge Galam is right when he says that all these people are *'climato-alarmistes'* who have lost their compass. They have also lost their tempers because Galam has been personally attacked by the white knights of CO_2 and other gases on moral grounds! Some *climato-alarmistes* have even demanded sanctions against him.

This is one more piece of evidence that, in a century hailed to be that of democracy, the free expression of one's thoughts is not permitted. Orwell's Thought Police seem to be on duty – and, with this, democracy has taken leave of absence, accompanied by personal responsibility and accountability.

7. The Able Management of Human and Natural Resources

Two syndromes often beset societies whose time is up. One is listlessness and drift as discipline crumbles, moral values plummet and ideas of how to come up from under dry up. The other syndrome is a desperate hyperactivity led by one-track minds and aiming to cure the symptoms of an illness rather than its underlying cause.

Of course, a good reason for focusing on symptoms is that they are easier to identify than the real cause. Many people today appreciate that dumping into the seas and burying in the earth well over 200 million tons of toxic, untreated industrial garbage per year is *not* environmental conservation. But:

- they don't go to the core of the problem, with the aim of finding out why this is happening, and
- they fail to appreciate that if we are unable or unwilling to take appropriate – though painful – measures, our environment will consume us.

28 So far, only Britain and some continental Europeans have been hot on implementing the Kyoto Protocol, putting both their companies and employment under stress.

29 Like Barack Obama and Al Gore, Rajendra Pachauri, the IPCC's *eminence grise* was awarded a Nobel Peace Prize in 2007– for reasons which remain obscure.

30 Indeed, as we will see in Chapter 14, the Hollywood connection (Gore's film) also lands right in the middle of this unsettled argument, exposing its sponsor(s) to the allegation that it was invented and orchestrated to draw attention away from the upcoming financial catastrophe of 2007–2011 – which indeed was man-made.

Virgil, the Roman poet and philosopher, said: 'You must know the causes to draw the right conclusions.' Knowing the causes is the first basic step in able management, but it is not necessarily what our society does. Causality analysis requires:

- examining causes and effects underpinning longer-term time series, and
- projecting the likely aftermath (taken as effect(s)), by using hypotheses which reflect on the causes.

In science, closely associated to the analysis of cause and effect is testing and experimentation – which are also part of the method that supports able management. By contrast, poor governance favours easy solutions based on seat-of-the-pants approaches rather than on fundamental thinking. (Another pitfall in management is ideas drawn out a hat and generalizations).

Those who believe that an easy way has been found to redress the environmental balances point to Brazil's large biofuel industry, which already provides 40 per cent of the energy consumed by its cars and should soon supply 15 per cent of its electricity. This argument, however, fails to consider that Brazil and the rest of Latin America, Africa and Asia are experiencing a baby boom which:

- puts all these countries' resources under great strain, and
- contributes in a big way to the overall pollution of planet Earth.

(Back in 1966 while carrying out research for the American Management Association on developing human capital, I met up with a senior official of the Brazilian government in Rio de Janeiro. Obviously, the issues of education and employment came up. The Brazilian said that to create decent employment for the new crop of youngsters joining the labour force, his country would have to invest its whole GDP year on year, which was not possible.)

Human resources and natural resources are two sides of the same coin – but they are not being considered in unison. This disregard for the larger picture, which includes the educational and employment needs of individuals and societies, is unsettling. We live in a world where *quantity* rather than *quality* is the goal, and this has significant implications for everybody because it means that:

- the management of human and natural resources fall short of the proverbial long hard look, and
- near-sightedness ensures that the seemingly simple imperative duty of a regime – caring for its citizens – turns out to be a big intractable problem.

Governments should have appreciated long ago that precisely because a consumer society changes the rules under which natural resources are employed, the cause and effect of relationships become stochastic, affecting their ability to forecast disasters. The general trend is one of rapid increase in energy usage as documented by per capita consumption of electricity in the United States. In the twentieth century this multiplied *more than 135 times* and kept on rising at a rapid pace, without any measures being taken to bend the curve.

European statistics exhibit similar trends. At the end of World War II and for more than a decade thereafter, car ownership was a luxury. Today it is commonplace. Motor vehicles have been actively promoted by the state because these industries and their suppliers provide more than 10 per cent of employment, as well as a wide basis for taxation. Little attention has been paid to the fact that:

- these millions and millions of private motor vehicles are big contributors to pollution, and
- great environmental damage has been caused by the massive switch from rail to truck transport (see Chapter 10).

The Organization for Economic Cooperation and Development (OECD) the former Marshall Plan and currently club of richer nations, forecasts that by 2030 total world energy demand will approach 350 million oil-equivalent barrels per day. As shown in Figure 1.1, an estimated 57 per cent of this colossal amount will be consumed by non-OECD members – that is, countries in the process of development. Nobody, let alone their own governments, seems to care that:

- these are precisely the countries suffering the most from population explosion, and
- their energy usage will skyrocket even if by 2030 they represent among themselves only a third of global domestic product.

The aftermath of such carelessness on both energy consumption and worldwide pollution is scary.

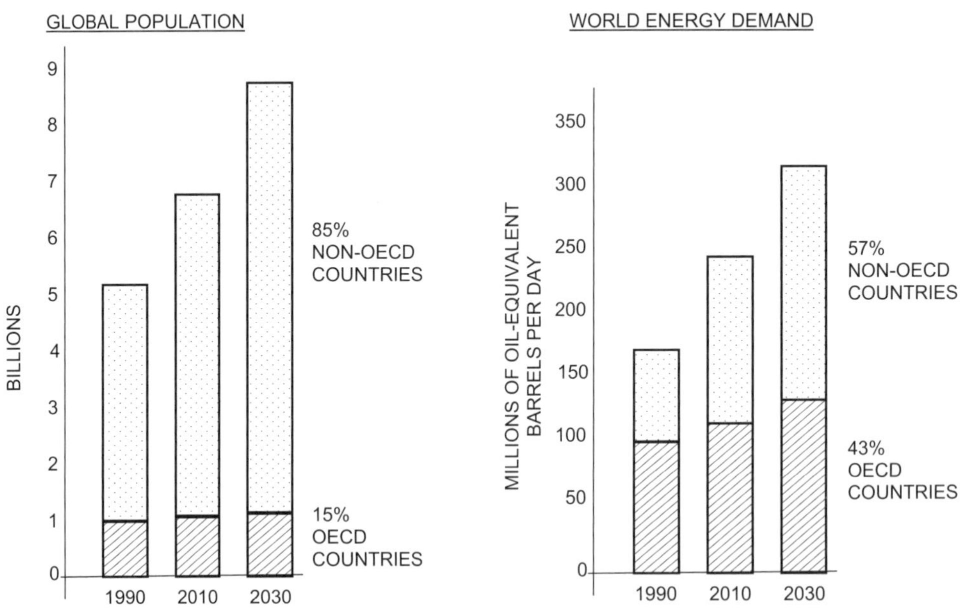

Figure 1.1 OECD forecasts on global population and energy demand

Source: Statistics from the Organization for Economic Cooperation and Development (OECD).

Optimists respond by saying that the world is endowed with huge resources, which is nonsense at best. Apart from the fact that the earth's mineral resources are being depleted, other energy sources like solar energy and wind power are exploitable only in certain parts of the globe (see Chapter 5). Nor does it make any sense for the current next generations to use up what is available, leaving little or nothing for those who follow.

A second message from Figure 1.1 is that energy demand is at an inflection point – and, among OECD and non-OECD member countries, the latter are far ahead of the former in terms of spoiling the earth's resources. However, this does not mean that the West does things right.

Nevertheless, in contrast to the European Union which tried to obtain at least a verbal commitment to move its energy production away from coal-fired plants (Britain being the only country to have done so), in China 79 per cent of all electricity is generated using coal[31] even though coal is not in abundance in that country. Atmospheric pollution aside, promoting coal rather than renewables ensures that, although estimates talk of the earth's coal supplies lasting more than a century, a tight coal market has developed as coal inventories at power plants fall.

If developing nations continue to depend on coal and if the West puts climate change ahead of energy supplies, it will damage the West's industries and employment opportunities and achieve only a minor positive environmental outcome – without really bettering the status of global pollution. Let's never forget that:

- economic growth and power generation correlate, and
- the fact that China has kept retail electricity prices low and stable since the 1960s provides incentives for wider energy usage, not for conservation.

The mining and extraction of minerals and metals is relatively low in energy intensiveness, but converting this ore into refined metals requires a large amount of energy[32] – and this is now taking place on a large scale. In the aftermath, the world is facing a new energy equation characterized by rather poor management and many unknowns. Apart from being a key driver of both rising energy prices and environmental pollution, the dissociation of human and natural resources poses complex questions which have not been addressed through conferences of Rio, Kyoto and Copenhagen vintage (Chapter 14).

8. The Priority should be Energy Conservation

The worries and shortcomings outlined in the preceding sections help to demonstrate that the best policy is *energy conservation*. Governments, as well as the public, should be keen to implement conservation measures. But there exist conflicts of interest with the result that the various parties concerned are not inclined to conserve energy. Had they

31 Hydroelectric power plants provide another 15 per cent of the Chinese electricity supply.

32 For instance, electrolytic reduction used to convert alumina into pure aluminium is the most electricity-intensive process among the base metals, and electricity is not the only form of energy used to process ore in metal. Smelting requires large amounts of heat to separate the metal from the ore.

been inclined to do so, they would have taken the initiative in applying measures to reduce an ever-growing consumption long ago.

It is not that governments are stupid, but rather that energy conservation has significant effects – a lower standard of living and greater unemployment being two of them. Yet there is no doubt that conservation is 'a must', and it must start with making the public aware of the need for it. That's the kind of action these various 'save the environment' love affairs, from Rio and Kyoto to Copenhagen, should have been keen to take.

For any practical purpose, the emphasis on energy conservation is a matter of self-preservation, as failure to establish a conservation strategy leads to unexpected consequences. Back in 1961 Indonesia was one of the founding members of the Organization of Petroleum Exporting Countries (OPEC). Today, five decades down the line, Indonesia is importing oil.

- Its wells did not run dry,
- but huge consumption and spoilage of cheap, local energy led to demand exceeding supply.

For its part, Saudi Arabia is not the world's bottomless oil depositor and economic giant it used to be. When it runs short of oil, it will become a rather poor country with a very high birth rate, unable to live off its oil wealth because this has been depleted. Rationally, though, oil-producing countries should be setting an example in terms of energy conservation. After all, for them, oil is a resource that which will become just a memory a few decades down the line.

Governments – all governments – have been not only consistently promoting energy conservation, but also appreciating that, though important and necessary for its own sake, the notion of 'conservation' is not enough. There are two other basic elements in implementing a conservation policy, along the line of Al Capone's dictum that 'You can get further with a kind word and a gun than with just a kind word'.[33]

1. *A ceiling on individual consumption of energy*, commensurate with pollution targets and with available supplies, so that they last longer.

   During World War II, in German-occupied Greece, a three-member family was allocated a ceiling on energy consumption of 14.5 kW *per month*. The first time this limit was exceeded, the subscriber paid a penalty; the second time around, electric power was cut for a month. Every morning, prior to leaving for school, I was responsible for measuring how much electricity we had consumed the previous day. The family was responsible for self-regulating the day's consumption.

   If the different head-of-state partygoers participating in the December 2009 United Nations climate summit in Copenhagen meant what they said, *then* they should have imposed limits on energy usage, with all its implications for the economy, employment and standard of living. Nothing like that took place, even though the time to bring

33 Spoken by Robert de Niro, playing Al Capone, in *The Untouchables*, dir. Brian Palmer, 1987.

energy consumption under control has come. (An early application of limits will help in avoiding wartime-type measures later on.)

Argentina offers a preview of what may happen when things come to a head and electric power is rationed in a big way. In the late 1990s financial analysts predicted that the consequences of the Argentinean government's populist energy policies would be painfully felt. During the country's 2001–2002 economic crisis the government forcibly converted all energy tariffs from dollars to pesos. With the peso in free fall, this represented a cut of nearly two-thirds in real value which hit the utility companies like a hammer blow.

Since then, only a handful of modest increases have been permitted in energy prices – a policy which has greatly discouraged investments in power production. To cope with peaks in demand, the government uses short-term fixes, most of which are highly costly. For instance, it imports liquid fuels from Venezuela and electricity from Uruguay and Brazil, which have proved to be extremely expensive, leading to forced rationing.

2. *A system approach to energy conservation*, which includes greater energy efficiency and penalties for not using high technology to improve transmission efficiency[34] and control consumption.

An example is provided by clear-eyed energy specialists, like Amory Lovins of the Rocky Mountain Institute (RMI), who have been arguing against trying to meet burgeoning energy needs simply by building more power stations. Instead, they say, governments should limit demand by matching an appropriate, efficient and cheap source of energy to each task.

This is an interesting out-of-the box idea. A lesson can be learned from the early years of the Industrial Revolution, when energy efficiency was improved by ceasing to attach machines to the same shaft and powering them individually instead.

Lovins not only argues for energy-efficient buildings, but is also concerned with:

- what people do in those buildings, and
- how they travel back and forth to them.

'Smart' buildings date back to the late 1980s. Sometimes the energy savings they have yielded have been significant, but so have been the investments. Therefore, although they were technically feasible and reasonable control over energy usage has been confirmed in practice, they did not multiply but remained, more or less, as prestigious examples.

For instance, at the Tour Lombard office building of the Union Bank of Switzerland in Geneva, a smart-building solution achieved 40 per cent reduction in the consumption of heating oil within five years. During the same period, the use of electricity was cut by more than 25 per cent. Note that:

34 The 'smart grid' argument (see Chapter 15).

- these results have been reached mainly through the incorporation of controls in the air-conditioning system,
- but much more can be achieved by knowledge engineering-based energy management, leading to intelligent building solutions.[35]

The point is that not only we should become conservation-conscious; we should also fine-tune conservation policies and put them into practice. Rather than being predominantly oriented towards the maintenance of status quo in energy consumption, new engineering designs should focus on how to bend the ever-rising curve of energy needs.

Examples abound all over business and industry. For instance, electric motors use about 60 per cent of the power generated in the United States, but experts say that this can be significantly downsized if they are more efficient, well-maintained and better suited to their task. A fertile domain for energy conservation is constraining the amount of energy used in transportation. Energy consumption by American motorists America provides an example:

- In 1980 the United States consumed 19 billion barrels of oil.
- In 2007 it consumed 142 billion barrels,[36] and transportation roughly accounted for 63 per cent of US energy demand.[37]

Much of this energy consumption was superfluous as major savings can be made by sharing transport facilities. This has been promoted in California through special lanes accessible to cars with more than two passengers, but there are few takers because such an approach infringes on individuals' travel patterns. People want to see energy savings without giving up their freedom of choice – an impossibility. Worse is the fact that this attitude, which also prevails in Europe, has been copied by citizens in developing countries in a widening horizon of:

- unsettled priorities, and
- wrong perceptions of after-effects.

Some of the people who want to reduce pollution and preserve the environment press for priority to be given to the notion of *cogeneration*, the combined production of heat and power. But so far there is little evidence to sustain the prediction that eventually many factories will generate their own power, using the waste generated by industrial processes. The bad news is that such schemes, already in use in parts of the world, have not raised total plant efficiency in a way that matches expectations.

Energy consumption is also encouraged by the wrong belief that electricity is a clean source and an 'obvious fuel'. Electrification is widely considered as a sign of progress.

---

35 D.N. Chorafas, *System Architecture and System Design*, McGraw-Hill, New York, 1989.

36 *Bloomberg News*, 5 June 2008.

37 Correspondingly, industrial usage takes slightly over 24 per cent, residential and domestic nearly 8 per cent and electric generation 5 per cent.

Up to a point this is true. But what is that point? Australia needs $92 billion in energy infrastructure investments in this coming decade to match demand.[38] Is this rational?

The bubble connected to electric cars is growing in the belief that they use 'clean energy' (see Chapter 10). People and governments, however, fall short of asking the most critical question: 'How will electricity be produced?' Not only private individuals, but also factories, offices, shows, lavish street lighting and a horde of other services overuse and waste electric power:

- People cram domestic appliances into their homes, as well as PCs, TVs and air-conditioners.
- And those responsible for offices are generally careless about turning off the lights outside working hours, and typically switch on air-conditioning and heating systems without giving a thought to energy wastage.

As for the estimated 600 million vehicles on the world's roads – three-quarters of them privately-owned vehicles – they waste petrol to a degree unparalleled in the history of transportation. All this serves as reminder that public awareness about for the need for energy conservation is plainly lacking. In the relatively wealthy OECD countries the energy gobbled up by transport now exceeds that consumed by the residential and services sectors taken together. As for the rapidly growing demand of developing countries for road transfer, this has become the main force behind the global rush for oil as well as a sensitive political issue – and therefore untouchable.

38 *Bloomberg News*, 19 March 2010,

CHAPTER 2 The European Energy Policy Dilemma

1. The European Union and Energy Supplies

Chapter 1 brought to the reader's attention that, during the nineteenth and twentieth centuries, harnessed energy became a pivotal point in the Western way of life, making a major contribution to development and the sustenance of a higher standard of living. At the same time, however, as people became more affluent and more numerous, the uses of energy proliferated and new sources had to be found – which continues to be absolutely true in the twenty-first-century globalized world.

Countries lacking energy resources are evidently at disadvantage. Many EU member states have significant coal deposits and they have been classically using coal-fired plants for power production. Coal, however, is on the black list with regard to CO_2 emissions, and, as we will see in Chapter 5, so-called 'clean coal' is still a mirage.

Several EU countries sponsor research into renewable sources of energy aimed at providing consumers and firms with new technologies as alternatives to the status quo. However, as already discussed, investment must be both forthcoming and profitable for a major changeover to actually take place. Nothing can be done if sufficient funding is not available.

In addition, issues connected to CO_2 and greenhouse effects – from coal- and oil-fired plants to other installations for energy production – cannot be seen in isolation. They have to be examined within the overall framework of decisions about *where*, in *which* media and at *what* pace:

- to direct public policy, and
- to invest significant amounts of capital.

One thing not so difficult to prognosticate is that demand for all sorts of energy is likely to rise, with some media substituting for others that are theoretically more polluting. Britain made a substitution, choosing gas as the energy source in the revamping of its power production infrastructure. Policy decisions must, however, be far-reaching. Roughly a third of British power stations will need replacing by 2015, and in environmental terms gas and nuclear power are better than coal – but nuclear energy is still controversial (see Chapter 4).

Although the various EU governments talk a lot about renewable energy, so far much of the new capacity looks likely to be more traditional. No effective solution can be based on the technical errors of the Greens and other environmentalists which include, but are not limited to, zero-point energy or 'solutions' like artificial photosynthesis, rooftop wind turbines in urban areas and other options:

- simply pulled out of a hat, and
- a waste of time and money above all else.

Like capacity expansion, replacement decisions are usually associated not only with implementing higher technological standards and cleaner solutions, but also with a favourable political climate which allows a fair balancing of opportunities and risks. In the context of a European electrical power policy the options mainly relate to the following sources:

- fossil fuels – carbon, oil (Chapter 3) and gas,
- nuclear power (Chapter 4) – with the associated problem of radioactive waste disposal,
- hydroelectric dams – often severely affecting the natural environment,[1]
- geothermal power – which its critics say it is not a hope but a curse,
- wind power – whose potential is subject to the stochastic behaviour of winds,
- solar power – practical only in some parts of Europe, and even there it faces opposition (more on this in Chapter 5).

Experience teaches that no single alternative has ever constituted on its own the best possible solution. Fossil fuels are a classical source of energy supply and the worst in terms of CO_2 emissions. Solar and wind power are much greener approaches, but when their proponents promote them as a steady source of energy, they promise something that they cannot deliver. It is nature, not man, which decides about sunny and windy days and hours.

Considered in watts per square metre of production facilities, nuclear has significant advantages over its competitors. But there is also the memory of Chernobyl and Three Mile Island, while over the years the Greens and other street demonstrators have ensured that the nuclear industry has fallen into the doldrums and its expertise squandered.

Nor can the nuclear industry alone solve all of the European Union's (or any other region's) power-generation problems. France, for example, was supposed to be well protected because 80 per cent of its energy supply comes from nuclear plants (see Chapter 4). It is not.

In 2009 Electricité de France (EDF), the country's state-held quasi-energy monopoly, had to import large quantities of electricity because almost a third of its nuclear plants went out of service at some point, either because of strikes or for maintenance. EDF warned that the country could even suffer power cuts during the winter of 2009–2010 and, to the distress of its clients, this scenario materialized.

Investments have been another challenge. Experts say that to fulfil the rising requirements for electrical power production EDF will need to extend the life of its reactors from 40 to 60 years – a big investment. A few years ago the company thought that the cost would be negligible, but it now reckons it will have to spend €400 million on each plant – a major budget. (As we will see in Chapter 4, a new nuclear plant today

1 Hydroelectric, geothermal, wind and solar power are discussed in Chapter 5.

costs between \$5 and \$6 billion, and decommissioning an old one costs more than \$1 billion.)

This is not written to discourage nuclear power production, but to warn that energy policy decisions should never be wholly in one direction. Another objective of bringing 'pluses' and 'minuses' to the reader's attention is to discourage silly talk about eliminating *a priori* some of the options (for instance, nuclear power).

Limiting one's options is always counterproductive and it will weigh heavily on future energy supplies, as well as on the cost of power to industry and consumers. It is also preposterous to pretend that one option is the earth's saviour, when it is well established that understanding temperature changes on this planet is still a challenge to science.

2. Big Lies Have Short Legs

Politicians, let alone the Greens, who promise an uninterrupted power supply and, at the same time, a cut in emissions – without considering costs and investments – are either morons, liars or big spenders of other people's money. There is plenty of irrelevance in such pronouncements and directives, like the one issued by the European Commission to airlines to cut emissions. Airlines don't design jet engines. The only way they can cut CO_2 emissions is by reducing air traffic.

- People must travel less and over shorter distances if they want less pollution.
- As long as energy requirements continue to grow, pollution will soar because science and technology does not know how to do miracles.

Short of draconian conservation measures, none of the European Union's member states, or any other country in the world, can prevent the situation from worsening. Saying the opposite is a big lie, and lies have short legs.

The exponential growth in energy consumption clearly has lasting implications, ranging from environmental pollution to power plays for the control of energy sources. Taking *world* electricity generation as a reference, in order of magnitude in 2009 the origin by source has been:

- 40 per cent coal,
- 19 per cent renewables (including hydroelectric),
- 16 per cent nuclear,
- 15 per cent gas, and
- 10 per cent oil.

In an aggregate sense, within each region as well as within branches of human activity, the distribution of peaks and valleys in energy usage changes over time, but the general trend is upwards. The exact percentages of surge in energy requirements vary by country, but among Western nations the pattern does not differ widely. Developing nations, however, have a changing pattern. Let's take the United States as an example of total energy consumption (not just electricity generation), starting with the nineteenth century when it was still a developing nation.

Statistics show that in 1850 less than 10 per cent of American energy was supplied by coal; the balance came from wood. By contrast, within a quarter of a century coal provided almost 30 per cent of energy, and this grew to about 73 per cent in 1900. By the time the United States had become a developed nation, the share of wood had shrunk to 20 per cent and there were two newcomers.

- About 4 per cent of energy came from gas, and
- at the dawn of the twentieth century another 3 per cent was provided by oil.

By 1950 the distribution of American energy sources had again greatly changed: oil accounted for nearly 38 per cent of energy production, coal for 35 per cent, gas for 14 per cent, with the balance shared between nuclear, hydroelectric and wood. By 2009 the share of oil and gas had ballooned, that of coal had come down to about 20 per cent, with wood making an insignificant contribution. On a percentage basis the share of hydroelectric power shrank, nuclear power production gained slightly, and a thin slice went to new sources of renewable energy.

Recent statistics[2] on the pattern of change in energy types in the European Union are not too different from those just mentioned. Today, still in order of magnitude, oil accounts for about 42 per cent and gas for another 22 per cent; taken together the two represent nearly two-thirds of energy supplies. Britain, Norway (not an EU member) and Romania are oil producers, the first two and the Netherlands have gas supplies, but, as a whole countries, in the European Union are far from being energy-independent.[3]

- Central and western European countries buy a huge amount of energy from somebody else, and
- in doing so they give their economies as collateral, which is not a wise policy and therefore should be discontinued (see Chapter 6 on the wisdom of betting on biofuels).

An eye-catching reference to remember is that the amount of energy consumed by the average European every year is enough to make a car journey one-and-a half times around the earth. As we have already seen, the rate of energy wastage is unprecedented (not just in the European Union but worldwide) and while there is plenty of talk about controlling pollution, no effort, official or individual, is made for conservation. Environmental concerns aside, this happens despite the fact that of the total amount of energy consumed in the European Union:

- only 42 per cent is produced within its borders, and
- 58 per cent is being imported, putting the old continent at the mercy of energy producers.

From the time of the 1997 Kyoto Protocol to the December 2009 Copenhagen UN Climate Change Conference the various governments and private initiatives have paid

2 From a meeting on energy investment attended by the author in mid-2009.

3 With the exception of France where about 80 per cent of electricity production is nuclear.

scant attention to the fact that wastage is widespread in both transparent and opaque ways. Only some sporadic, and relatively minor, measures have been recorded.

The EU directive on changing light sources is an example. An incandescent bulb comprising a wire filament encased in glass, which dates back to Emerson and the late nineteenth century, emits as light only 5 per cent of the energy it consumes; the balance is wasted as heat. Fluorescent bulbs which consist of tubes filled with mercury vapour, are roughly four times more efficient, but they still waste as heat 80 per cent of their energy. Switching from incandescent to fluorescent lamps is an improvement, but no big deal. Besides:

- the European lamp industry has not been prepared for the changeover, with the result that the Chinese will cream off the European Union's fluorescent lamps market, and
- this will happen while EU consumers (and companies) will be paying a high multiple of the cost of incandescent lamps.[4]

Lies and empty words left aside, over the years the Greens and their political allies on environmental protection have offered no technical solutions providing a roadmap for greater energy efficiency *and* for quality of life. The environmental zeal of other political parties, too, is nothing but smoke and mirrors, and this exacerbates the energy problem. Indeed, their, energy-consuming initiatives, like promoting global tourism and bending over backwards to foster world trade, contribute greatly to:

- soaring energy usage, and
- widespread environmental pollution

One of the hoaxes being perpetrated at the beginning of the twenty-first century is that some countries keep to their emission-reduction targets under the Kyoto Protocol, while others don't. If they miss them, those who supposedly want to keep to their pledge can buy up offsetting CO_2 'credits' from countries that have cut their emissions by more than their target. Yet, today, with a couple of exceptions in the West, it is the developing countries that make the biggest contribution to greenhouse gases. Ironically, for example, China sells to Britain a lot of CO_2 'credits'.

This sort of trade is irrational, as China has overtaken the United States to become the world's largest energy consumer and polluter. The International Energy Agency claims that from 2000 to 2010 China's energy demand has doubled, even if its consumption per person is still only one-third of the OECD average.[5] Such a sharp increase in energy usage is not without consequences.

According to the World Bank, China's economic loss due to pollution is estimated at nearly 6 per cent of its GDP, mainly from coal-fired power plants which produce 80 per cent of the country's electricity supply. To address the challenges posed by associated pollution, China plans to increase its nuclear power generation capacity to more than 70 gW by 2020, seven times more than current levels.[6]

4 This does not mean that incandescent lamps should stay, but it is a good example illustrating that our society is unprepared for changeovers.

5 *The Economist*, 24 July 2010.

6 Crédit Suisse, 'Thematic Energy Research – Uranium', 29 November 2010.

Neither are CO_2 emissions the only challenge connected to pollution. Disposing of computers, monitors, printers and mobile phones and other devices is a large and growing environmental challenge to which no international conference on environmental damage has addressed itself. An estimated 30–50 million tons of e-waste is produced each year, and, according to experts, this has become the fastest-growing pollution source.

Indeed, the Rio, Kyoto, Copenhagen and numerous other environmental conferences are themselves part of the pollution problem, rather than part of its solution. The same is true of the money thrown by the West to the various less developed countries to control *their* pollution. No better aphorism can be found for this situation than the one by Montagu Norman, the former governor of the Bank of England, in connection to the first Great Depression: 'We achieved absolutely nothing, except that we collected a lot of money from a lot of poor devils and gave it to the four winds.'[7]

3. Energy Policy in the European Union

Right after the end of World War II, in many European countries that are now EU member states, all sorts of energy supplies, from gas to electricity and oil, were either nationalized or brought under overall government control. This was done mainly for political reasons, although investment considerations, too, were present. The public has generally looked favourably on state ownership of the energy industry (and of the transport and communications industries as well).

Cheaper energy, which was the stated purpose of nationalizations, can be obtained in either of two ways. The better one is higher efficiency in regard to its usage. As has been explained, however, this requires investment, takes considerable effort and is not easy to attain. The easier way is that of taxes and subsidies which abound in the realm of socialist policies and through government intervention – but have the nasty habit of contradicting one another:

- first, by keeping prices at consumption level lower than the market's, thereby inviting wastage, and
- then by exercising discretionary tax measures on the use of energy resources, making money available for selective subsidizing.

The long decades of treating energy production as an appendage of the state explain the wariness of EU member states about any attempt by the EU Commission to extend its competencies on energy and related tax matters.[8] Many EU plans and projects have contained rather diverging value added tax (VAT) provisions – initiatives taken in the name of environmental protection such as a combined energy/carbon dioxide tax. This is a poor indication of a uniform and well-coordinated energy policy. The same is true of the fact that, to keep their economies going, member states have thrown caution in energy consumption to the wind. In terms of directives targeting energy efficiency, the EU Commission does no better than member states – notwithstanding the Green Paper on energy issued by Brussels in March 2006 (see section 4).

7 Liaquat Ahamed, *Lords of Finance: The Bankers Who Broke the World*, Penguin Press, New York, 2008.

8 D.N. Chorafas, *The Business of Europe is Politics*, Gower, Farnham, 2010.

There are many ironies attached to this sorry state of affairs, one of them being that the lack of a determined EU energy policy has a dual negative effect: first, on the price of electric power and, second, on the amount of pollution. Based on statistics by EWS (the central Swiss electricity company),[9] Figure 2.1 clearly shows that the price of energy and resulting pollution correlate.

- French tariffs are lower because electricity production depends on nuclear power and, relatively speaking, the cost is lower and so are unwanted emissions.
- Likewise Switzerland – as well as Sweden (see section 6) has capitalized on hydroelectric power.
- By contrast, Germany's power production has been cornered by the Green opposition to the nuclear option, with the result that both the cost per kWh and CO_2 emissions are high.

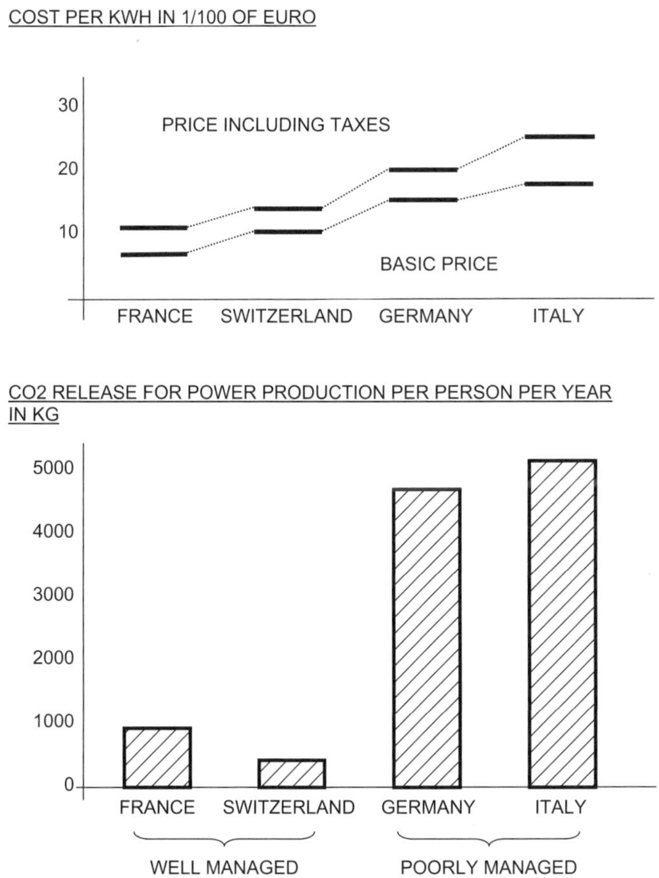

Figure 2.1 Well-managed and poorly-managed electricity production and environmental protection

Source: Statistics from Verband Schweizerischer Elektrizitätsunternehmen VSE.

9 The statistics come from the brochure 'Geht uns ein Licht auf, wenn der Strom knapp wird'. Available at: http://www.stromzukunft.vh/stromzukunft/downloads/.

This is precisely what happens when political eggheads are allowed to dictate energy policy. The French decision to bet on nuclear power was taken in the early 1970s under a right-wing government. It is to the credit of the French Greens that when they participated in the Jospin government in the late 1990s, they opted to eat their words rather than reverse the French policy on nuclear power production.

By contrast, because they have simultaneously promoted both high energy prices and high CO_2 emissions, the anti-nuclear policy followed by the Greens in Germany has become highly counterproductive. However, the worst offender of the four countries in Figure 2.1 is Italy where the cost of electricity per kWh is the highest with concomitant high levels of associated CO_2 pollution.

This sharp contrast in terms of cost-and-CO_2 results between France, Germany and Italy – all EU member states – highlights the fact that the European Union's plans for an energy policy are still unsettled, despite frequent pronouncements to the contrary. It is indeed curious that although, as individuals, the Greens and other environmental vigilantes are not that stupid, together:

- they choose the wrong way to attack environmental issues, and
- cling to that wrong way, tooth and nail.

Nor has there been any change towards better coordination of energy policies among EU countries since 23 January 2008 when the European Union proclaimed a 'new era' in the fight against climate change – including supposedly sweeping measures to cut greenhouse gases and boost renewable energy sources. José Manuel Barroso, president of the European Commission, called on the world to join forces in 'the great project of our generation' (whatever that means).

In contradiction to official pronouncements, in the years since then, little, if anything, has been achieved. Part of the reason – but only part – is that to follow the announced course of action the 27-member European Union would face costs well above €60 billion ($90 billion) a year, or roughly 0.5 per cent of gross domestic product. This is the theoretical price-tag of that policy; in practice, the cost is about double that level, and unaffordable.

- The finances of EU member states are weak and cannot support grandiose schemes.
- Not only are some of the newly acceded countries teetering on the edge of bankruptcy, but even the coffers of the European Union's major economic powers are empty.

Another key element that favours talking rather than action concerns the fact that politicians cannot make up their minds about what takes precedence: a policy on assuring uninterrupted energy supplies *or* environmental issues. Back in January 2008, in the middle of a full-blown economic and banking crisis, Barroso contradicted himself by stating that his proposals would simultaneously control emissions *and* give Europe more energy security, making it less dependent for oil and gas supplies 'on regimes that are not our friends'.[10]

As if to underline that what he had to say contained incompatible assertions, the president of the EU Commission also promised favourable treatment for heavy industries

10 *Financial Times*, 24 January 2008.

such as iron, steel, aluminium, chemicals and paper. Everywhere in the world these three industries are environmental polluters, and prior to Barroso's 'new talk' they had made it plain that higher energy costs would drive them out of Europe to areas with laxer controls.

'We want to create jobs, not destroy jobs,' said Barroso, contradicting yet again his pollution control aims. This is not written as a criticism of Barroso, who is a nice chap, but as evidence that there are plenty of conflicting aims as the member nations and the EU Commission try to:

- keep the energy market functioning,
- improve job prospects, and
- control emissions – all at the same time.

In the absence of a firm and non-contradictory EU energy policy, conflicting aims hold the upper ground. The European Commission and EU member states seem unable to sort out their priorities and set out on a path of diversification in energy supplies. Yet, it is not difficult to see that the European Union should first and foremost focus on safeguarding its security of supply:

- by exploiting what it has in abundance, which means biofuels (see Chapter 6), and
- by forming appropriate alliances that will assure an enviable degree of diversification.

Experts suggest that by 2030 Russia will provide about 50 per cent of the European Union's gas imports. This might be alright *if*, and *only if*, the European Union invites Russia to be the next member state[11] (rather than Turkey with its empty pipeline to be built sometime in the future). In addition, the European Union should negotiate with Russia as a body. Currently rather than working together, EU countries are bent on striking bilateral deals.

4. What the European Union's Energy Plans Have Missed

If environmental pollution controls are as important as they are considered to be, *then* everybody should abide by them, and trade sanctions are a good way of instilling some discipline worldwide. This does not mean that Europeans are not polluters. Like the Americans, EU citizens are so used to believing in 'limitless energy supplies' as a sort of God-given right that they, their governments and companies are unable to imagine what life might be like when the fuel and other energy sources are rationed or really start to run out. As will be demonstrated in subsequent chapters:

- for the time being, many of the alternative energy and renewable-energy solutions (they are not exactly the same thing) are either half-truths or half-baked, and

11 And even that will be a partial solution, as there are fears that Russia may be physically unable to produce enough gas to satisfy EU demand.

- as consumption of energy continues to increase, with no EU policy in place there is nothing to decrease dependence on oil (Chapter 3) besides gas and nuclear energy (Chapter 4).

Even if the demand for 'more energy' grows more slowly in Europe than in China, India, Brazil and other developing nations, a policy to provide direction, coordinate efforts and avoid having to make repetitively similar decisions is still required. At the same time, policy-makers should appreciate that the one-sided imposition of pollution-control rules makes absolutely no sense: it does not really help the global environment and also increases the cost base.

In a nutshell, this explains the European Union's dilemma and subsequent inertia regarding the growing energy crisis. This applies all the way to the so-called Green Paper, published by the European Union in March 2006,[12] which, although it went into oblivion, is useful to review because it helps in revealing the lack of seriousness as far as an EU energy policy is concerned. Simply stated:

- this Green Paper's guidelines are hollow, and
- notably absent is a plan for energy conservation, which should have been its central feature.

As Chapter 1 showed, conservation is a 'must'. Environmental reasons aside, energy resources within and around Europe's borders are being depleted. This is true of North Sea oil and it is also the case with aging nuclear and other power stations. Politicians may say that in a free economy investment is surely the responsibility of private companies. But the dual effect of public opposition to nuclear energy and economic nationalism demolish this argument.[13]

The European Union is currently poor in big transborder energy companies, and there is no reason to believe this will change in the near future. Energy companies continue to merge within their national borders – which they should not do, because this creates monopolies. At the same time, companies cannot grow 'at home' to global size even if political will guarantees that they don't run into trouble with anti-trust authorities. By contrast, cross-border mergers of energy firms within the European Union would signal the start of a long-awaited consolidation, in which:

- four or five big players would be able to lead the entire EU energy utility market, and
- would have enough capital and clout to renew Europe's decaying energy system.

For a trained engineer, like myself, what is advanced as a 'solution' by the EU Commission, such as encouraging new 'interconnections' between national electricity grids, is just a joke. And the suggestion of setting up stockpiles of gas is nothing more than an ill-conceived, inefficient and dangerous measure. Specifically in regard to the argument about interconnecting energy grids, a report by the EU Competition Directorate complained that monopoly suppliers:

---

12 European Commission, *A European Strategy for Sustainable, Competitive and Secure Energy*, Brussels, March 2006.

13 D.N. Chorafas, *The Business of Europe is Politics*, Gower, Farnham, 2010.

- still carve up markets,
- stifle competition, and
- block new entrants' access to gas and electricity.

Here we have a body, the EU Commission, in which the left hand does not know what the right hand is doing. The best way to interconnect power grids is to establish pan-European technical standards, rather than create a power distribution monopoly that stifles competition. Some people say that the hand of EDF can be seen behind this EU interconnection statement, but it really does not matter whose hand it is. What matters is that:

- the European Union must come up with a technically valid solution at an affordable cost, and
- it should not let movers and shakers push it to corners – a practice that will ultimately damage the competitiveness of European industry.

In addition, if they want their words about energy solutions to be taken seriously, EU commissioners and bureaucrats must explain where the money for 'this' or 'that' proposed action will come from. At present, this is far from clear. For instance, in the Brussels energy policy meeting of EU prime ministers on 23 March 2006, it was said the network interconnection pipedream will cost €1 trillion over ten years – but nobody cared to explain who is supposed to pay this €100 billion per year.

Investment, state control and associated economic nationalism aside, from a cost-effectiveness standpoint grid interconnection makes sense *if*, and *only if*, there is a significant cross-border energy market with flows of power unimpeded by artificial barriers. This is not the case. The EU landscape is characterized by national power production plans and national grids, with few bilateral links because there is no overwhelming reason for them. For instance:

- Spain's market is almost entirely self-contained, and
- the one in the Baltics is also more or less self-contained, as it currently stands.

Gas stocks is another domain where an EU policy is glaringly absent. At present gas stocks vary from about four months' supply in Austria to practically zero in Ireland, Finland and Sweden. The consequences of a Russian gas cut-off, like that exercised on occasion against Ukraine, would ripple right through the European Union whether or not a country had stockpiled gas supplies. With economic nationalism on the rise, each EU country with a gas surplus would keep it for itself. To my knowledge, the EU heads of state have never really approved guidelines on:

- how many months of gas and oil supplies each EU country should keep, and
- whether these supplies should be EU resources to be equally shared by all member states, or be reserved solely to the country that owns them.

There is a similar lack of policies, plans and standards in regard to the proliferation of gas pipelines, which cost a fortune to build but in some cases have no guaranteed source of supply.

- Nord Stream starts near the Russian–Finnish border and ends in Germany.
- South of Nord Stream, Yamal crosses Russia (starting at the Yamal peninsula in the Arctic[14]), Belarus and Poland and also ends in Germany.
- South of Yamal is Brotherhood, which again starts in Russia, crosses Ukraine, Hungary and Slovakia, and goes to Germany and Italy.
- There is also Blue Stream in southern Europe, and Nabucco. The latter is planned to transit through Turkey and is distinguished by the fact that its source of gas supply is at best uncertain.

There is a reason for this Russian concentration of point of departure. Almost 60 per cent of the world's known gas deposits can be found in just three countries: Russia, Iran and Qatar. On the other hand, as already explained, overdependence on Russia is not a wise policy. The output of Gazprom's current three giant gas fields, which account for 75 per cent of its production, is declining at a rate of some 6–7 per cent a year, and output from a new gas field brought on stream in 2001 has already peaked.

As a matter of policy, the European Union is wrong to place all its bets on pipelines. Experts say that, beyond 3,000 km, it is cheaper to liquefy and ship natural gas than pipe it. The European Union has failed to produce the right product mix in terms of pipelined gas supplies and liquefied natural gas terminals, because the latter have not received the attention they deserve – and this works against the principle of diversification.

5. Patterns of Energy Usage: An Example from Britain

Britain's electricity demand peaked at 58.9 gW, within 3 per cent of deliverable peak simultaneous demand, on 7 January 2009.[15] A little over 45 per cent of that power came from power plants fuelled by gas from the North Sea, another 35 per cent from coal, somewhat less than 15 per cent from nuclear power, with the remaining 5 per cent or so from different sources.

The 58.9 gW was considered a temporary peak, yet unquestionably the trend in demand is for more electric power production. Experts say that *if* the 2007–2010 economic crisis becomes a memory and economic growth resumes, as a reasonable guesstimate Britain will need 65 gW to cope with rising requirements, and even that would have to be revised upwards. This poses two questions:

- Where will that electricity supply come from?
- For how long will a rise in the provision of power be assured?

The same questions are being heard all over the European Union (with the possible exception of France where a long tradition of nuclear power production provides some assurance).[16] In their background lies the fact that Western nations are running short of power supplies so quickly that some economists claim the 2007–2011 financial crisis

14 In 2006 Gazprom decided to develop a massive field in the Yamal peninsula – a frozen and barren Arctic landscape – but that will take years.

15 Claverton Energy Research Group at: http://www.claverton-energy.com/is-wind-power-reliable.html.

16 See also the discussion in sections 1 and 7.

and associated economic slowdown has been useful in that it has given governments a breathing space in which to re-examine their energy policies and look for a solution. However, critics add that there is no evidence this has been done.

As far as trends in energy needs in the European Union are concerned, there is not even agreement on how big the gap is between increases in power production and what may be reasonably available in the coming years. This is difficult to ascertain because Britain, France, Germany and other EU member states have to close some electricity factories for system maintenance and also phase out the older ones on account of their age and inefficiency. However, what is planned in terms of new facilities falls short of projections about rising demand.

In 2009 a survey of experts conducted by Mitsui Babcock, a power-station builder, found that some 75 per cent of those who participated expected blackouts in Britain by the time of the London Olympics in 2012. Electricité de France (EDF), the French power company that wants to build nuclear plants in Britain, puts the size of the power production/consumption gap at 32 gW, while E.ON, its German competitor, says it will be 26 gW – not too different.[17]

Clearly, the British Labour government has not been the only EU government to fail to take account of the steadily rising demand for power, the multitude of the forces pushing it upwards and the misconception that electric power is the most environmentally-friendly power alternative. Also, nobody seems to be asking how all this electricity can continue to be produced on an exponential scale without negative effects on the environment.

(In a way, this inability to challenge the obvious resembles the false assumption by American law-makers at the beginning of the subprime crisis and alt-As[18] huge mortgage crisis in July 2007 that Fannie Mae and Freddie Mac, the two giant government-sponsored mortgage institutions, would solve the problem of foreclosures single-handedly although everyone knew that they were themselves overleveraged. The pipedream of *Deus ex machina* ended in early September 2008 when both Fannie and Freddie teetered on the edge of bankruptcy, and the US government had to take them over using an inordinate amount of taxpayers' money.[19])

Complacency in government quarters intent on avoiding making tough decisions does not last forever. Tough decisions have to be made, and there is no lack of warnings. Britain's energy grids, for example, have shown signs of stress, and fuel supplies have experienced price spikes. In the winter of 2005–2006, a cold snap led to a spike in demand for natural gas, used both as a heating fuel and to generate electricity. Prices shot up, but despite several pipelines linking Britain to continental Europe, no extra gas was forthcoming.

- There was speculation that big factories might be instructed to stop work, and
- the National Grid, which runs the electricity network, faced the possibility of cutting supplies to homes as its gas reserves plunged to just two days' supply.[20]

---

17 'Dark Days Ahead', *The Economist*, 6 August 2009

18 Subprimes are the mortgages (wrongly given by US banks and S&Ls (building societies) to people who did not qualify for them. The term 'Alt-As' stands for mortgages given to families or individuals on their simple declaration they have an 'A' (good) credit rating, without providing any evidence for it.

19 D.N. Chorafas, *Capitalism Without Capital*, Palgrave Macmillan, Basingstoke, 2009.

20 'We're in for Nasty Weather – And No Gas', *Guardian*, 22 February 2006.

In the background was not just shortage in capacity but also (if not mainly) a changing pattern in power usage. Statistics from central Switzerland[21] provide another example. Consumption figures indicate that in nearly five decades, from 1960 to 2008, electric power usage per head per year rose:

- from 3,000 kWh in 1960,
- to 7,600 kWh in 2008.

Behind this wholesome increase of 253 per cent in less than 5 years lies the changing pattern of electricity consumption in Western countries, including new devices ranging from all sorts of electronic gadgets (TVs, mobiles, personal computers and Internet surfing included) to the introduction of electric and semi-electric cars and other replacements for fossil fuels. These increases far exceed electricity savings generated by more efficient household appliances (refrigerators, ovens, washing machines and so on).

One of the ironies is that in developed countries many people have been lulled into thinking that computers are almost synonymous with energy-saving. That's totally wrong. Computers are hugely hungry users of electricity, and 'communicate, don't commute' proved to be a cloud-cuckoo proposition as far as energy consumption is concerned. The Internet devours billions of kWhs.

Here is a recent example. Nicknamed 'cloud computing', because it is represented as a cloud in many charts, this new strategic inflection point in information technology (IT) promises to be a huge problem in terms of energy consumption and environmental damage. In 2000 IT's worldwide energy consumption stood at about 66 billion kWh per year. In 2010 it stands at an estimated 350 billion kWh, a 510 per cent increase in a decade, roughly divided between:

- servers,
- cooling,
- networking and other gear.[22]

Information technology is gulping up electricity at an unprecedented scale. Google operates a global network with over 1 million servers. Microsoft is making huge investments in giant data centres and is also looking for a site in Siberia where its servers can chill. In Iceland, Hitachi and Data Islandia have built one of the largest underground data storage facilities, while the country is marketing itself as prime location for global data centres.

Britain is one of the topmost users of information technology worldwide, and, as the aforementioned facts suggest, this poses challenges in terms of energy supplies and associated environmental pollution. Being ahead in technology is highly important in terms of global competitiveness, but there is now evidence that this is accompanied by unintended consequences.

Without doubt, matters are worse in developing countries which are trying to leapfrog the West in IT and other areas like home appliances and private motor vehicles. Experts say that in most cases in the developing world the new capacity for electrical power

---

21 From EWS, the power authority in Canton Schwyz and part of Canton Luzern.

22 D.N. Chorafas, *Cloud Computing Strategies*, Auerbach/CRC, New York, 2010.

production already in the works is probably the wrong sort. With no official energy policy, power firms go for the easiest and least-cost option, constructing more coal-fired power stations which are cheaper and quicker to build – but also the most polluting.

That drive is further promoted by the fact that carbon emissions are not penalized in developing countries, even though CO_2 is already assigned a price through the European cap-and-trade mechanism (Chapter 10) and Britain pays a great deal every year in this trade. These inconsistencies make up a system that is unstable and vulnerable to gaming on global basis, because politicians have failed to take the proverbial long, hard look.

6. Phasing Out Oil: Case Study on Sweden

As a French saying has it, 'Oil is like your girlfriend who you know will leave one day. Therefore, you should leave her first.' That might be more easily done with girlfriends than with a product with which literally everybody is intoxicated – in the north and south, in the east and west, and in rich countries and poor ones.

Two other quotes may be worth mentioning. Paraphrasing Louis XIV, the last drop of oil wins the war. As for George Clemençeau, the French prime minister at end of World War I, he got it right when he said: 'Every drop of oil secured to us saves us a drop of human blood.'[23] To get out of the shadow of Clemençeau's dictum, a country must be prepared to relinquish its dependence on oil. To date, only two EU member states have seriously tried to do it: France and Sweden. Time will tell:

- whether Sweden's has been an attainable goal, and
- if attainable, whether it is sustainable in the longer run.

For a start, it should be appreciated that Sweden currently gets about 40 per cent of its energy from hydroelectric power. On the basis of this statistic, the country has set an objective of obtaining at least half of its electric power from renewables. A plan based on these premises is not pushing into uncharted territory with unknown roadblocks and costs. So far, Sweden has been able to position itself against the world's thirst for energy without ruining its finances.

Back in June 2006 no less an authority than the International Energy Agency (IEA) said that investments totalling \$45 trillion might be needed over the next half-century to prevent energy shortages and greenhouse gas emissions from undermining global economic growth. In one brief statement, reported on *Bloomberg News*, the IEA brought into perspective the fact that the combination of growing demand for energy, the dangers of climate change and scarcity of resources was going to require huge shifts in the way the global economy was organized.

In the same news bulletin, the IEA also added that to meet those challenges, nations would have to overcome objections to building nuclear power plants, storing large amounts of carbon dioxide underground or beneath the ocean floors (see Chapter 5 on 'clean coal'), as well as take other measures which at first sight are not necessarily to their liking.

23 Jean-Pierre Wauquier and Jean-Pierre Favennec (2001), *Petroleum Refining. Volume 5: Refinery Operation and Management*, Editions Technip, Paris, p. 3.

More than most EU nations, Sweden has appreciated that it is very vulnerable to the interruption of its imported oil supply, and moreover its dependence on oil has a significant effect on its foreign and economic policies. To regain its freedom of decision and action, its government decided to adopt a credible plan to reduce its dependency on foreign oil even if this meant a reasonable increase in energy costs.

Swedish politicians can be credited for having understood that when a country depends on imported energy resources, the normal laws of supply and demand do not apply. Supply is determined by how many barrels of oil producers are willing to pump and sell at any given time, and this is affected as much by international politics as by internal economics. Therefore, through a cross-industry effort Sweden set a 2020 vision of phasing out of the oil habit. Details of the initiative were thrashed out by a committee which featured:

- car manufacturers,
- other industrialists,
- academic researchers,
- the country's farmers, and
- civil servants working on energy conservation.

According to Mona Sahlin, the minister of sustainable development, 'A Sweden free of fossil fuels would give us enormous advantages, not least by reducing the impact from fluctuations on oil prices'.[24] Even spiking oil prices, however, are the lesser evil when compared to the fact that addiction to energy on the part of every consumer, every company and a country as a whole carries a much higher risk.

The risk Sweden faces is that it has more or less fully exploited its hydroelectric power resources. Its escape into the future is that it currently generates a good deal of its electricity from nuclear plants while its fossil-fuel consumption is mainly in the transport sector, where it plans to increase the development of biofuels derived from its forests (see also Chapter 6). The government has been working with car-makers Volvo and Saab to develop cars and lorries which effectively use various types of biofuel.

Sweden's initiative is a most commendable example, but people's attitudes, too, must change. Both consumers and producers should understand that not only are fossil-fuel resources limited, but we also cannot suffer an energy crisis every decade or so. This means avoiding self-indulgent pronouncements about theoretically plentiful oil and falling oil prices which cause the search for alternative energy sources to be largely abandoned.

Some critics say that while Sweden is a good example on switching away from oil in terms of energy supplies, the government has been less successful in changing public attitudes from energy consumption towards energy conservation. The oil addiction made sure that although energy consumption today is much less popular, it is still just as necessary. Swedes still buy SUVs.

The trouble with changing public attitudes towards oil is that we are accustomed to it and consider it to be so much more attractive than other forms of energy. Next to natural gas, it is both cleaner and cheaper than coal, and it is the most widely used energy source for vehicles, with a network of service stations tuned to it, providing plenty of points

24 *European Automotive Design*, March 2006.

of supply. (Biofuels, however, can compete with that and, most importantly, they can successfully exploit oil's already existing network of service stations.)

The message to retain from this brief case study is that a change in attitudes is not impossible. Just as Sweden has done with electrical power production, the European Union needs a far-sighted and precise plan of action on energy sufficiency. This is a salient problem, not an option. Like Sweden, in order to assure continuity of energy supplies to businesses and consumers, the European Union must have in place an elaborate strategy to counter that of oil producers now and in the future.

7. Biofuels Can Assure Continuing Delivery of Energy to the European Union

During World War II General George Marshall, chief of staff of the US armed forced and military advisor to President Roosevelt, received in his office the cream of nuclear scientists, all of them solidly convinced about the unprecedented power of the atomic bomb as a weapon-to-be. Having listened to their story, Marshall explained to them that the power of the military lies in its ability to continue to deliver. The same is true of power supplies and their impact on consumers and business.

It is one of the worst-kept secrets that if there is a major oil crisis the European Union will be in the front line of shortages, and it will pay dearly for lack of a prudent, well-studied policy on *energy reserves*. Worst hit will be Germany which is by far the largest consumer of energy in the European Union. In terms of energy consumption among the European Union's 27 member states, Germany is followed by France, Britain, Italy, Spain, Poland, the Netherlands and Belgium (in that order).

Germany alone uses more energy than Austria, the Czech Republic, Denmark, Estonia, Finland, Greece, Hungary, Ireland, Lithuania, Portugal, Slovakia, Slovenia and Sweden taken together. Despite the country's insatiable appetite for energy, successive governments have done nothing to take care of the energy shortfall. Yet back in 2006 statistics compiled by the Deutsche Bundesbank showed that in the previous year:

- of total German exports valued at €786.2 billion, energy represented only 2.1 per cent, and
- by contrast, of total German imports of €625.6 billion, the share of energy was a wholesome 11.1 per cent.[25]

Germany's and the whole European Union's energy challenge brightens up when the Union's first-class agricultural industry is taken into account and, as Chapter 6 explains, *biofuels* provide a solution to the shortfall, particularly in combination with nuclear energy. Many EU member states have a first-class agricultural system whose production Brussels unwisely tries to curtail rather than fully exploit (see Chapter 8).

- Biofuels like ethanol, butanol and biodiesel are made from agricultural crops.
- Biofuels are also biodegradable.
- The global biodiesel industry is ranked as among the fastest-growing markets.

25 Deutsche Bundesbank, *Monthly Report*, March 2006.

Biofuels are the European Union's best bet for kicking the oil habit; they provide the old continent with a competitive advantage and they radically reduce its dependence on imported oil and gas. Successfully exploiting the biofuels option requires thinking out of the box, as well as focused and well-funded research projects to significantly improve the current rudimentary manufacturing processes.

The omens are good, particularly as intensive biofuel research can work in synergy with biotech and cleantech – two other major forces of the new decade. In addition, a return to the predominance of agriculture is consistent with European farmers' cultural roots and provides plenty of employment. In turn, this makes feasible an energy programme that is both rewarding and realistic.

Those who say that they doubt the rationale of using food for energy should be reminded that food *is* energy. Moreover, in the past the exact opposite procedure has been practised. In the 1960s when a barrel of oil was dirt-cheap:

- Exxon and other oil companies were making steaks out of oil, and
- ranchers were feeding their veal with food derived from oil.

An even more powerful motivation for the European Union to bet its energy future on biofuels is sound management. A question every good manager should always ask is: 'What are my alternatives?' From an EU perspective, apart from biofuels and nuclear energy – followed closely by hydroelectric, wind and solar power which alone do not fill the rapidly growing electricity needs – the alternatives are not on call. Gas is a case in point.

Even if Germany has the largest storage gas capacity (in billion cubic metres) in the European Union, followed by Italy, France and Britain (in that order), this storage capacity has to be filled not just once, but steadily. Germany's and the European Union's energy vulnerability was illustrated in January 2006 when a dispute between Moscow and Kiev disrupted gas deliveries to Ukraine and, by extension, to central and western Europe.

Political risk associated with instability in the countries through which the pipelines pass has not so far received the attention it deserves. Yet it is not only Ukraine, but also countries like Bolivia, an energy producer, that offer examples on what can happen overnight in the aftermath of an election leading to regime change

On 1 May 2006 Evo Morales, Bolivia's (then) new president, announced the renationalization of the country's hydrocarbons industry, effective immediately. This was done without compensation to companies that had invested in developing the sites, including European and Latin American firms. The measure was enforced by the deployment of Bolivian troops, primarily the engineering battalions, to:

- oil and natural gas installations, and
- fields across the country where energy resources were found.

The opening clause of the nationalization decree declared that the general welfare of Bolivians takes precedence over other interests. 'Our hydrocarbon wealth returns to the hands of the nation, and will be used for the benefit of the country,' the decree began.[26]

26 *EIR*, 12 May 2006.

Hardest hit by the nationalization was Brazil's Petrobras. Also significantly affected were energy supplies to neighbouring Argentina, Brazil and Chile.

Another example is provided by Ecuador. There Occidental Petroleum was producing around 100,000 barrels of oil per day, about a fifth of Ecuador's total output, having invested $1 billion in the country since 1999. A newly elected government accused it of having breached its contract by failing to register its sale of a stake in another field – a claim rejected by the company – and its assets were expropriated.

The irony with wild investments in unstable energy-producing countries is that they violate the principle of making big oil and gas investments: these presuppose political stability which, in many cases, does not exist. Theoretically, giant firms like BP of Britain, Royal Dutch Shell of Britain and the Netherlands, Total of France (all of which suffered at the hands of the Russian government) and ENI of Italy (which suffered in Kazakhstan) should not be throwing billions of dollars into oil and gas exploration when a 'no-matter-who new president' can builds his political capital on expropriating *national wealth*. In practice they do so because there are no alternatives. This issue raises the following questions:

- Is the European Union able to protect its companies and their energy investments?
- Can Brussels be in charge of the European Union's energy supplies?
- How can the European Union can exercise its power to help its energy companies overcome the growing financial nationalism of different countries?

It is not only the exploration and extraction of mineral resources which are at risk that must be covered under a protective umbrella, but also the security of transport facilities. Both are part of the European Union's energy strategy. The big pipelines enumerated in section 6 are wide open to terrorist attacks, and oil tankers are hijacked by pirates – as has happened time and again with the Somalis.

Concomitant to this is the fact that a shortage of finance has led to underinvestment in central and eastern Europe's aging energy infrastructure, raising the prospect of severe capacity constraints when something happens to the big pipelines. In June 2009 this question was brought up by the European Bank for Reconstruction and Development (EBRD) in connection with energy shortages when growth returns to the region, but its bearing is much wider. The concern is about:

- a whole series of projects that have been delayed because of a lack of funds, and
- projects targeting alternative paths of energy transfer, which can serve as back-ups and channels in case of crisis.

From origin of supplies to transport channels and distribution networks, an assessment of the European Union's energy needs in the coming decade gives the impression that current policies are inadequate at best and disastrous to the European economy at worst. It cannot be repeated too often that critical issues such as energy conservation are way down the list of priorities, and there are reasons to doubt whether there will ever be a fair market in the European Union for private major trans-European firms.

Another issue waiting for EU energy policy decisions, and solutions is the unbundling of energy production, transmission and distribution. Some countries, like Britain, the Netherlands and Spain have gone further than others, such as France, Germany and

Italy, in unbundling their energy companies. Yet, unbundling is a sound approach which permits specialization and greater efficiency.

In October 2008 the European Commission announced a new 'level playing field' clause which it inserted into its draft energy liberalization package to allow countries that have fully unbundled to block acquisitions by more vertically integrated companies. But between what Brussels says and what member states are doing there is often a wide gap – in time and in actions. At the time of writing, in January 2011, the 'level playing field' is nothing more than unkept promises.

CHAPTER 3 Managed Oil Markets

1. The Power of Power Cuts

In mid-August 1990 the computers, lights, lifts and air-conditioners were turned off in several New York buildings when a fire destroyed a Con Edison power substation in lower Manhattan. The inconvenience lasted for a couple of days, and commentators said that the experience helped them appreciate why President George Bush had committed US military forces to the defence of Saudi Arabia and the liberation of Kuwait.

'We are over there to keep the lights on over here' was the opinion expressed by political commentators. Others, however, said: 'We are over there to challenge the act of naked aggression by Iraq' – but the odds were that the US president would not have risked engaging almost half a million soldiers if there was not so much oil under Kuwait, and even more under the sands of Saudi Arabia.

As Chapter 1 explained, oil means power – electrical power, transport power and political power. As Harold Ickes, US secretary of the interior, wrote in the 1930s, 'We have passed from the stone age, to bronze, to iron, to the industrial age, and now to an age of oil. Without oil, American civilization as we know it could not exist.'[1] And not only American civilization.

When they conquered the Persian Empire, Alexander's armies passed through what are now the oilfields of Mesopotamia, and history books say that they looked at the crude black stuff with disdain. But that is history. The world's first oil well was bored in Titusville, Pennsylvania, in the late nineteenth century. It produced 50 barrels a day. Within a few decades – which means with extraordinary speed – this viscous, black, smelly liquid became the world's biggest business.

- Oil played havoc with the world order.
- Fortunes were won and lost because of oil.
- Individual transportation on a mass scale became possible.

Very important in terms of oil's efficiency as an energy source is the ratio of *energy return on energy invested*. This ratio has not been stable. According to analysts, oil discovered in the 1930s delivered about 100 units of energy for every unit invested. This dropped to 30-to-1 in the 1970s, while present-day oil and gas finds, like undersea reserves, are thought to offer a return in the range of 16-to-1 to 20-to-1. Oil sands might lower this ratio to a single digit.[2] Let's keep these references in mind when we discuss biofuels (Chapter 6).

1 Daniel Yergin, *The Prize: The Epic Quest for Oil, Money, and Power*, Simon and Schuster, New York, 1990.

2 *The Economist*, 24 October 2010.

During the first years of its history, oil primarily powered light sources. But it was soon discovered that the first well in Pennsylvania gave the right fuel for the motor car. It did not take long for this discovery to set the pattern for oil's future into the twentieth and twenty-first centuries.

One of the vast domains where oil has been widely used is in the military. The apparatus of mechanized warfare became established during World War I, powered by oil. When in August 1914 the British Expeditionary Force (BEF) landed in France, it possessed only 827 motorcars and 15 motorcycles. By the end of the war, oil was needed for 56,000 lorries, 23,000 cars, 34,000 motorcycles and thousands of airplanes.

Oil's impact has also been felt on the oceans. At Winston Churchill's insistence, the British fleet was converted to oil. With this, it gained extra speed and range which translated into naval superiority dented only by Germany's diesel-powered submarines.

At the outbreak of World War II both Britain and Germany, the main belligerents, realized that military supremacy would depend on manoeuverability. Consequently, their forces were powered by oil. Since then, national security has demanded that foreign policy should secure reliable oil supplies – or else.

Statistics from civilian motor transport also tell an interesting story. In 1916 there were 3.4 million cars registered in America; by 1930 there were 23 million, and in the 1950s the yearly production reached half that number. Today the world production capacity is in excess of 90 million cars per year and may soon reach 100 million with the new factories being built in China.[3] The following statistics illustrate the steep rise of oil addiction:

- Oil consumption in America more than doubled during the 1920s.
- Between 1950 and the first oil crisis of the 1970s, the world's energy consumption grew by about 350 per cent.
- Over the same timeframe in Europe it increased fifteenfold and in Japan a hundredfold.

Oil consumption fell slightly with the oil shock of the 1970s, as hefty price increases made people and companies conscious of the need to conserve energy and increase the efficiency of its utilization. This, however, was soon forgotten and in the mid- to late 1980s it zoomed up again, eventually reaching an unprecedented level in terms of demand and supply.

Experts blamed that 1980s spike to the massive increase in mass transportation – by land, air and sea – but the transport industry is not the only big user of oil. Power production is a challenger for the top position and the same is true of a myriad other processes, the manufacture of plastics being one of them. It should not be forgotten that cheap energy and plastics has transformed life over the last 60 years throughout what was formerly known as the First, Second and Third Worlds.

As long as the United States was a main oil producer and exporter, the pipeline was never shut. But as consumption approached (and later exceeded) US-based supplies, new

3 Prior to the 2007–2010 economic crisis about 60 million cars per year were produced, leaving idle about a third of installed capacity. With the crisis, production levels dropped to 45 million cars per year, with the result that car manufacturers have been working at half-capacity.

resources had to be found in other countries – particularly the Middle East – and things don't always go according to plan.

Since the end of World War II the flow of oil has frequently been threatened by conflict. The first major adversity came in 1951 when Mohammed Mossadegh nationalized the Anglo-Iranian Oil Company's (later BP's) Iranian holdings. In 1956 the Suez Canal was blocked. In 1967, as Israel attacked and overran Egypt's massing armies, the Arabs tried to embargo supplies to the United States.

- Another Arab–Israeli conflict caused the first oil shock in 1973, and
- promoted by Jimmy Carter and Giscard d'Estaing,[4] the Iranian revolution led to the second oil shock in 1979.

On each occasion the oil price spiked, and the world economy shuddered. Sudden price shifts were nothing new for the oil industry. In a free market the price of any commodity can go up or down. Nearly half a century prior to the Middle East wars, the Great Depression of the 1930s caused oil prices to fall from $1 a barrel to 6 cents, precisely at the time the enormous East Texas Oil Field was discovered by Columbus Marion ('Dad') Joiner, a wildcatter.

The surge in oil prices in the 1970s and fear of wider conflict in early 1991 caused far-reaching responses. Oil means power as long as it keeps flowing. When the flow stops, the dynamics of the equation change. That is what has happened when the Iraqis torched hundreds of Kuwaiti oil wells during the Gulf War of 1990–1991.

The Iraqis did not just occupy Kuwait; they also put everything on fire. They set ablaze the oil heads, the pipelines, anything where the oil comes out or goes through. This was a totally indiscriminate offence against the whole world and its oil supplies – yet not only were there no public protests about the destruction and accompanying pollution, but the pledges to bring those responsible to justice have also been forgotten, which led to the Iraq War, which started in 2003.[5]

2. The Insatiable Public Demand for Oil

Decades have passed since the events described in section 1, but their memory still lingers on and the problem of oil supplies remains. In mid-July 2007 the International Energy Agency (IEA) issued a pessimistic report on the supply of oil and natural gas, predicting an oil supply crunch after 2010 because of slow production among non-OPEC countries. The IEA projected that demand for oil will rise by 2.2 per cent per year,[6] with China at the kernel of such projections despite the severe 2007–2011 economic crisis.

More and more energy supplies are necessary because the global demand for energy is huge. At end of 2009 the world's population consumed about 16 terawatts of electrical power[7] – the ultimate downstream energy commodity whose worth translates to the tune

4 For ill-conceived reasons they both promoted the return of the ayatollah and handed him Iran as a gift.

5 I have deliberately chosen not to include references to the Iraq War even, if according to many opinions, oil was by far its number one motivation.

6 *The Economist*, 14 July 2007.

7 A terawatt is 1,000 gigawatts, and a gigawatt is the capacity of the largest available coal-based power station.

of about $6.5 trillion a year or a tenth of the world's economic output. What's more, energy's share of global GDP is rising.

Figure 3.1 uses 1990 to 2010 statistics and 2010 to 2030 forecasts for global energy consumption, dividing them by tranches corresponding to the six main consumers. It would not be easy to satisfy this rise in demand, and, as we will see in sections 6 and 7, as far as oil supplies are concerned, a growing share will come from *unconventional* sources.

It needs no explanation that pollution is conjoined to this huge increase in energy consumption like a Siamese twin. According to some predictions, *if* the current shaky commitment to contain pollution materializes, *then* by 2050 power consumption is likely to have risen to 30 terawatts, a 187.5 per cent increase over current levels.

The call for birth control as a cornerstone of the effort to swamp pollution can find no better justification. The projected demand for food and energy cannot be otherwise satisfied. Estimates suggest that about a third of this increase in energy requirements will be due to global population explosion and two-thirds due to higher global living standards.

- In both cases this means people, and
- the after-effects will inflict expensive consequences on both the environment and the remaining energy resources.

Optimists say that transition from an economy based on fossil fuels to one using alternative green energy is most urgent precisely for these reasons. Leaving aside the fact that alternative energy is not that 'green' (see Chapter 5) this argument totally forgets that, under current conditions, alternatives have not yet proved themselves as massive providers of energy resources, with some exceptions like:

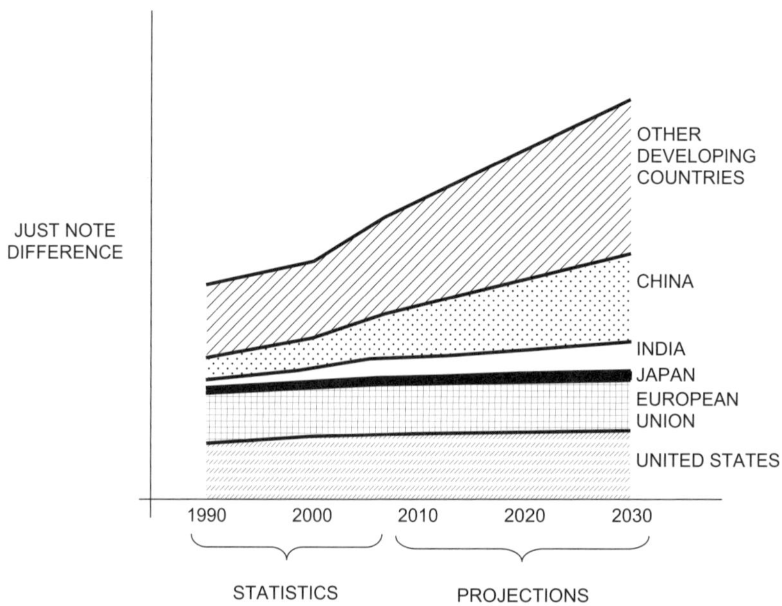

Figure 3.1 Demand for energy by developed and developing countries

- hydroelectric power and
- more recently, biofuels (see Chapter 6).

The 'alternatives' to oil, coal and gas are still far from moving into the mainstream. Therefore, realists answer that the hopes placed on alternative energies are overinflated and one-sided, failing to account for the impact of both the population explosion and higher living standards. It is not only motor vehicles owned by families which demand large energy supplies – particularly oil. Consumer appetite for washing machines, freezers and other household appliances is also rising rapidly.

- A higher consumption of durables brings a substantial increase in energy use.
- This pushes up both pollution levels and prices, the latter because of resurfacing supply constraints.

In addition, each of the three most important defining trends of today's global economy – *urbanization, industrialization* and *globalization* – are huge energy consumers. People moving from the countryside to the cities basically turn from low-intensity solar-power consumers into high-intensity thermal fuel users.

For all practical purposes, industrialization and urbanization correlate between themselves and with globalization – requiring further major boosts in electricity production. The trend towards urbanization and industrialization in China, India, Africa and South America will continue to grow over the next decades, leading to unstoppable requirements for more energy.

Each one of these trends, and the three in unison, ensures that demand for energy outpaces supply by a significant margin: global oil production is reaching its peak, and improvements in efficiency fall way short of closing the gap. Projections talk of global energy efficiency increasing by 1.8 per cent year on year, while, as we have seen, demand will increase by 2.2 per cent per year – a 20 per cent gap.

Clearly, over and above that gap comes *realpolitik* – the fact that, under the best conditions, peace in the Middle East is precarious and the eventual possession of atomic bombs by Iran will dramatically change its political landscape. In his published memoirs, Alan Greenspan writes: 'I am saddened that it is politically inconvenient to acknowledge what everyone knows: the Iraq war is largely about oil.'[8] The next wars, too.

3. Markets for Energy and the Macro-economy

One of the positive side-effects of power generation is that it provides a good snapshot of the macro-economy. This is one, but only one, of the reasons why the slack in investments should be taken seriously. According to some opinions, capital markets are jointly responsible for the current disparity between energy supply and demand. Their failure has been a global misallocation of capital. As far as the macro-economy is concerned, they have channelled too much money into real estate and too little into energy.

8 Quoted in 'Alan Greenspan Claims Iraq was Really for Oil', *The Sunday Times*, 16 September 2007.

- *If* saving rates were high and capital had been efficiently allocated to the most productive sectors in the global economy,
- *then* both developed and emerging economies would have seen a higher investment rate in the energy sector, which could have allowed a certain self-sufficiency in energy resources.

As all this suggests, there are many good reasons for an intense debate over the future of energy investments and their impact on the macro-economy – particularly that of the European Union. Investments also impact on the right-pricing of energy supplies where exchange rates, too, play an important role. Oil and gas have a dollar-denominated global market price,[9] which is not true of electricity.

In the European Union, business customers in different countries pay prices that vary by as much as 100 per cent across the so-called 'single European market'. Moreover, because energy is one of industry's most critical inputs, especially in manufacturing, rising prices are putting pressure on already tight margins – placing European companies at disadvantage as they try to compete with emerging low-wage manufacturing in Asia (particularly in China). According to the EWS statistics to which reference was made in Chapter 2:

- In the European Union the most expensive power is in Italy, at 9.5 cents per kilowatt hour (kWh).
- The least expensive is in Sweden, at just over 4 cents per kWh.
- Mid-range is Britain, with 6.7 cents per kWh, while other EU member states gravitate around that price.[10]

Quite obviously, these pricing differences impact on the macro-economy, as does oil-price volatility (as well as spikes). In addition, the European Union is a patchwork of national energy policies and regulations, while the power of EU authorities in Brussels to intervene as an equalizer is limited.

The EU Commission can fine individual companies, but it is up to national governments to decide on issues concerning the behaviour of energy companies, their ownership and prices. Furthermore, EU energy directives are unwieldy, and governments have been unwilling to work together towards equalization solutions when the price of oil goes through the roof.[11]

The need for relative stability in energy prices by operating through the fat tax margin of EU governments is not the only one escaping the European Union's watch and jurisdiction. There are other issues, too For instance, how should European society respond to the changing and increasingly alarming interaction between energy consumption and well-being? How can the ongoing energy transition be steered consciously towards a more supportive and sustainable relationship between energy, health, the economy and the environment?

9 Although there is a bifurcation between West Texas Intermediate and light Brent (more on this in section 4).

10 Statistics from Verband Schweizerischer Elektrizitätsunternehmen VSE.

11 And it can reach the stars when military expenditures are added, According to Joseph Stiglitz, of Columbia University, the cost of the Iraq War may total $5 trillion.

The fact that some of these issues are discussed in international tourist gatherings called 'UN environmental conferences', should have prodded Brussels to run faster to reach an EU consensus. Let's face it, the European Union and its citizens never had the required open debate on the connections between energy and well-being, supported by a greatly expanded research programme to clarify the evolving pattern of benefits, downsides and costs of an ever-greater use of energy. Instead:

- large uncertainties attend many issues, including those of macro-economic importance, and
- poorly documented optimistic scenarios for energy are based on relatively low population growth and progress in energy efficiency – both divorced from reality.

As far as oil supplies and prices are concerned, a great deal of the reality is political. Much of the prevailing conditions is the effect of cartels, which have deep roots, and governments have been unable or unwilling to eradicate them. Since the late nineteenth century, oil producers have worked through monopolies.

- The greatest of these was the Standard Oil Trust which was the subject of anti-trust action by the US government.[12]
- Post–World War II came the 'seven sisters', a group of Western (mainly American) oil companies (three of them formerly part of Standard Oil),
- Then, the early 1960s saw the creation of the Organization of Petroleum Exporting Countries (OPEC) which, as a cartel, violates the basic rules of the free market.

Little appreciated is the fact that, for oil producers and users alike, the managed market of monopolies and cartels provides only temporary relief and, sometimes, the wrong guidance. Even without oil's vexed politics, managed markets mask price signals, trading a false sense of stability 'now' for greater volatility 'later' and making investment in energy sources much more uncertain than they would have been otherwise (see also the discussion on the effects of speculation in section 4).

In addition, because the profitability of alternative greener energy sources is questionable, banks are not forthcoming in loaning money to renewable energy projects. For reasons that will be discussed in Chapters 4 and 5 respectively, nuclear and renewable power are not attractive investments, and the motive power behind less polluting energy sources is not enough on its own to change people's habits.

4. Financial Speculation and Oil Markets

At least in statistical terms, the economic crisis of 2007–2011 has had a salutary effect on energy consumption. There has, however, been no repetition of events in the 1970s when the after-effects of the oil shocks guided the hand of machine designers and users. Since practically everyone knows that the cost of oil became detached from the laws of

---

12 Standard Oil was broken up by trust-busters, but eventually each one of its pieces became bigger and more powerful than the parent company had ever been.

supply and demand because of speculation, very little attention has been paid to energy efficiency.

By the 1980s, a decade after the oil shocks, the pace of improvements in energy conservation had already slowed down. In the 1980s cheap energy short-circuited the drive to use energy resources more efficiently. Two decades later, in 2008 in the middle of a major economic and banking crisis, the price of oil skyrocketed, but the answer to the question 'Why?, rather than efficiency, attracted most attention.

(There are different grades of crude oil commanding different prices, but these tend to move more or less in unison, even if at any given time they vary and may even diverge. The different grades of crude oil are traded by density (light or heavy) and by their sulphur content (sweet or sour). In North America, the light and sweet West Texas Intermediate (WTI) is the benchmark; in Europe, the North Sea Brent dominates; while Asia uses the relatively heavy and sour Dubai Fateh.)

As the price of all grades of crude oil soared, opinions about the reasons for this have been divided. Financial experts said that, with very low oil inventories, the contraction of global crude oil production and bottlenecks at refineries, it was little wonder that petrol and diesel prices reached record highs or that the potential for a further surge was omnipresent. These factors had more influence than the fact that oil prices themselves experienced significant upward pressure. However, other commentators, including several economists, had a diametrically opposed idea about the reason for the price surge of oil and its derivatives. They viewed soaring oil prices as purely the work of financial masterminds – a game played at the expense of the large majority of oil and petrol users.

As will be recalled, despite a severe economic crisis which started in developed countries and spread worldwide, the price of crude oil continued to rise steeply until mid-2008 and then underwent a sharp correction because speculation alone could no longer sustain it. This provided further evidence that energy is not only subject to temporary price volatility, but is also a component in speculative financial gaming.[13]

It needs no explaining that higher prices in the international crude oil markets negatively affect the economies of oil-consuming nations – from enterprises to households. They also have a snowball effect because, by and large, the prices of other key sources of energy are pegged (or almost pegged) to crude oil prices. In addition, since oil and other basic commodities are notified in dollars, changes in exchange rates impact on their quoted prices in each currency or currency zone.

Rising energy prices, initially paid by consumers, amount to a transfer of real income from oil-consuming to oil-producing countries, since the vast majority of primary energy used domestically has to be imported. But the intermediaries and speculators also make hefty profits and so do governments because of high energy taxes they receive.

A measure of the total burden on households and enterprises is the economy's bill for imported energy, which obviously rises in proportion to the price of oil. The effects of the interplay between business sectors and households always tend to be to the latter's disadvantage, but the economy also suffers.

---

13 Now-defunct Enron is 'credited' with having been the first big derivatives player to significantly push up energy prices. The ploy took place in California, spiking electricity prices, bankrupting one of the two main Californian electric companies and finally bringing Enron itself into bankruptcy . See D.N. Chorafas, *Management Risk. The Bottleneck is at the Top of the Bottle*, Palgrave Macmillan, Basingstoke, 2004.

- The financial means of households are reduced due to higher expenditure on energy.
- This unwarranted extra taxation impacts on household demand and thereby weighs on the business sector.
- In turn, this leads to rethinking return on investment and can also bring about changes in the structure of economic forces.

Such considerations have been among the reasons for Sweden's strategy of phasing out oil by 2020 (Chapter 2), which helps in reducing the impact of oil-price fluctuation on its economy, as well as in the promotion of biofuels as a sustainable EU energy solution (see Chapter 6). They also provide evidence to those who maintain that energy at large, and oil in particular, is not a free market.

Lobbyists make sure that speculators have a hold on parliaments and governments. Speculation adds to market uncertainty by increasing opaqueness. Were it not for speculation, free markets would have been more effective in managing the price of a barrel of oil in the global economy through longer-term commitments. Indeed, until the 1970s more than 90 per cent of oil was traded, and priced, by long-term contracts. During that period spot markets were much less important than today, and they had little influence on the worth of contracted barrels. They were used mainly to:

- balance the books, and
- pass small quantities of oil between companies which relied on long-term contracts for most of their supplies.

Nevertheless, even before the second oil shock in 1979, more than half of internationally traded oil was priced according to spot rates, although everyone knew that under long-term contracts big quantities of oil frequently changed hands. After 1979 certain events suggested that violent price changes might occur rather frequently, and buyers began to favour exchange-traded contracts. With that, a *futures market* was born, built around:

- the New York Mercantile Exchange (Nymex),
- the London-based International Petroleum Exchange (IPE), and
- Singapore's International Monetary Exchange (Simex), as the Asian junior partner.

The futures market, which attracted both investors and speculators, was propelled by the leverage offered by derivative financial instruments. And it has been supported by an ever-increasing demand for oil which, by the late 1980s and early 1990s, was rising at or above 2 per cent per year.

The chemistry of trading also made its contribution. Oil producers and traders expected the price of oil to rise, and this suggested not only greater profits on the barrel, but also an opportunity for long-term gains through the (then) relatively low stock prices for oil service and drilling stocks.

It is also worth noting that during the late 1990s lower equity prices ignited a mergers and acquisitions (M&As) movement among oil companies. Before this, M&As among oil companies were only really active at the peak of the oil boom in 1981–1982 when buyers were attracted by scarcity and high prices. But in the late 1990s the opposite circumstances prevailed: oversupply and low prices caused oil companies to merge.[14]

As the futures markets for oil skyrocketed and oil barrels started to rapidly change hands (see section 5), new trading houses came up with one-off deals that hedged years into the future – in contrast to hedging on stock exchanges, which extends only up to 18 months. Experts say that today a barrel of oil may be traded up to *50 times*, and this passing from hand to hand starts even before the oil is extracted from the earth. A financial contract, however, is one thing; physical delivery of the commodity is another.

- Physical delivery requires real oil reserves, not just a trading pattern.
- Nevertheless, gambling on a pattern of reserves shapes the way in which oil is traded.

This is eventually reflected in oil markets and national policies, with the tip of the iceberg being the asked oil price. In late May 2008, after oil hit the (then) record price of $135 a barrel, the US Congress approved a bill to prevent the government from adding to its strategic stock of oil and contemplated another to enable US prosecutors to sue the governments of OPEC countries for market manipulation.

Other governments, like that of India, said that they were sure that speculation makes commodities dearer. India banned the trading of futures contracts for some of the commodity prices,[15] but ironically not the oil futures.

5. Paper Barrels

In answer to the critics who maintain that the oil market is wide open to financial speculation, the pros argue precisely the reverse. They say that oil prices are bound to keep rising since supplies are running short, the sizes of new fields are diminishing and new finds are largely *unconventional* (see section 6). They also add that investment can flood into the oil market without driving up prices because they are not buying any actual crude oil. Instead, they buy contracts for future delivery.

This argument is hollow because the lion's share of the money which goes into derivatives trading is *not* investment but speculative capital looking for a fast buck. That said, for the sake of examining both viewpoints let's follow the hypothesis that when the derivative contracts mature, they are either settled with a cash payment or sold to genuine consumers.

This is the notion of so-called *paper barrels* which says that oil products (and other commodities in the futures market) can be bought and sold at will, without consumers always becoming poorer and despite speculators becoming wealthier.[16] To appreciate the basis on which the paper barrels theory stands, we have to:

---

14 The announcement of the biggest merger in history between Exxon and Mobil highlights how strong that M&A movement was.

15 Crédit Suisse, for example, paid a hefty penalty because one of its traders in India imprudently dealt in futures contracts.

16 Governments, too, profit because of windfall tax revenue when oil and other energy products are sold at much higher prices to businesses and consumers.

- briefly review the role of derivatives, and
- examine the relationship between the paper barrel and the spot price for oil.

For a start, *derivatives* are financial instruments whose value changes in response to a change in an underlying spot price, such as an interest rate, index, price of a barrel of oil (or other commodity). Typically, a derivative instrument requires no initial investment, or calls for one that is smaller than would be needed for a classical contract with a similar response to changes in market factors.

- Generally, derivatives contracts are settled at a future date, and
- they either require (or permit) net settlements, or provide for delivery of an asset that in practice puts the buyer in a net settlement position.

The barrels of oil specified by a derivatives contract will notionally, but not physically, change hands every time that financial instrument is bought or sold. Therefore, no oil is hoarded or somehow kept off the market. In short, these oil contracts are really *a bet* on which way the price will go:

- In principle, the number of bets does not affect the amount of oil available.
- What it affects is the price of the contract through the interplay of offer and demand.

The aforementioned notional, as opposed to physical, characteristics give rise to the term *paper barrel*. The spot price of oil, however, is the *underlying* factor in the derivatives transaction. (Usually, the underlying factor may be a specified commodity price, interest rate, currency exchange rate, index of prices, or something else. It may also be a variable applied to the notional principal amount to determine the cash flows or other exchange of assets required by the derivatives contract.)

Important in drawing up a derivatives contract is the *notional principal amount*. This term is borrowed from the swaps market, where it signifies the quantity of money on which the transaction is based. This money is never actually intended to be paid or received, but it serves a purpose. In interest-rate swaps, for example, it is used as the basis for calculating periodic payments of:

- fixed interest, and
- floating interest.

Derivatives instruments may be basic or exotic.[17] The former are those that have become institutionalized – for example, options. An *option* is an agreement between a buyer and a seller that, when exercised, gives the former the right, but not the obligation, to require the option *writer* (seller) to perform certain specified obligations. For example, an option on a barrel of oil gives the buyer the right, but not the obligation, to purchase that barrel from the writer during (or at) a stated period of time, or at a specific point of time at a stipulated price.

17 D.N. Chorafas, *An Introduction to Derivative Financial Instruments*, McGraw-Hill, New York, 2008.

- *If* the buyer decides to exercise his or her option to purchase,
- *then* the seller is obliged to turn over the barrel of oil at the agreed-upon price.[18]

The price a buyer pays to a seller for an option is its *premium*, meant to compensate the latter for his willingness to grant the option and the risk assumed with it. The price at which the option can be exercised is the *strike price*. The last day on which an option can be exercised or offset is the *expiration date*.

Futures are current commitments that can be exercised, as their name implies, in the future. They are traded in exchanges and have a market, except, of course, in the case of panic. Futures take the form of contracts in which the quantity of the underlying factor and expiration date are standardized. Paper barrels are futures derivatives contracts.

Forwards resemble futures but they are not traded on exchanges; they are over-the-counter (OTC) instruments, essentially bilateral agreements that have no active market. *Swaps* typically involve two parties who enter into an agreement that for a certain period they will exchange regular payments. A standard *swap* involves period receipt of a predetermined fixed amount, and corresponding period payment of the spot value of a unit of the asset concerned.

Hence investors and speculators may hold either long or short positions, depending on their market insight. Index funds (a rather recent development) create additional demand for oil futures, conceivably pushing up oil prices, or at best increasing market volatility.

In this chapter what interests us in connection to paper barrels is futures. These can involve major risks because of the leverage that derivatives instruments make possible. Hence, they are suitable only for investors and speculators who have sufficient liquid assets and, moreover:

- are familiar with this type of instrument, and
- are able to absorb any losses that may arise if the market moves in the opposite direction to that which they forecasted.

This is true whether the price of oil (or any other commodity or index) is rising or falling. The snowball effect of geared deals comes from the fact that, whether they are rushing in or out of the door, leveraged market players create a snowball effect. Derivatives are, in principle, useful instruments, but this snowball effect poses a major risk to the economy and to the players themselves – whether individuals, banks or other entities. Remember this if and when you plan to play in paper barrels.

6. Unconventional Exploration of the Earth's Oil Resources

According to current estimates, over 60 per cent of the world's proven oil reserves that can be recovered from known oilfields using existing technology are in the Middle East. If Russia and Venezuela are added, the share rises to nearly three-quarters. Saudi Arabia alone accounts for a little less than a third of that sum, followed at a distance by Iran, Iraq and Kuwait.

18 In contrast, after an originally stipulated period of time, an option that is left unexercised expires as worthless.

Compared to the heavyweights, the European Union's oil reserves – located mainly in the North Sea and in Romania – are as light as a feather. American oil production is roughly one-fifth of that of Iran[19] whose oil output is slightly over 40 per cent of Saudi Arabia's. China's, Brazil's and Algeria's annual production is each roughly half that of the United States – which means not much.[20]

Even worse, paper barrels or no paper barrels, these resources are being rapidly depleted. The United States, Britain and China will be producing very little by the end of this decade, as their reserves will be almost exhausted by then. By contrast, Kuwait's oilfields are expected to last for more than a century and Iraq's about 150 years. These prevailing guesstimates are based:

- partly on measurements, and
- partly on hypotheses which may be revised upwards or downwards in the coming years.

Barring unexpected major finds within the reach of existing technology, annual oil production from known oil reserves is travelling south (see Figure 3.2). Nobody can tell for certain where oil will come from in the future. Old statistics give no guidance. The United States was the world's second-largest oil producer in the first half of the twentieth century. Now, although it still is a producer, it is also a big importer. Indonesia is next in line to join America in that twilight status.

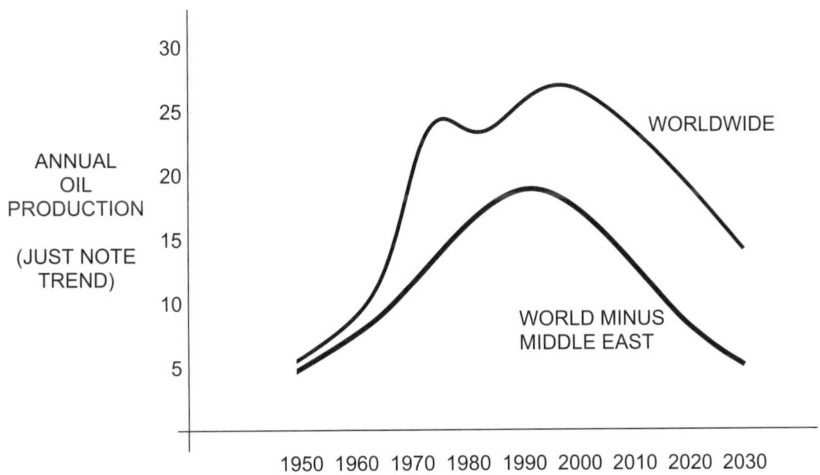

Figure 3.2 An 80-year trend in oil production from known reserves

19 And it has been falling by roughly 5 per cent a year while consumption soars.

20 The new finds in Brazil are still a long way from full-scale exploration and further still from exploitation. Besides, they are highly unconventional and the trouble BP encountered in the Gulf of Mexico raises doubts (to my mind) as to whether it will be practical to exploit them.

While at present a handful of developing countries are lucky enough to export oil, and the rest of the world is pleased to import it, every day their black-gold resources are becoming depleted. Speculation aside, this depletion of reserves is the reason why financial markets are nervously raising the price of oil. According to current estimates and assuming present-day consumption habits, the earth can only provide the following:

- 30–40 years for oil,
- some 50–60 years for gas,[21]
- up to 150 years for coal.

The 40 years for oil and 60 years for gas will pass in no time. Therefore, despite all environmental 'commitments' by heads of state, coal may come back with a vengeance (and forget about 'clean coal' – Chapter 5 explains why). These horizons are, of course, subject to the accuracy of current estimates of reserves (which are only partly known), any acceleration in energy consumption rates and the possible advent of new technologies that would make deeper exploration feasible.

Back in the 1990s people spoke of the vast Middle East oil reserves as dwarfing those anywhere else. With a torrent of oil running out of wells, no expert is saying that any more. But there is a growing interest in oil and gas reserves which, to be exploited, require very advanced, or not yet invented, technology. These are termed *unconventional* sources, with shale oil and gas hydrates, as well as very deep-sea reserves, accounting for most of the newer finds.

Furthermore, according to informed opinion, all of the 'extra' oil supply to be found anywhere in the world is likely to come from *expensive* and *environmentally-damaging* unconventional areas. These include Brazil's deep offshore, Canadian oil sands (tar sands)[22] and Venezuela's Orinoco tar belt.

Among the oil majors (in alphabetical order), Chevron, ExxonMobil, Shell and Total have been investing heavily in Canada and Venezuela (though in the latter they are facing political problems). Chinese energy companies are considering extracting heavy oil from Madagascar. Petrobras and Statoil are active in Tanzania which has sizeable reserves of natural gas, and so on.

Added to the urge to establish new sources of oil supply before the current ones are exhausted is the fact that most of OPEC's current spare capacity is of a quality for which there is only limited demand.[23] All these reasons, as well as the narrow spare-capacity cushion all along the presently-known oil supply chain and resulting high sensitivity to unanticipated changes in the supply–demand balance:

- are pushing towards more intensive exploration, and
- like it or not, unconventional resources are at the centre of that effort.

---

21 An estimate revised upwards because of shale gas which lies in vast quantities in many parts of the globe.

22 It takes up to four barrels of water to generate one barrel of tar-sands crude. Mining the sands also strips forest and creates vast ponds of toxic by-products. Moreover, 20 per cent of Canada's natural gas is used to produce tar-sands oil. *The Economist*, 7 August 2010.

23 Occasionally it is suggested that Iran and other heavy-oil producers will eventually be obliged to drop prices far enough to make processing their stuff worth the refiners' while.

How critical the situation is can be appreciated through reference to a late 2005 study by the US National Commission on Energy Policy (NCEP), a group of leading energy forecasters and politicians. NCEP simulated the likely impacts of a natural calamity and supply disruption to *just 4 per cent* of the world's oil trade.[24] The conclusion was that prices would probably jump from the assumed baseline of $60 per barrel to over $160[25] even without:

- any massive al-Qaeda attack on Saudi Arabia's oil infrastructure, or
- any other major debacle or event that could be considered oil-sensitive.

Oil-sensitive debacles considered by that study did not include the damage to the energy infrastructure in the US Gulf of Mexico following Hurricane Katrina, which had exercised pressure on the oil supply chain. (As it will be recalled, crude oil prices rose to a new all-time high in Katrina's immediate aftermath, but returned to pre-hurricane levels the following week, partly in response to the release of strategic oil reserves by the US government and the release of other reserves coordinated by the International Energy Agency.

However, the hurricane's effects were considerably more pronounced on the prices of refined oil products. They gave rise to higher netback margins (spreads between refined and crude oil, minus processing costs) in the United States, opening up arbitrage opportunities and attracting plenty of oil cargoes from other regions of the world – pushing up prices there.

In conclusion, complacency about the (undocumented) inexhaustibility of the earth's oil reserves has joined the population explosion and widespread energy wastage as a scourge of standards of living. The stakes are all the higher because conservation has taken leave of absence and the pipedream that alternative energies (Chapter 5) will fill the gap remains just that. But biofuels might make the difference (see Chapter 6).

7. Challenges Associated with Unconventional Deepwater Exploration

The move into deeper and deeper water has been the oil industry's response to the progressive depletion of known resources. To a significant extent, this has been enabled by a succession of technological innovations, many of which were pioneered as part of a focused, long-term commitment to the development of deepwater exploration and production skills.

Persistent demand for oil has underpinned this development. Another major factor which has lent weight to the push for deepwater exploration has been the scientific challenge connected to unconventional geology and its associated technical infrastructure. Explorers have been upbeat even if:

24 *The Economist*, 8 September 2005.

25 Which almost materialized in mid-2008. In May 2008 as oil prices soared, Goldman Sachs reckoned consumers were handing over $1.8 trillion a year to oil producers.

- the ability to locate and exploit deepwater unconventional reserves in significant volumes is still non-existent, and
- in many cases it takes vast quantities of scarce and valuable potable water, as well as natural gas, to turn unusable heavy oil into heavy low-quality oil.

Energy and exploration companies face massive problems in extracting unconventional oil – and deep water is not the only challenge. Other hurdles (many of which have yet to be overcome) include the remoteness of the fields and (in the case of gas) a quickly eroding ground that makes construction difficult. Most evidently There are clearly also numerous environmental implications.

Experts say that oil prices would have to rise in order to induce investment, since financing is far from forthcoming. Nevertheless, there is plenty of evidence that oil and gas majors like ExxonMobil and Chevron (in America), Royal Dutch Shell, Total, ENI and BP (in the European Union) are moving ahead with challenging projects. Reportedly, they predict that more than 30 per cent of their production will come from unconventional sources as soon as 2020.

A whole new terminology has arisen with offshore deepwater projects which are stretching the frontiers of know-how in oil exploration. Terms which in the future might become commonplace include:

- subsea (SS) exploration,
- early production systems (EPSs),
- floating production, storage, and offloading vessels (FPSOs),
- tension leg platforms (TLPs), and
- deep draft caisson vessels (DDCVs)

Exxon Mobil says that the new technology is paying dividends. In West Africa the EPSs have added value through their positive contribution to development costs.[26] The FPSOs contributed to the Kizomba A and B development projects, the company's largest in the world. Successive EPS and FPSO implementations benefited from the concept of 'design once, build multiple', which allowed the company's broad deepwater portfolio to be leveraged in order to:

- achieve design and construction efficiencies, and
- improve project cost-effectiveness in a domain of frontier exploration

How deepwater technology progressed over a 60-year timeframe, from 1947 to 2006, is dramatized in Table 3.1. From 16 metres to 1,500, 2,000 and 3,000 metres,[27] technology's big leaps have covered two orders of magnitude in deepwater exploration. This is by no means a minor feat, and the engineers who achieved it should be congratulated. The problem is that risk management did not follow up in terms of sophistication. In deepwater technology, the challenges are threefold:

---

26 ExxonMobil participated in 54 deepwater discoveries in West Africa estimated to contain more than 16 billion gross oil-equivalent barrels of recoverable resources.

27 Pedido oil well, in 2010.

1. Operating costs are high, given the difficulty of working under kilometres of water to develop reserves trapped below several kilometres of rock.
2. The complexities of these potential reservoirs require complex modelling and steady detailed measurements to assure that the right well is drilled and in the proper way.
3. The risk of a catastrophe from a profound lack of expertise, as unconventional technology is always a 'first', is unprecedented – a fact which neither companies nor governments are keen to admit.

Table 3.1 The increase in the notion of deep water over six decades

Year	Location	Depth in meters
1947	Grand Isle, Gulf of Mexico	15
1962	West Delta, Gulf of Mexico	50
1976	Hondo, California	260
1983	Lena, Gulf of Mexico	300
1989	Harmony, California	365
1993	Zinc, SS, Gulf of Mexico	450
2000	Hoover/Diana, DDCV, Gulf of Mexico	1,460
2000	Marshall/Madison, SS, Gulf of Mexico	1,500
2003	Xikomba, EPS/SS, Angola	1,350
2004	Kizomba A & B, FPSO/TLPs, Angola	1,100–1,250
2006	Erha, FPSO/SS, Nigeria	1,100

Source: Statistics from ExxonMobil's *Annual Report.*

There is no better example of the risks associated with these challenges than the deep trouble BP got into in the Gulf of Mexico in 2010, and the fact that the oil company agreed to demands that it should set aside $30 billion for an independent compensation fund and cancel its shareholder dividends for 2010.

As successive attempts to stem the torrent of oil gushing out from the damaged deepwater well failed, estimates of the volume of escaping oil continued to rise, reaching between 35,000 and 60,000 barrels a day by mid-June 2010.[28] Finally, in late July BP got a grip on the problem, but the financial and public relations damage was done.

Critics said that BP already had a miserable safety record in America. As an article in *The Economist* put it:

> *In 2005 an explosion at one of its refineries in Texas killed 15 people. In 2006 corrosion in its pipelines led to a sizeable spill on Alaska's North Slope. Since then, regulators have often fined it for breaking safety standards. There are indications that BP's approach to the drilling of the Macondo well was similarly slapdash.*[29]

28 'BP Well Gushing as Much as 60,000 Barrels a Day', *Business Week*, 16 June 2010.

29 For instance, engineering measures that might have prevented the calamity were not carried out, and tests of safety equipment delayed. Quote from 'Obama v. BP', *The Economist*, 17 June 2010.

But this is only part of the story.

The real issue is that such risks are unavoidable when the mass society wants its thirst for oil satisfied with no 'ifs or buts'. What President Obama should have said to the American and international public is that the demand for more and more oil has a huge cost on the environment. Instead, he said is that new regulation will ensure that such a disaster will not happen again. This makes no sense. The basic problem is not regulatory, but political and social.

- Do we want pristine shorelines and a little conventional oil?
- Or do we want torrents of oil, dirty beaches and all the other consequences associated with such massive flows of oil?

Responding to the consequences of increasingly deep deepwater oil extraction is not the only challenge. Typically, though not always, unconventional oil and gas reserves are a long distance from the nearest port, pipeline or airstrip, and, by all evidence, extracting and refining unconventional oil and natural gas uses large amounts of energy. But, as already made clear, most of the earth's easily accessible oilfields are in decline.

It's a fair bet that the search for unconventionals will intensify, particularly as several experts estimate that the world still holds vast amounts of difficult-to-extract oil and gas. Indeed, according to those in the know, the transition to unconventional hydrocarbons is approaching faster than many had expected, and the shift will be by no means a smooth one.

Also known as *frontier resources*, unconventionals require a whole stream of technological breakthroughs for their exploitation. Developing such resources calls for simultaneous advances in exploration, extraction, transport and refining, as well as (most probably) changes in the ways in which the output of unconventional oil and gas will be used. By all likelihood it will involve politically tough decisions – not only technical and financial ones. The stakes are high since experts say that the exploration and exploitation of:

- heavy oil might represent over 100 billion barrels,
- extra-heavy oil, another 450 to 500 billion barrels (hypothetically), and
- shale oil up to 2,800 billion barrels (tentatively).

Frontier energy resources attracting the developers' attention include (estimated) large amounts of coal-based methane, tight gas sands and shales reserves, hydrates and more. But, as with deepwater exploration, their exploitation is by no means assured, nor are current guesstimates a dependable guide to what the earth might still hide in the form of energy reserves.

Due attention should also be paid to unexpected consequences (the theme of section 8). High technology aside, the economics of exploration of unconventional energy sources is itself a problem, at a time when the marginal cost of oil production is ballooning. Oil sands, offshore or Arctic projects will all probably require oil prices above $90–100 a barrel to maintain a certain level of profitability – and even this may be an underestimate.

For example, to get a 10 per cent internal rate of return, some greenfield Canadian oil sands projects require $100 a barrel,[30] suggesting that lower oil prices may well be crowding out new investments. (On reflection, the combination of higher exploration costs, dollar weakness, inflation and the higher cost of money for risky deals could push prices structurally above $150 a barrel during the next years *if* economic activity starts to recover.)

In conclusion, finding and developing new oil and gas fields is a necessary but highly challenging, expensive, time-consuming and by no means 'sure' business. The engineers, survey ships and seismic rigs that oil firms need to find and produce new deposits are not available at the required level of sophistication, and, when they are, they will cost big money.

If the past is used as a way of foretelling the future, the costs of finding oil have doubled over the last few years because everybody is looking for new oil resources and skills, as well as machines for doing the work. In addition, the giant new discoveries, like the one in deep waters off Brazil, are unlikely to produce oil for a decade or more.

8. The Case of Financial Consequences

There is no reason to believe that governments, oil companies, experts, financial wizards and environmentalists are fully considering the unexpected consequences of massive exploration of frontier resources. Not only the physical variables but also the operational characteristics are significantly changing. Heavy investments can pay dividends if the quantities to be extracted are significantly increased. We no longer talk of relatively small conventional volumes that are easy to develop and exploit, but of large volumes:

- with (quite likely) new characteristics, and
- operational challenges that are difficult to monitor, measure and manage.

In the early years of exploration of unconventional resources, accidents, even major ones, can in no way be excluded. In addition, new product characteristics like extra-heavy oil and methane will require, respectively, the redesign of refineries[31] and of engines using the new energy resources. It goes without saying that new designs cannot be as stable as the old ones because of a mare's nest of as yet unknown problems, costs and (most importantly) environmental effects.

With new fuels coming out of frontier exploration, at least some of the consequences will be unexpected. However, once colossal investment has been made in frontier energy, and national governments have chipped in, it is difficult to see how the process can go into reverse gear – no matter what the Greens and their friends may be saying. In other words, in addition to purely technical and environmental issues, there are going to be unexpected financial consequences, and these are likely to involve not only private capital, but also large sums of taxpayers' money, making it difficult for governments

30 Merrill Lynch, *Global Energy Weekly*, 1 October 2008.

31 In the Unites States and the European Union, refineries have been designed for West Texas Intermediate and light Brent, not heavy oil. That's why Westerners are trying to persuade the Chinese to build heavy oil refineries – which, at their end, may be counterproductive.

to back out. On top of that come financial speculation and the power of the lobbyists working for the speculators.

Section 5 gave an example with paper barrels. There are also paper ships[32] and other commodity futures, which have caused formerly real assets to become financial instruments opening up new opportunities for profits (and losses) alongside stocks and bonds. In principle:

- the more leveraged and more complex is the art of finance,
- the less reversible are the transactions done in its name.

Complexity aside, the volume of these transactions is also a barrier to changing into reverse gear, as the 2007–2010 financial crisis has shown. By 2009 the number of transactions involving oil futures on the New York Mercantile Exchange, the biggest market for oil, more than tripled in comparison to five years earlier. Not unexpectedly, the price of oil (up to mid-2008) followed a fairly similar path. Speculative oil trading and geared deals have yielded examples of unexpected financial consequences. With the popularization of paper barrels:

- no oil market player needs to hoard oil;
- instead, he trades points and quarter-points, and
- buys a contract if the oil price is rising and sells it three minutes later to someone convinced it will go higher.

Along with buying and selling their contracts, traders spread rumours, suggesting increased shortages of the commodity, lower than expected inventories, increased guerrilla activity in Nigeria or elsewhere and other events, real or imaginary. That's not too hard to do, and when the price moves upwards the trader unloads while lobbyists take care to freeze government action.[33]

Although what has been pointed out in the preceding paragraphs is a direct consequence of the globalized financial market and not of unconventional oil exploration, it has created a perpetual-motion machine which directly affects all frontier projects. Much has to do with the fact that derivatives permit market players:

- to separate the ownership of commodity price risk,
- from the ownership of the physical commodity.

The silver lining of this process is that the huge risks associated with frontier exploration can themselves be traded. Dealing in futures (section 5) enables not only quick gains to be made (if one guesses right), but also the transfer of risk from the producers or holders of a commodity to agents willing to accept different types of exposure. Typically:

- insurers hedge against natural and man-made catastrophes,
- commercial players hedge against wider price fluctuations, and

32 D.N. Chorafas, *An Introduction to Derivative Financial Instruments*, McGraw-Hill, New York, 2008.

33 In Brussels alone the European Union features 15,000 lobbyists making a living by influencing government decisions.

- non-commercial players, who have no interest in any physical delivery of oil, exploit their (supposedly) better information by anticipating market movements in prices.

The pros say that while it is possible that speculative activity leads to great oil-price fluctuations in the short term, this also facilitates price determination, which leads to a more efficient functioning of the market. In addition, whether real or imaginary, the transfer of exposure stimulates people's appetite for risk, making possible the financing of projects which, otherwise, would have been starved for money.

Critics point out the damage done to the market by high leveraging, helped by the fact that the margin required to buy, for instance, an oil futures contract is small compared to the commodity's price. They also contend that the fact that derivatives greatly facilitate position-switching leads to asset bubbles. To appreciate this argument, the reader should know that:

- by and large derivatives transactions are not between producers and consumers, to which apply classical economic rules, and
- not only are such trades unregulated, but also the role of the players is not that of intermediaries but of making a good profit by getting the market on the move.

At the same time, lack of transparency concerning oil-market fundamentals – more specifically, regarding the prospects for supply, demand and level of oil inventories – generates significant trading uncertainty. This can be easily manipulated in the case of frontier explorations by means of rumours like those detailed earlier.

Because of their longer-term horizon, frontier projects will very likely exercise a countertrend to rapid buying and selling, turning attention to trends spanning a given timeframe. (This change was first signalled a few years ago by record-high oil prices for contracts extending to 2016 – therefore, not for immediate delivery – as institutional investors and end users started to look further ahead in their investment plans.)

One question is whether the futures market for oil and other commodities, originally intended for the short term, could reflect processes and events associated with unconventional energy exploration and other longer-term projects. Is the futures and forwards oil market able to assist traders, who have been concentrating on short-term transactions, in analysing more distant supply and demand balances? More precisely:

- Is the back-end of the demand curve a proxy for the long-term sustainable price?
- Do market players understand the longer-term fundamentals by extending in time their short-term bets?

Far from being academic, these questions are practical and important, particularly because the longer-term perspective in trading has developed following the arrival (around 2000) of institutional investors, like endowment policy managers, pension fund managers and insurers. These parties have a longer investment horizon and also want to diversify their portfolios. For them, commodities:

- can offer a hedge against inflation, and
- tend to perform relatively well when asset classes like equities are weak.

Because the aforementioned investors do not seek immediate returns for all their money and instead use financial tools that lock them into contracts for later delivery, temporary changes to the supply–demand picture have a limited impact. (Their thinking is that the short term is volatile, while longer-dated contracts feature returns based on more fundamental supply and demand imbalances.)

It will not be long before the market sees futures and options in exploration and exploitation of frontier resources with underlying contractual rights and an index based on milestone progress, realized costs and the spot price of the targeted energy resource. Modern markets are much more complex, esoteric and speculative than ever before. Oil no longer means 'oil' in the classical sense of the term.

CHAPTER 4 *Nuclear Energy*

1. Past, Present and Future

In 1942 Enrico Fermi, the physicist, led a group of scientists who built the first nuclear reactor as part of the Manhattan Project. Although this first attempt was simple in design, many of its features are still part of almost every nuclear power plant today, which speaks volumes about the lost seven decades in nuclear engineering.

The core of Fermi reactor's core consisted of pellets of uranium fuel inside bricks made of graphite, serving as a moderator. Then, as now, the moderator's mission was to reduce the speed of the neutrons, maximizing their ability to cause further fission. Presently, most reactors use water as the moderator.

Another feature of Fermi's reactor was control rods made of a material that absorbed neutrons. If necessary, these could be inserted into the core to slow or shut down the chain reaction. This, too, has become one of the classical processes characterizing reactor design. With these facts in mind:

- superficially, it looks *as if* a 70-year-old technology is still in the mainstream,
- pragmatically, great strides have been made in reactor operations, including safety measures.

Accidents do happen, however; and they happen with all systems, natural or man-made. The safety of nuclear energy installations got a bad name because of a near-disaster at Three Mile Island in America in 1979, a real disaster in Chernobyl in Ukraine in 1986 and several other smaller accidents or incidents. Added to this is the problem of disposal or recycling of nuclear waste (see section 8).

While these worries are present and real, the greater risk from a proliferation of power production by nuclear plants around the globe is a proliferation of opportunities for building nuclear weapons. Iran provides an example. It is therefore ironic that various campaigns against nuclear energy have focused on the accidents issue and not on the weapons, despite evidence concerning the diversion of nuclear fuel to make bombs.

Another real (as distinct from political) problem is that many nuclear plants built back in the 1960s and 1970s are nearing the end of their projected lifespans. This was the case with the Soviet-era plant in Lithuania which was shut down in December 2009 under pressure from the European Union. Still another problem, in my view, is that while Fermi's reactor was made by a genius, those who designed the thermal transmission process lacked vision. They copied too much of the workings of coal-fired electrical power-production plants.[1]

On top of the old technology issues,[2] and a reason for them, has been adverse public reaction which has discouraged many countries from either going ahead with nuclear plants or increasing their original investment in them. Nevertheless, according to fairly recent statistics, an estimated 440 nuclear reactors in 31 countries supply part of the world's electricity, guesstimated by various sources as lying in a wide range between 9 per cent and 13 per cent. This is projected to increase to over 20 per cent by 2030.

The United States has a little more than 110 nuclear plants which produce nearly 20 per cent of its power needs. According to the World Nuclear Association (WNA), as of January 2011 there is just one reactor under construction: the Tennessee Valley Authority's Watts Bar unit 2. The project began in 1973 and may be completed in 2012. That compares poorly with eight reactors being built in South Korea[3], four in India[4] and 23 in China.[5]

Experts suggest that many US plans for new nuclear power plants have been slipping in their timetables. Critics say that the Obama administration has failed to take a strong position on nuclear power production.[6]

In the European Union, France has 59 nuclear plants, contributing 80 per cent of France's electrical energy production; in Britain 23 nuclear plants master a 20 per cent share; and in Germany 17 plants produce about 30 per cent of the country's electricity needs. Eastern Europe still uses Soviet-era reactors, which are unsafe and should be decommissioned.

An interesting case is Australia. It has 38 per cent of the world's low-cost uranium reserves, but has never made its own nuclear power. In 'down under', coal fuels 80 per cent of the country's electricity, with all the implications in terms of CO_2 pollution, and gas provides the rest. Australia's priorities in energy production may, however, be changing.

In other countries the number of nuclear plants lies in single figures. Austria and Switzerland have given up their nuclear programmes – stopped from further developing their nuclear energy by the Greens who were instrumental in turning public opinion against the atomic energy option. Other countries have been fence-sitting to see which way the chips fall (but Finland joined France in developing and testing a new-generation nuclear reactor – more on this later).

Today it is China and India that are building, or planning to build, the most nuclear plants with the intention of increasing their nuclear energy capacity several times over in the next decade. Iran is widely thought to be concealing a nuclear weapons programme

1 In 1953 when I went to UCLA as a young electrical/mechanical engineer and Fulbright scholar, Professor Conally offered me an opportunity to work with him in nuclear power production. But, after examining the Palaeolithic way in which heat transmission was handled, I declined and chose computers instead.

2 Even the project to build and run an International Thermonuclear Experimental Reactor (ITER) is 25 years old. It was proposed in 1985 by Mikhail Gorbachev who put the idea of working together on fusion to Ronald Reagan.

3 South Korea currently has 25 nuclear plants, contributing roughly half its electrical energy needs.

4 India presently features 16 reactors producing an estimated 3–5 per cent of its electricity needs – peanuts.

5 Statistics from the World Nuclear Association at: http://www.world-nuclear.org/info/inf17.html.

6 However, in February 2010, a company building two nuclear-power reactors in Georgia was granted $8.3 billion in federal loan guarantees, the first to be awarded under the American government's clean fuel initiative. 'U.S. Supports New Nuclear Reactors in Georgia', *The New York Times*, 16 February 2010.

under the guise of nuclear power generation. Turkey and Vietnam are considering starting nuclear power programmes. Argentina, South Africa and other countries plan to expand their existing nuclear power base.

All in all, however, these are small potatoes compared to the world's rapidly rising electricity needs and the stated objective to cut CO_2 emissions and other pollutants. This may be changing as the world is moving towards universal curbs on greenhouse gases, but not fast enough to kill two birds – more power and less CO_2 – with one well-placed stone.

The companies that will benefit from this charge are the current nuclear reactor builders who include (in alphabetical order) Areva of France, the GE-Hitachi nuclear joint venture, KEPCO of South Korea, Siemens of Germany and Westinghouse (owned by Toshiba of Japan). South Korea's KEPCO, beat French, America and Japanese rivals to win a $40 billion contract to build and run four nuclear reactors in the United Arab Emirates.[7] All these companies have been working to benefit from participation in the widely proclaimed *nuclear renaissance.*

2. Who Will Pay for the Nuclear Renaissance?

In an article published a few years ago in the *EIR*, James Muckerheide envisioned nuclear energy growing from supplying the aforementioned 9–13 per cent of the world energy to 33 per cent of global energy demand in 2050.[8] That's a tall order. At the core of his argument have been some interesting questions:

- Can fossil fuels continue to provide the amount of energy it takes to cover demand in 2050?
- Is it likely that hydroelectric, wind, solar and other alternatives can provide electricity to a level practically equivalent of 100 per cent of today's total energy use?[9]

The answer to both questions is evidently 'No!'. One has to keep in mind, however, that the barriers to new nuclear construction are not technical even though technology has not been moving as fast as it should. There are two barriers: the first is political and the second is economic.

Responding to this wholesome demand for energy is not going to be a kid's game. In addition, figures advanced by current projections are sure to surge over the coming four decades. By 2050 meeting such a demand would require a level of nuclear power production roughly equal to 18 times its current level. According to Muckerheide's calculations this is equivalent to 6,000 nuclear power plants.

- Theoretically, this sounds like a way towards the solving the global energy problem.
- In practice, it involves a mare's nest of problems from economic feasibility to security and systems management.

7 Which will form part of the first civilian nuclear energy project in the Arab world. The UAE decision became known in December 2009.

8 James Muckerheide, 'How to Build 6,000 Nuclear Plants by 2050', *EIR Science & Technology*, 24 June 2005.

9 See also Chapter 5.

The complex issues involved in such a massive deployment of nuclear power plants around the globe are the best evidence that no single solution can answer the world's energy problems. The best approach is a range of energy sources, and the challenge in establishing an optimal mix is that it must be simultaneously:

- realistic,
- affordable,
- sustainable, and
- manageable.

Political issues come under the heading of 'realistic', and they must be solved first because otherwise they will obstruct the other criteria. Assuming that this is done, the next challenge is economic. Somebody must come up with the money – with a huge amount of capital, in fact.

A modern nuclear plant currently costs between \$5–\$7.5 billion to construct – say, \$6 billion on average. Even assuming that prices do not increase over the 20 years or so it will take to build these 6,000 reactors (which, of course, leads to underestimating the necessary capital expenditure), it will cost \$36 trillion at the minimum. Finding the money to finance such a massive project cannot therefore be taken as foregone conclusion.

Wishful thinking is an enemy of sound governance. Moreover, any plan we adopt must be affordable and sustainable. This automatically means that the building of the aforementioned nuclear power plants will not be the only cost. Training – more precisely lifelong learning – will require a major budget – much of it upfront. There is a dearth of trained nuclear and systems engineers. Let's never forget that:

- some 95 per cent of all accidents are man-made, and
- with nuclear plants there should be zero tolerance for accidents – a wholesome task.

Another imperative requirement is first-class operational management, including secure administration of nuclear fuel and of waste (see section 8). Nuclear waste cannot be dumped into landfill sites, as per usual practice. Not only is it radioactive with a long half-life, but it can also fall into the possession of hotheads and other terrorists, and from there find its way into 'dirty' atomic bombs.

As can be seen, the notion of a sustainable global network of nuclear stations poses a long list of requirements – all the way from conception to designing, building, operating and, finally, decommissioning. Each of these activities demands first-class management, and system managers with vision, skills and courage are rare birds.

System managers are required not only for operating and controlling nuclear power plants, but also for decommissioning – which is a real challenge. Crucial decisions start with the act of going ahead with it, because this will deprive the electricity grid of lots of input. Decommissioning is also a costly, risky and long-term enterprise.

After long years of EU pressure on the Lithuanian government to decommission an old and not-so-secure nuclear plant of Soviet-era vintage, the Lithuanians decided to close it down. This was by no means an easy decision because this plant supplied about 80 per cent of the country's electric power. Note that:

- the dismantling and disposal of that nuclear plant is planned to take 25 years(!), and
- it is scheduled to cost €1 billion ($1.5 billion) – and probably double that amount by the time it is completed.[10]

Therefore, to the average cost of $6 billion per nuclear power plant should be added another $1.5 billion for decommissioning, and the best way to budget for it is upfront. Otherwise, the excuse will always be found that there is no money available for nuclear decommissioning. This will raise the budget for the 6,000 nuclear energy plants to a cool $44 trillion (a conservative estimate).

It might sound as if James Muckerheide has been way overoptimistic in his projection, but it is not necessarily so. His proposal is not based on a love of nuclear plants, but on his estimate of forthcoming global energy requirements at current rates of increase – something the various UN environmental conferences and other government 'initiatives' have failed to do. Quite rightly, Muckerheide's projections don't just account for electricity use in the areas that are the most popular today, but also for other upcoming uses: for example, to produce fresh water by desalination of seawater; electric cars and transportation in general; process heat for industry; and more.

- *If* the reader thinks that 6,000 nuclear plants are 5,500 too many,
- *then* he or she should notice that, according to the aforementioned study, even that number does not displace coal, oil and gas usage.

By 2050 about 200 per cent of *current* energy use would still have to come from fossil fuels and alternative sources, according to Muckerheide. *If* oil and gas are dwindling, *then* this would require an even greater commitment to nuclear energy – particularly if nuclear energy is needed to extract oil from tar sands, oil shales and other unconventional sources of supply (Chapter 3).

These facts are not yet seeping down through the decision processes among those who govern. But several countries are experiencing an awakening, and this is one of the basic reasons why, despite opposition from the Greens and other eggheads, nuclear power production is no longer seen as an outcast. Reactors present many challenges, but have the following advantages:

- They emit almost none of the greenhouse gases responsible for global warming (if CO_2 is the source).
- They are fuelled by uranium, which is relatively abundant and available from many sources.
- Although the capital investment is high, once the plant is operational, the cost of production is relatively low, which will allow subsidization of expensive sources of alternative energy for some time to come.

Nevertheless, as we have seen, capital investments and other constraints may prove a greater deterrent than the eggheads. For their part, vendors of new nuclear plants

10 Marielle Vitureau, 'Soviet-era Lithuanian Nuclear Plant Shut Down under EU Deal', AFP, 31 December 2009 at: http://www.google.com/hostednews/afp/article/ALeqM5gSm1FWVzrPcijnsHTUSKzxzorUcQ.

technology such as (in alphabetical order) Areva, General Electric (GE), Hitachi, KEPCO, Siemens and Toshiba-Westinghouse, will have to put up large capital budgets to retool their know-how and incorporate demands from utilities and operators with decades of experience in power services. None of this is easy to accomplish, but it is doable, albeit at much smaller scale than that proposed by Muckerheide.

3. Nuclear Energy: Love it or Hate it

The biomass track (discussed later in Chapter 6) is a commendable enterprise, turning plants, grains and waste into power. But it is also wise not to expect one source of energy to solve all problems. As for renewable alternative energy sources, Chapter 5 makes the point that they should be definitely exploited while always keeping in mind that:

- their cost is way above the current cost of electrical power, and
- their energy input will be insufficient to cover a big chunk of rising power needs worldwide.

Like it or not, this leaves atomic energy not only in the list of serious answers to the challenge of increasing electricity production, but also as the best medium-term alternative to fill the breach between demand and supply – *if* coal and oil, the environmentalists' nemesis, are downplayed. Nuclear power produces no carbon dioxide, and even anti-nuclear groups admit that, as an option, it should be on the table. But habits, too, including engineering design habits, must change by adopting:

- a measured and cautious approach to development and usage of reactors,
- intensive training to fill the current huge gaps in nuclear and system engineering skills,
- extensive precautions to limit the consequences of control mechanisms failure,
- accurate projection of types and risks of accidents (which cannot be eliminated altogether), through reliability studies.[11]

Rigorous maintenance standards must include safety protocols, steady reliability monitoring, advanced instruments, orderly preventive procedures and, obviously, stress testing.[12] Deteriorating equipment, untrained workers, and inept management undermine safety at nuclear power plants (and everywhere else) to the point that the likelihood of a disaster can grow exponentially. This, however, is not the fault of nuclear energy; it is the fault of people.

The point should not be missed that risks exist with anything we do, and each of the alternative processes of electricity production has its negatives and its critics. But it is not only Chernobyl and lesser accidents that keep nuclear energy unpopular in the public mind.. There is also the psychological warfare conducted by the Greens who, at the same time, see no problem in using highly polluting cars and jet travel.

---

11 D.N. Chorafas, *Statistical Processes and Reliability Engineering*, D. Van Nostrand Co., Princeton, NJ, 1960.

12 D.N. Chorafas, *Stress Testing for Risk Control Under Basel II*, Butterworth-Heinemann, London and Boston, MA, 2007.

In the meantime, the slack in energy is taken up by coal-burning, with its consequences of acid rain, pollution and CO_2 emissions, and by importing energy generated by nuclear power from other countries – *as if* nuclear clouds released after an accident recognize state borders! Downplaying nuclear power for more than three decades has seen to it that, with a new generation still in development, the technology being used is not exciting – or so supporters of fusion research say.[13] So far, however, the results are not positive, as explained in the Appendix to this chapter.

In the 1950s when the American, British and Soviet governments set up laboratories to develop peaceful uses of fusion, many physicists believed that fusion power stations would be delivering cheap electricity in the 1990s. As it turned out, scientists achieved controlled fusion for the first time in November 1991, in a doughnut-shaped experimental reactor, the Joint European Torus (JET) at Culham near Oxford.

The JET experiment produced great newspaper headlines and it was indeed a major technical breakthrough, but the output was only 10 per cent of the power put into the reaction. It was said at the time that several more decades of development would be needed, costing tens of billions of dollars. This negatively affected research funding and therefore its pace.

The scientific case for fusion power seems irresistible, but even the pros do not envisage a full-scale demonstration fusion plant operating before 2030, with wider commercial application in the 2040s. As a result, we will have to make do with improved versions of current nuclear power plants. The silver lining is that:

- we know them better, and
- can significantly improve their reliability as well as their cost-effectiveness.

Note that in the late 1990s the Jospin government in France had Green ministers, but did not soft-pedal on nuclear power. Like most Scandinavian countries, Sweden and Finland have a powerful Green movement opposed to nuclear power, but both have gone ahead with nuclear power stations because their governments:

- persuaded the electorate that there was a serious energy problem,
- announced that parliament would take a decision on this matter, and
- promised to respect the people's representatives' decision, whatever it was.

Having provided assurance of a meaningful debate, the Finnish government put forward a nuclear power proposal and proceeded to win the argument. This made it possible to resolve part of the country's energy bottleneck. Fundamentally, the Finnish government's argument was:

- What are our alternatives?
- Do we have something better to offer?

There is no doubt that nuclear plants involve risk throughout their useful life, and even after they have been decommissioned (section 2), over and above the problem of

13 The process that fires the sun and the hydrogen bomb.

nuclear waste disposal. Studies which followed the painful experience of Chernobyl taught lessons on how risk can be managed. These can be summarized as follows:

- *The high impact area is 50–75 km around the plant.*
- Theoretically speaking, this should be a no-man's land. In practice, such a wide radius is not feasible, so only an inner (still to be defined) circle should be preserved for security reasons. However, people living in the outer circle must be informed of the risk they assume by being there.
- *Operational, emergency and risk control teams should be thoroughly trained* (see sections 7 and 8).
- Six Sigma's overriding concept of smaller standard deviation makes it an excellent quality-control methodology to apply with nuclear plants.[14] There should be a policy of lifelong learning for engineers, technicians, doctors, nurses and all personnel involved in nuclear power plants, as well as in support work on nationwide basis – with reserves ready for an emergency.
- *An international inspection force should have the right to visit and control any nuclear power plant, anywhere.*
- This must be done both regularly and by selecting inspection sites using a table of random numbers, leading to unexpected inspection visits. Moreover, the plants themselves, and their national inspection teams, should exercise worst-case drills. The responsibility for international inspection could be assigned to the International Atomic Energy Agency (IAEA) or some other transborder institution under the UN.
- *Quality histories should be maintained by all nuclear energy plants, no matter where they are located.*
- Today, we don't build simple nuclear plants like Chernobyl any more. Our plants are sophisticated with thousands of crucial components whose behaviour must be recorded on a database, as well as subjected to comparisons, tests, analysis and simulation. Statistical quality-control charts and qualitative quality histories must be maintained by the plants themselves and controlled by inspection teams.
- *An exit strategy should be established, obeying international norms; the latter must cover every nuclear plant and the system as a whole from cradle to grave.*
- Like any other natural or man-made system, nuclear plants have a life cycle. Today, this is believed to be about 30 years. The exit strategy must foresee, and account for, all issues connected to the dismantling and disposal of the nuclear plant – according to international standards (to be developed). This would add considerably to the dependability of life-cycle management.
- *Last but not least, sound management will require a far superior reliability than currently available.*
- While nuclear power plants are currently little better than 99 per cent reliable, the reliability threshold for an energy network of nuclear power plants must be better than 99.99 per cent – which means a two orders of magnitude improvement.

---

14 D.N. Chorafas, *Integrating ERP, CRM, Supply Chain Management and Smart Materials*, Auerbach, New York, 2001.

4. Eighty Per Cent of Power Delivered from Nuclear Plants: Case Study on France

Since the end of World War II the world's economy laboured to provide the energy necessary to meet growing requirements, first in the developed world and then in developing countries. Among the latter's needs were the transmission of electrical power to rural areas and providing a transportation infrastructure by building roads, railroads, canals and airline networks – all of which are power-consuming.

Other energy requirements which nowadays interest both the developed and the developing regions are water supplies and irrigation systems, wire and wireless telephony, medical and hospital services and other programmes, with quality of life playing a pivotal role. The problem is that few countries have a dependable plan on how their current and future energy requirements should be met.

- By default, everybody looks to fossil fuels as the solution, but this clashes with the policy of controlling CO_2 and other gases.
- France is one of the rare countries that avoided cornering themselves in energy supplies by betting successfully on nuclear power, since the 1970s.

Developing countries can learn a great deal from what has been achieved in French electricity production, particularly in the ability to sustain the system and make it profitable. (By contrast, in some other countries costs run wild, and billions have been spent bailing out loss-making nuclear power companies, making this sector of industry a byword for mendacity, secrecy and profligacy with taxpayers' money.)

The French model of power production proves that properly managed nuclear energy can be a successful undertaking, and this helps the government promote it. Good management and documented positive results can be as effective as the fear of climate change in boosting nuclear energy. All governments should appreciate, as the French did, that under a sound programme:

- nuclear power offers large quantities of baseload electricity that is cleaner than coal, more secure than gas and more reliable (in delivery) than wind, and
- if new electric loads are added, like cars switching from oil to electricity, the demand for power generated from carbon-free sources will be at a premium.

One of the interesting lessons from France is that nuclear power can be tuned to provide up to 80 per cent of the country's electricity needs.[15] This fact has divided its enemies. Some environmentalists retain their antipathy to nuclear energy, but others have changed their minds and embraced it; even public opinion has begun to come round.

- France decided to go nuclear in 1974, and
- today there are 59 nuclear power plants across the country.

15 And 90 per cent of EDP's production capacity in France.

This has also produced employment, as the French ingeniously exploit in other countries the know-how and experience acquired over 36 years with their own nuclear plants. Areva uses as a reference the series of successfully operating reactors it has built in conjunction with EDF, in France, its argument being that by capitalizing on these skills:

- it can save time and money by using as many identical parts as possible for the different nuclear plants it builds abroad; and
- it can reduce construction time to four years, so that a new reactor contracted in 2010 will enter service in 2015.

Areva is also active in developing the next-generation nuclear power production factories. The first of its EPR designs (jointly developed with Siemens) is under construction in Olkiluoto, Finland.[16] But it has teething troubles and is two years behind schedule, as well as over budget. To avoid such surprises, Areva also offers older, already-proven designs.

According to information in the public domain, Areva offered the modern ERP technology to the United Arab Emirates (UAE). Also, as already mentioned, on 27 December 2009, a South Korean consortium led by Korea Electric Power (KEPCO), a state-controlled utility, won the contract to build four nuclear reactors in the UAE, of traditional design. With this $20 billion contract the Koreans will also be running the four nuclear plants for six decades – their estimated life cycle. What tipped the balance was the Korean bid-price, reportedly billions of dollars cheaper than those of GE/Hitachi and Areva.

- The French offer was high-tech; hence more expensive.
- KEPCO[17] has focused on building reactors quickly and rather cheaply, but of older technology.

Other EPRs are under construction at Flamanville in Normandy and Taishan in China. Areva forecasts that by 2020 demand for nuclear capacity could bring it orders for 60 reactors, or one-third of the total market – each with a price-tag of around €5 billion ($7.5 billion).[18]

As section 2 has already highlighted and these figures further document, the high cost of building new nuclear power plants could significantly delay that option. One way to rein in costs is to amplify the installed base. In September 2008 Electricité de France (EDF) bought British Energy, a utility in which the British government held a big stake.[19] British Energy plans to build four new reactors in Britain.

In December 2008 EDF also launched a defensive $4.5 billion bid for half of the nuclear business in Constellation, an American utility, planning to use Constellation as a platform for expansion in the United States. (In February 2008 UniStar, the two firms' joint venture, agreed to propose Areva's EPR for construction in America. But in September

16 Third-generation EPR has been jointly developed by Areva and Germany's Siemens, but the collaboration of the two companies ended in 2009.

17 KEPCO's nuclear subsidiary runs similar plants in South Korea and plans to build 20 more.

18 'Power Struggle', *The Economist*, 4 December 2008. Westinghouse has orders from China for four of its new AP1000 reactors, and GE Hitachi's ESBWR design is being considered by several American utilities. Toshiba-controlled Westinghouse is Areva's major competitor.

19 'EDF to Buy British Energy for £12.4 bn', *Guardian*, 24 September 2008.

Constellation agreed to be acquired by MidAmerican, a unit of Warren Buffett's Berkshire Hathaway Inc.)[20]

What happens with the Finnish EPR plant may either promote or damage Areva's image. Competition is tough, and the die is not cast on who will dominate the global market. Some experts say that Areva will most likely hold the upper ground in Europe, and GE in America, but the big market for nuclear reactors will be in developing countries.

Moreover, since the mid-2000s, there is also a new kid on the block – the fourth-generation nuclear plant. The way *Le Canard Enchainé* relayed it in January 2006, when plans for the next-generation EPR nuclear plants – the touted fourth-generation – became more or less concrete, the former president of the Republic acted as chief nuclear engineer, announcing that this newest-generation prototype will be functional in 2020, eventually overtaking the Franco-German EPR.[21] However, professional nuclear engineers disagreed, maintaining that the fourth generation will be operational only in the second half of this century. According to some expert opinions, this division of attention between two advanced projects delayed the EPR.

5. Britain, Germany and Italy at the Crossroads

The way in which nuclear power production is being viewed in the European Union is changing. But, with the exception of France, this is happening by default, not because of a well-thought-out plan and general agreement among EU members about what should and should not be done with regard to energy-related challenges.

It is not unlikely that the issues outlined in sections 1 to 3 of this chapter, or at least some of them, have been in the minds of British, German and Italian political leaders as they started to lean towards nuclear energy production for their countries' future needs. Historically, Britain was the first nation to exploit nuclear power for electricity, but:

- its politicians were unable to decide whether to build a network of power stations to meet the nation's future energy needs, and
- opponents to the nuclear alternative disregard the fact that using more coal, or even gas, was contradicting their own aims.

The first fundamental official questioning of the inevitability of nuclear power in Britain took place in 1976 and was the mission of a top-level commission led by Sir Brian Flowers, president of the European Science Foundation. The commission recommended that the government should avoid, for as long as possible, a major commitment to nuclear power as a main energy source.[22]

This report was mainly concerned with the (then) new fast-breeder technology, a rather efficient energy producer because it generates more fuel than it consumes. Suggesting that fast-breeder reactors would not be needed for at least 50 years, the Royal Commission

20 *The Economist*, 6 December 2008.

21 *Le Canard Enchainé*, 11 January 2006.

22 In a 200-page report released in late September 1976, the Royal Commission on the Environment asked for more research into alternative power sources, such as the sun, wind and waves, in the hope that a nationwide system of nuclear generators will never be needed. *Nuclear Power and the Environment*, ('The Flower's Report'), Sixth Report of the UK Royal Commission on the Environment, Her Majesty's Stationery Office, London, 1976.

report came three days after Sweden's socialist government lost power, partly because of opposition to its plans to commit Sweden heavily to nuclear power. It should, however, be noted that the Commission did not advocate abandoning nuclear fission altogether.

- It just stated that this solution would neither be wise nor justified, and
- it questioned the Department of Energy's projections of future energy demands, as well as projected costs.

In contrast to the Royal Commission's thesis, Sir John Hill, chairman of Britain's Atomic Energy Authority, had strongly backed a fast-breeder programme, calling it 'vital to our future well-being'. Hindsight from the last 34 years documents that Hill, and not Flowers, was on the right path.[23] Although 2009 saw a rebirth of interest about nuclear power in Britain, it still has its opponents, their main argument being that the electricity produced is uneconomical – an argument contradicted by the results obtained in France (section 4).

Supporters of nuclear energy argue that simplistic comparisons with other electricity sources are silly. Coal-fired power stations pay too little for their high carbon emissions. Because nuclear power plants must recover their high construction costs over decades their balance sheets are lop-sided in the early years. But, with carbon pricing which forces fossil-fuel plants to pay for the environmental cost of the carbon they generate, the income figures radically change.

In fact, the German Institute for Economic Research (RWI) has shown that nuclear power is the cheapest form of generating electricity at *2 cents* per kWh versus *40 cents* for solar. Germany had different reasons from the Flowers Commission to be shy about nuclear power: the Greens. But by 2009 German citizens had warmed to the idea, and support for building new nuclear power plants is growing.

Still in July 2009 the emergency shutdown of a nuclear reactor in northern Germany did thrust worries about atomic safety back on to the political agenda. Although the failure of a transformer at the Krümmel nuclear reactor did not endanger the public, revelations that Vattenfall, the Swedish utility company, failed to install a monitoring device or promptly inform its regulator caused alarm. This incident – which was 100 per cent a management failure, not a nuclear failure – has been seen as a setback for the energy industry after a brief period when pro-nuclear arguments on prices and climate change had begun to carry weight in Germany. The incident provides evidence confirming what was explicitly written in section 3:

- that the reliability of nuclear plants has to be rigorously managed at a 99.99 per cent level of confidence, and
- the norms regulating design, operation and maintenance should be universal, with a supranational authority having full powers to inspect, control and bring to justice.

For its part, the Italian government has plans to generate a quarter of its electricity from nuclear power, which industry sources say would take a dozen 'gigawatt-plus' reactors. For Italy, which has long relied on others to provide 86 per cent of its energy needs, this is a new departure. By default that policy has left it more dependent on imports than any

23 In the mid-1980s Britain built the Sizewell B reactor which is still its newest – but it is already 14 years old.

other member of the Group of Eight countries – a fate due to worsen as its own supplies of natural gas have fallen sharply over the past decade and they covered only some 12 per cent of power consumption in 2008.[24]

Hydroelectricity used to be Italy's biggest domestic source of energy, but this has been exceeded by imports of electricity from its neighbours, a situation made worst by a 1987 referendum in which the people voted to close down the three nuclear power stations then in operation. (Italy also produces geothermal power at Larderello in Tuscany – see Chapter 5.)

Obviously, Italy has plenty of wind and sunshine, but as David Tabarelli, chairman of Nomisma Energia, was to comment, forecasts of a boom in wind and solar are optimistic, and Italy will not be able to meet the goal set by the European Union that renewables should provide 12 per cent of energy consumption by 2010.[25] (This '12 per cent' goal is one more indication that bureaucrats in Brussels and elsewhere live in cloud cuckoo land.) By all evidence the only way for Italy to take its energy future in its hands is to reverse its ban on nuclear power.

In May 2008 the then newly elected government said that within five years Italy planned to resume building nuclear plants. 'By the end of this legislature we will put down the foundation stone for the construction in our country of a group of new generation nuclear plants,' said Claudio Scajola, Minister of Economic Development. 'An action plan to go back to nuclear power cannot be delayed anymore.'[26] Let's wait and see.

For their part, other EU member states are afraid of taking the initiative. An example is Spain whose government reassessed its anti-nuclear stance by agreeing in mid-July 2009 to extend the life of its oldest commercial nuclear power plant – which carries more risk than building a modern nuclear facility. It is quite clearly ridiculous to be anti-nuclear energy and keep the old nuclear plants in operation.

It is also surprising that Sweden, the country which planned to end its dependence on imported oil (Chapter 2), has suddenly become ambivalent in its attitude. The reason seems to be that the four government coalition parties have long been split over nuclear power. So they agreed in a manifesto to keep all matters atomic off the agenda until their term expired in 2010. But a combination of tight climate-change targets, energy security worries and a wobbly economy has now caused a rethink. Drift is what happens when people are unable to decide.

6. A Nuclear Power Take-off in the United States?

In a timetable not too dissimilar to that of France, a major push in nuclear power production in the United States begun in the 1970s, eventually creating a surplus of electricity. Twenty years later, however, the construction of new power plants levelled off, while American dependence on electricity continued to grow. During those two decades electricity use increased by a whopping 92 per cent.

At the time of ambiguity (1990) the United States had 112 commercially licensed nuclear plants generating 20.6 per cent of its electricity. These reactors were of various

24 Italy's 5.2 million tons of crude pumped from local oil sources satisfies just 7 per cent of its needs.

25 *The Economist*, 6 June 2009.

26 *International Herald Tribune*, 25 May 2008.

sizes and types. Environmental activists said that they generated some 20,000 metric tons uranium (MTU) of spent fuel, temporarily stored in water pools at over 100 commercial nuclear power plants and awaiting final disposal in geological repositories. By and large, they are still waiting.

As we will see in section 7, as far as nuclear waste disposal is concerned, the opponents of nuclear power have a point. An amazing aspect of this waste (which, as some claim, will remain dangerous for many, many years) is that all over the world it is being stored 'temporarily'. It is not that the authorities have no plans for it. They have lots of plans and that's part of the problem because:

- no one has any clear idea about which handling method will be the best longer-term solution, and
- the practice of sending it abroad for reprocessing is nothing more than a way of avoiding a responsible decision.

Opposition from environmentalists' aside, another reason why nuclear power in the United States was put in mothballs for some time was that the higher-than-expected cost of many of the earlier projects, and state regulators prevented utilities from passing on cost overruns to consumers. As a result, most utilities simply dropped out of the nuclear plant construction business, as there was just no incentive for them to take the risks.

This is another way of delaying important decisions, but it can only last up to a point as several older US nuclear plants are about to be forced into retirement. This is also true of plants using fossil fuels. For instance, the average age of coal-fired plants in the United States is more than 25 years, and new laws on air pollution are beginning to bite with the result that the dirtiest polluters are being (or have to be) shut down.

With CO_2 emissions in the spotlight, it does not make great sense to build new coal-fired power plants, and America's nuclear industry, which has long been in hibernation, now sees an opportunity for a comeback. Some experts are expecting a rush of applications to build new reactors – the first in the United States in almost 30 years.

According to the Nuclear Regulatory Commission (NRC), *if* all these applications for new US nuclear power facilities are successful, *then* the number of reactors in the country will increase by roughly a third, while the output of electricity would grow sharply as new reactors would be more powerful than older ones. (America's most recent nuclear plant, at Watts Bar Unit 1 in Tennessee, started operations in 1996 but it took a surprising 23 years to complete. Watts Bar Unit 2 is now under construction. Another plant, at Shoreham in New York, was completed and tested in 1989, but never allowed to start commercial operations because of local opposition.)[27]

All this demonstrates that nuclear power's rehabilitation in the United States can in no way be taken for granted. Neither are government incentives on call. It may be no coincidence that the Watts Bar Unit 1 reactor built and commissioned in the United States and the second unit now under construction have been backed by the Tennessee Valley Authority (TVA), the huge federally-owned hydroelectric public utility of the Roosevelt years.

27 This, too, is part of the problem. Everybody wants to have and use plenty of electricity, but nobody likes to have a plant – nuclear or otherwise – in his vicinity.

By contrast, there is no shortage of initiatives leading nowhere. In June 2009 several US congressmen launched a plan for 100 new nuclear electricity factories over the next 20 years. But that support has so far been of little practical benefit to the industry. A $50 billion loan guarantee fund for new energy technologies was cut from the stimulus package in February 2009, yet the loan guarantees were essential for nuclear projects to go ahead.

For private-sector companies with shareholders to account for, the economics of nuclear power are not that attractive without loan guarantees because cost estimates have been soaring. After having calculated the cost of nuclear generation at about $2,000 per kW in 2003, in 2009 the Massachusetts Institute of Technology (MIT) doubled that figure to $4,000 per kW.[28]

The Obama administration is in a squeeze. *If* it really pushes a low-carbon energy strategy and decides it wants to embrace clean energy, *then* it has little choice other than to back nuclear power plants. The problem of rapidly growing electricity usage is ever present. If it is ignored, it will come back with a vengeance.

The United States could switch to gas, as Britain did, for electric energy, but oil and gas experts project that during the coming decades production cannot be maintained at the massive levels required to feed such power plants. For their part, electric power production experts respond that the projected shortfall would require an even greater commitment to nuclear energy. However, the following is certain:

- No industrial country can afford the business-as-usual policy of the last four decades.
- A downward shift in the supply of more traditional energy sources will lead to economic collapse.
- It is *this* generation's duty to assure adequate energy supplies for the coming generations.

In addition, because nuclear power plant construction has a long lead-time, decisions have to be made now – not in ten years' time. Practically every country in the world will need a substantial amount of electricity before 2030 and will have to install a power infrastructure that would produce a stable electrical production rate for the future.[29]

If it is not nuclear, it will be coal with far worse effects on the environment. China is reportedly building roughly one new coal-fired power plant per week now, and the United States has about 100 coal-fired plants on the drawing board. The irony of this investment is that these plants and hundreds of others will need to be replaced after 2050 and, by then, nuclear power may well be the only acceptable option. The challenge is how to effectively manage the transition and the downside that unavoidably accompanies any major change.

28 *Financial Times*, 23 September 2009.

29 An out-of-the-box idea currently being studied in the United States is to design and build new-technology nuclear plants which burn, as fuel, the waste of classical nuclear plants.

7. Train, Train and Test

In 1890 Pablo Casals, a 13-year-old Spanish boy who became a musical prodigy, was looking through bits and pieces in a second-hand sheet-music store in Barcelona when he came across a tattered copy of six cello suites by Johann Sebastian Bach, written in the 1720s. These had long been overlooked by better-known musicians, but for the young Pablo, their melodic beauty was audible.

- Casals spent the next *12 years* practising them *every day* before he would perform them in public.
- Eventually, his superb rendition ensured that the cello suites attracted a mass following, and they became the hallmark of the virtuoso cellist's fabulous career.

There is no better example of the sort of training required for nuclear engineers, system engineers, nuclear power station managers and their personnel. They must always live by the rule that their skills are perishable, new risks can always come up, and, even if they are confronted by many unknowns, their know-how should enable them to be in charge.

Intensive programmes should be the order to the day, aimed at training engineers, technicians and managers all the way from design to operation and maintenance of a nuclear plant. Such training must include case studies on what has gone wrong with nuclear reactors and other units of the plant, as well as solutions adopted to solve emergency problems. The best approach is:

- to include myriad medium and small incidents,
- rather than just concentrate on the big incidents, such as Three Mile Island and Chernobyl.

For example, in the early hours of 24 March 1992 radioactive gases and iodine escaped from reactor no. 3 in the nuclear power plant at Sosnovy Bor, 100 kilometres west of St Petersburg. It was an accident measuring only 2 on the scale of 7 used by the International Atomic Energy Agency (IAEA), but it was also a reminder that accidents are not confined to the big ones. The accident occurred when contaminated steam escaped from one of the 1,690 fuel channels in the reactor's core, either because the channel ruptured or (more likely) because a seal on the channel gave way. The level of radioactivity in the turbine hall rose to 60 micro-roentgen an hour, three times its normal level – although there was no danger of meltdown because the reactor's safety system worked.

This is an excellent case study which is worth analysing in detail, with the lessons learned and used in textbooks about nuclear reactor safety. An interesting hindsight is that two months prior to the accident Swedish engineers had visited Sosnovy Bor and recommended a raft of improvements. Instead, the number of Russian safety inspectors was halved, because of lack of money. According to Vladimir Sukhoruchkin of the Kurchatov Nuclear Research Institute in Moscow, the fuel channel at Sosnovy Bor failed because of poor testing in manufacturing.[30]

30 *The Economist*, 28 March 1992.

Quality control in manufacturing is one of the areas where plenty of attention should be paid to training engineers, technicians and managers working, or to be working, with nuclear plants. Quality-control standards established in the early 1940s for the Manhattan Project are the best guide I can think of in this connection.[31] Case studies on what has happened – and on what did not happen but might have happened – should include comparisons between:

- the older technology in nuclear power design, such as the KEPCO product line, and
- new technology like Areva/Siemens EPR, where greater efficiency may be accompanied by not-so-well known (or studied) operational problems.

Closely connected to these two bullets, and a 'must' for training programmes, are arguments revolving around simpler versus more complex designs of nuclear power plants. The pros say that simpler designs cut maintenance and repair costs. A training programme should examine whether this is a strong or weak argument, given that cutting maintenance costs is the worst possible decision because:

- even older nuclear reactor designs now include features which have become more sophisticated, and
- the very act of cutting down maintenance may well open a Pandora's box of minor and major accidents.

Some experts state as evidence of nuclear plant safety that shutdowns are now far less frequent, to the extent that a typical station in the United States is currently online 90 per cent of the time, up from less then 50 per cent in the 1970s. They also point out that new passive safety features can shut a reactor down in an emergency without the need for human intervention.

A training programme should provide its participants not with opinions, but with factual and documented research findings – for instance, hard data on whether or not, and under which conditions and within which limits, it is safe to let computer models decide on the most critical issues pertaining to the plant's operations.

Equally important is explaining that one of the principal problems with the safety of nuclear power plants is not an originally unsafe design (although this, too, exists), but the fact that the companies and agencies that operate them worldwide simply do not have the personnel or the capital to maintain and run the plants properly. In several countries one reason for this is that electricity is priced so cheaply that utilities are short of the capital margins necessary to finance:

- intensive training, and
- desperately needed safety improvements.

For better or for worse, there is a plethora of minor, but significant, nuclear plant accidents which, properly compiled as case studies, can provide an excellent real-life basis for intensive training. For instance, about a year prior to the Sosnovy Bor accident, there

31 MIL-STD-105A, published on 11 September 1950.

had been press reports about a long chain of mechanical failures at one of Tokyo Electric Power's big nuclear plants in Fukushima.

- Several of its reactors had to be shut down, and
- one of them remained out of action for more than a year.

The Japanese incidents show that technologically advanced countries are not immune to nuclear plant failures. Such failures had left Japan's 37 nuclear power stations operating at only 70 per cent of capacity in 1989, at the time of one of my visits. Then in February 1991 Japan narrowly escaped a major nuclear accident when a pipe apparently broke in a 19-year-old nuclear facility and touched off a series of events that ultimately forced the emergency flooding of the plant's reactor to cool its nuclear fuel.

Government officials and Kansai Electric Power, which owned the nuclear reactor, said at the time that no radiation escaped from the plant itself and that no one was injured. But in the opinion of outside experts, that accident was the worst since Japan began operating nuclear power plants two decades earlier. Some experts further suggested that:

- *if* the emergency cooling system had not worked properly,
- *then* the result could have been very similar to the kind of damage that took place at Three Mile Island in 1979.

An interesting case study can be developed around the theme that the trouble at the Kansai Electric plant had abnormal radioactivity readings in the secondary cooling system, while the 500 mW reactor was operating at maximum capacity. The timely response by operators is also an eye-opener. When they detected the higher-than-usual readings, workers in the control room started to reduce the power output. The reactor automatically stopped after the water pressure in the core had dropped precipitously. Experts said that:

- *if* the core was left exposed, *then* the result could be a meltdown, and
- *if* it were left uncontrolled, *then* this would have led to a large release of radioactivity.

In a case study on potential failures, all these *ifs* must be analysed in detail and become scenarios for training nuclear engineers, safety personnel and plant managers worldwide. Also in Japan, a chain reaction at a uranium processing plant in Tokaimura exposed three workers to lethal doses of radiation and irradiated hundreds more with smaller amounts. This 1999 event was the world's worst nuclear disaster since Chernobyl.

- Investigation revealed that the company had hired unskilled labourers to do many technical jobs, providing little in the way of training, and
- the factory workers had unknowingly dumped six times more uranium oxide into a mixing tank than they should have done.

Most pertinent to a training programme is to emphasize that these unskilled workers were, moreover, under instructions to break safety rules in order to save time and money.[32] Some of the equipment was also outdated. While none of these occurrences is by no means unheard of, attaching to them a name of country, company and factory gives them so much more impact.

It is also pertinent to emphasize that if this sort of mismanagement can happen in Japan, which after World War II became renowned for its quality-control standards and procedures, it can happen anywhere in the world. And it can occur with higher frequency in less developed countries in any industry – nuclear being only an example.

Another more recent case comes from Slovenia, a member of the European Union. On 4 June 2008 the Slovenians blamed a faulty valve for a water leak at a nuclear plant, seeking to play down the incident as EU ministers questioned the safety of the country's nuclear reactors.[33]

After the leak, the Slovenian authorities shut down the plant in Krsko, near the Slovenian–Croatian border, saying that it was confined to the plant building and caused no damage. But details have not been released, and statements not backed up by credible evidence are the worst that can happen with any accident, anywhere. As Louis Brandeis, the US Supreme Court justice once said, 'Sunshine is the best disinfectant'.

8. Managing the Downside of Nuclear Energy

Let's face it. In many countries, management and maintenance of technological installations (not just nuclear plants) is substandard. While it is true that good design facilitates maintenance, it is no less true that it does not substitute for preventive maintenance which must be consistent, focused and done by highly qualified technicians.

The safety methods we have available must be continuously improved – for instance, to guard against mechanical failures that could interrupt the transfer of heat, most reactors employ multiple back-up systems, a strategy known as *defence in depth*. The problem is that:

- these back-ups are not always kept in tip-top condition, and
- there is a lack of worst-case drills to test their preparedness.

No matter how many layers of safety are built into a conventional reactor, it can never be 100 per cent safe against a meltdown. Stress tests and worst-case drills reveal weaknesses which otherwise go undetected until it is too late.

Plant size is another cause for concern. Economic factors incite nations and companies to opt for bigger nuclear power plants, a policy which steeply increases the need for first-class management and well-trained technical experts (section 7). Therefore, even though their power outputs are modest (hence less economical), smaller units could be regarded as safer plants and could prove to be more cost-effective in the long run. In addition:

- they can be simpler in design,

32 *The Economist*, 4 March 2000.

33 'Slovenian Nuclear Reactor Leak Forces Emergency Shutdown', *Daily Telegraph*, 4 June 2008.

- their components can be built on an assembly line,
- it is easier to exercise rigorous quality control, and
- if put into operation one module at a time, they permit operating capacity to be matched with demand for power.

In addition, smaller units tend to pose (at least individually) less of a problem with one of the most controversial issues with nuclear power: the disposal of radioactive waste. Ironically, while this is indeed a major challenge, it will not be made much worse by building new reactors – because of the mountains of nuclear garbage from weapons systems.

Let's start with the hypothesis that the various bureaucracies have a strong incentive to safeguard nuclear waste and assure that security is maintained. After all, if radioactive material were ever to be used by unauthorized groups, it would most likely be used against its former owners. It also appears that those in charge of weapons in the United States, Britain, France, Russia and China are in close touch with their counterparts in an effort to assure that an accident does not occur.

But accidents do happen, and many are man-made. In a prophetic article published two decades ago, Mark L. Melcher stated:

> *[In a] nuclear story, I think, much more important than the possibility of an East–West weapons confrontation is the corruption, misrepresentation and incompetence associated with the production of nuclear weapons in both the Soviet Union and the United States ... Both nations have shown a strong propensity in the past 45 years or so to put chowderheads in charge of their nuclear defense establishments. This problem hasn't gone away.*[34]

Add to this the chowderheads of all other nuclear nations – most particularly, the North Koreans and, tomorrow, the Iranians.

Melcher went on to say that (up to that time) the clean-up of nuclear waste generated by the US military could cost taxpayers upwards of $130 billion, adding that the Russians may not be able to clean up their mess as thoroughly as needed, and rely more heavily on the use of a 'national sacrifice zone, permanently sealed off from public access' (which is one of the American answers to the problems at Hanford).[35]

The reason why Melcher's reference to military nuclear waste is so important is that, ironically, this is one of the aspects of the so-called 'temporary' storage of nuclear waste by electric utilities. It is also an issue that has done great damage to the image of the nuclear power-production industry.

The non-temporary history of 'temporary' solutions for disposal speaks volumes about the problem of nuclear waste produced by electricity plants. According to the US Council of Energy Awareness (USCEA), all operating nuclear factories worldwide have been generating tons of nuclear waste material, virtually all of which is in 'temporary' storage, awaiting the creation of a long-term solution to the problem, *as if*:

34 Prudential Bache, *Strategy Weekly*, 11 September 1991.

35 Among other sleeping beauties in a deep pit carved out of the poisoned earth of the Energy Department's Hanford nuclear weapons reservation lie 21 huge, grey steel cylinders. They used to be parts of nuclear submarines. When the subs were scrapped by the navy, the bow and stern were scuttled, but the centre units housing the radioactive reactor cores were barged up the Columbia River and dumped in this nuclear graveyard. *International Herald Tribune*, 3 December 1991.

- governments don't really care to seriously study and solve the nuclear waste problem, and
- even prefer to keep it that way, for reasons proper to their own interests but contrary to public safety.

Britain considers nuclear waste reprocessing as a big business and has lots of plans for future waste-treatment facilities and processes. But there is plenty of smoke and mirrors. Supposedly, the siting and construction of a long-term repository for intermediate-level and future low-level waste 'is being addressed', sites 'have been identified' and detailed geological investigations 'are in progress'. Yet as of December 2010 ministers talk of a renaissance in nuclear power, involving the construction of up to eight nuclear reactors in the coming decade.[36]

Despite all these assurances, however, progress is not visible. In the meantime, thousands and thousands of metric tons of spent, but still highly radioactive, nuclear fuel pile up at power plants located in countries that use nuclear material.

Philosophers may say that every generation leaves both assets and liabilities to the next. The way we are going will leave our grandchildren the appalling problem of a huge world population and plenty of waste of all sorts with an ever increasing level of pollution. Are we also going to leave future generations an insoluble problem about what to do with nuclear waste?

In theory, in the United States the Department of Energy is in charge of looking after the nuclear waste problem. It requires utilities to set aside a tenth of a cent for each kWh of nuclear power they generate to help defray the costs of transporting nuclear waste to a safe repository and storing it there in a supposedly permanent way. The hitch is that no such repository so far exists.[37]

By and large, countries have decided that the best place to store their nuclear waste is underground, in watertight containers. The United States got as far as selecting a site for one, at Yucca Mountain, a ridge in the middle of a former nuclear-testing ground in Nevada. But in 2009 the Obama administration cut all the funding for the Yucca Mountain project – first proposed in 1978 and, since then, the subject of plenty of lawsuits.

The important point is that the nuclear waste problem is serious, general and global. It is not specific to the civilian power-production industry or the weapons industry alone. It is not just an American, British, French, Russian Chinese or any other nuclear nation's problem. It is everybody's problem, and each nation should contribute its best people to analyse, test and comprehensively study it.

Out-of-the-box solutions are called for, and no hypothesis should be eliminated *a priori*. 'We are asking the wrong question,' says an American expert about the handling of nuclear waste.[38] Rather than asking how spent nuclear fuel can be kept safe for 10–100,000 years, we should worry about keeping it safe for only 100 years. Because nuclear waste still contains an enormous amount of energy, future generations may be able to harness it as an energy source through tomorrow's better technologies.

36 *The Economist*, 11 December 2010.

37 *The Economist*, 8 September 2007.

38 Personal communication.

Appendix: Fusion Power is a Far-out Event

In 2006 the European Union, United States and (in alphabetical order) China, India, Japan, Russia, South Korea and Switzerland signed an agreement to test the possibility of harnessing the power of nuclear fusion. Known as the International Thermonuclear Experimental Reactor (ITER), this project includes the constructing a huge testing facility at Cadarache, France.

At the time of writing, the ITER project is five years old and its deliverables are not stellar – as it is faced with major cost overruns as well as delays. The initial estimated construction costs of €10 billion ($13 billion) have tripled, which has naturally made European policy-makers and their international partners quite nervous, and the extension of project timetables does not help.[39]

Scientists joke that fusion is 30 years away and always will be,[40] as the new reactor's first power-producing experiments have been pushed down the line. While major scientific projects cannot and should not be rushed, the question of their cost-effectiveness is always present. Both goals and their financing must be subject to steady evaluation through design reviews based on more recent findings. This does not seem to have taken place, and its absence is a pity.

It comes therefore as no surprise that several ITER partners and a score of technical experts (who were initially full of enthusiasm) have begun to question whether this project really worth the effort and investment. Others wonder how much more are various governments prepared to pay. As these reactions are suggesting, not everyone is convinced about the ITER project's viability. Some even say that this project is tantamount to throwing money at the CO_2 problem, with little hope of a return.

The lack of expected results with high technology saw to it that rather than continuing to rely on huge reactors costing billions, energy executives are now looking at small, inexpensive ones. These are mainly based on proven designs from nuclear submarines and other warships. Russia has developed small floating reactors to deliver energy to its Arctic regions, though primarily because of the problem of building power plants on permafrost.

Scientists at the Nuclear Energy Institute, the industry body in Washington, DC, estimate that more than 20 countries have expressed a serious interest in buying mini-reactors.[41] Thousands of nuclear engineers are said to be working on eight different technical approaches, in several countries, but it is too early to tell what these will deliver. There is also the disquieting fact that too little attention is being paid to the training of engineers able to install, operate and maintain these nuclear mini-reactors.

39 'Expensive Iteration: A Huge International Fusion-Reactor Faces Difficulties', *The Economist*, 22 July 2010.

40 *Financial Times*, 28 July 2010.

41 *The Economist*, 11 December 2010.

CHAPTER 5 Alternative Energy Supplies

1. Sustainable Development or Alternative Energy

The phrases 'sustainable development' and 'green (or alternative) sources of energy' keep popping up every day – often in unexpected places. Usually these and similar labels are simply shorthand for anything environmental. Many people nowadays see them as a better basis for taking decisions which, by some miracle, are balancing energy costs, providing well-rounded benefits and controlling pollution. They don't.[1]

Finding a better mousetrap is laudable, but running after lots of hares at the same time guarantees that none is caught. Sustainable provision of energy supplies able to reduce the costs associated with them and controlling pollution seem to be the right targets. In reality, however, they contradict each other. Solar and wind energy, for example, cannot compete with coal and oil without hefty subsidies.

We can better appreciate what comes under the term 'alternative energies' by looking at them as processes promoting *decarbonization*. Their opposite is reliance on continuing to produce energy by carbon compounds – coal, oil and gas – which today dominate energy sources.

- Theoretically, it is not really difficult to set alternative energy targets.
- In practice, the challenge is to meet them in a consistent, affordable and sustainable way.

For example, the US Department of Energy has set a target for 30 per cent of the 2004 petrol demand for vehicles to be met by biofuels by 2030 (biofuels is the theme of Chapter 6). The European Union wants 25 per cent of transport fuels to be derived from biofuels by the same date.

These targets sound great, provided that technology delivers the better mousetrap. Today's ethanol does not pack a particularly powerful punch, although that seems to be changing. A biofuel that did not suffer from the current stuff's limitations would be welcome. Plants are rich in carbohydrates (a group of organic compounds based on carbon and water). By contrast, an ideal fuel should lack oxygen (which exists in abundance in water) and feature molecules that contain few carbon atoms.

1 On 10 June 2010, Bill Gates and other CEOs of big US firms pledged funds for clean energy. At the same time, however, they asked the government 'not to be near-sighted' in funding. In other words, they want the taxpayer to foot the bill.

- Several ongoing projects are addressing these prerequisites, and eventually they will come up with a solution.
- To be sustainable, however, the solution they provide must be cost-effective in comparison to today's carbon-rich energy sources.

Cost-effectiveness is a prerequisite with all alternative sources of energy. The cost of the better mousetrap must be attractive to invite wider adoption. If it is higher, people, companies and nations will find a thousand reasons for not choosing it and, characteristically, none of those reasons will include the word 'cost'.

Assessing all costs and benefits that come into an energy algorithm is not easy. Apart from what the kWh or barrel of oil bears as price-tag, other costs – such as greenhouse effects and polluted waterways – have to be estimated in ways that are difficult and frequently controversial. Decision-makers don't truly care if the accumulation of pollution may deliver nasty and irreversible surprises until it is too late (the lack of any global agreement during the Copenhagen circus of 7–18 December 2009 being the best proof of this statement).

A great deal of the challenge of estimating costs and benefits is that the latter are often lightly spelled out, whereas most of the former are in the more distant future. As if to add to the difficulty of measuring and estimating comes the fact that the science of economics is badly equipped to deal with:

- CO_2 measurements,
- environmental after-effects, and
- timescales that extend way beyond short-term benefits and damages.

It goes without saying that this has a direct impact on investments for alternative energy. The most widespread principle is that these must earn a return high enough to compensate the investor for uncertainties, including those connected to sustainability. If not, the taxpayer is the only other source of financing – an often abused source, for that matter.

Private investments have not been promoted by the fact that companies specializing in alternative energies have been performing poorly during the last few years. Here is a practical example. According to one of the brokers I have spoken to, in the 2008–2009 timeframe the wind and solar sectors constituted respectively 42 per cent and 32 per cent of global *cleantech* market cap. They did so because they had attracted the most investor interest.

Solar energy, however, has been by far the worst performing (–26 per cent) of all energy generation processes and wind the second worst. Investors who have lost plenty of money in solar ventures (Bill Gates among them) are not likely to come back with plenty of capital.

- Financial losses are hardly the stuff to induce private investors into putting up the money.
- Instead, the taxpayer is asked to contribute billions, but subsidies are poison in the longer run because they feed inefficiency.

Nothing can match past experience, and experience teaches that subsidized industries would rather be followers than leaders. That's too bad for alternative energies, because breaking out of the current production straits requires real leaders not 'me toos'. As long as return on investment (ROI) remains unappealing, sustainable development from alternative energy sources is little more than an empty slogan. This will change *if*, and only *if*:

- projects are dimensioned the right way,
- everybody, including *all* developing nations, accept the concept behind environmental protection, and
- there are dependable metrics available to monitor the deliverables, and measure all of their consequences.

Fulfilling these requirements takes study, documentation and concerted action on a global scale, because neither the use of energy nor the resulting pollution recognize man-made frontiers made. *If* there is no global agreement – and so far there is none – *then* there is no solution whatsoever, no matter what the different politicians and environmentalists may be saying.

2. Alternative Energy and Decarbonized Energy are not Exactly the Same

The way it has been generally defined, by and large, alternative energy means renewable energy. This, however, is not necessarily synonymous with a decarbonized or environmentally-friendly energy solution. It does precious little good to confuse the different goals, terms and processes, taking one as meaning the other.

Gas is an alternative energy to coal and features a significant degree of decarbonization, but it is not renewable. Biofuels provide the advantage of decarbonization, when compared to coal, but they are renewable. Energy based on biofuels is also more sustainable than that of wind turbines or solar cells, which are idle when the raw materials – wind and sun – are not on call.

Hydroelectric power, also known as *white coal*[2] (see section 3), is renewable energy and it is sustainable, provided that rainfall continues at an expected rate. By contrast, it is not necessarily environmentally-friendly, no matter what the Greens say, because large hydroelectric projects affect the environment – including human habitat and wildlife.

The brick which hit the head the government of Ecuador in 2009 provides an example of what can happen when a country depends on a couple of energy sources, one of them hydroelectric. During that year, oil production fell by 4.9 per cent (in the 12 months to November) as threats against foreign companies led them to cut investments. At the same time, drought cut output from Paute, the country's biggest hydroelectric plant,

2 The term 'white coal' is used here in contrast to black coal. The same term has also traditionally been used for fuel based on dried chopped wood. However, although the dual meaning might create initial confusion, 'white coal' is a currently used term for hydroelectric power.

which contributes 35 per cent of the country's total electricity. The result was months of frequent power cuts.[3]

Is '*white coal*' environmentally-friendly, as it is mooted to be? Not always, and not necessarily. Huge water dams have radically altered the environment and ecosystem over distances of many miles. People, indeed whole villages, have been uprooted and other animals chased away (if not drowned) as the waters of the artificial lake rose.

- In the construction of many water dams salmon and other fish lost their traditional sites for reproduction.[4]
- The Aswan dam altered the silt and traditional fertility of the Nile Valley.

Moreover, even the impressive numbers regarding power delivered by water dams can turn on their head when electricity usage explodes. When the Hoover Dam was built in Nevada during the Roosevelt years, it provided more than 90 per cent of Los Angeles' power needs. This was in the 1930s, but in the 1950s, when I was a graduate student at UCLA, the contribution of the Hoover Dam was a mere 3 per cent of electricity requirements of greater Los Angeles.

- The Hoover Dam's power production did not change.
- What changed is that power consumption increased by nearly two orders of magnitude, in about two decades.

These days, hydroelectric power generation supplies only a relatively small fraction of American electricity. The time when electrification greatly depended on 'white coal' as a better alternative to black coal is past. True enough as the United States moves to reduce its dependence on carbon and oil fuels, alternative energy sources are back in vogue – but 'vogue' is not making the limits disappear by magic.

In the 1930s the Tennessee Valley Authority, Hoover Dam, Columbia Dam and other epoch-making projects (for their time) were looked up to as the solution for the future. In 2008, according to the US Department of Energy, hydropower production accounted for about 7 per cent of America's total energy consumption – a figure dwarfed by energy supplied from oil, natural gas and coal.

Another constraint of hydroelectric power generation is that dams negatively affect both agriculture and wildlife. The Aswan Dam has reduced the fertility of the Nile Valley because it is blocking silt from travelling downstream. In fact, experts and environmentalists are now asking for Aswan, the formerly glorious dam of Nasser years, to be dismantled.

In addition, past success stories with hydroelectric power are no guarantee for continued performance. Negative reactions from environmentalists aside (see section 3), there are simply not enough good locations to build dams to satisfy galloping electricity

3 *The Economist*, 16 January 2010.

4 Though in Canada the engineers created platforms which returning salmon could jump to reach their traditional breeding sites. Likewise, to help migrating juvenile fish circumvent the spinning turbines at hydroelectric plants, five major dams on the Columbia River, in the north-western United States, have been equipped with submerged travelling screens. In fact, the federal dams on the Columbia are among the best in gaining operational experience using downstream fish passages.

needs. There is no second big dam built near to Hoover or Bonneville because geography does not offer that opportunity.[5]

This in no way minimizes the contribution of 'white coal' to electricity power production and in proving a better (albeit rather limited) alternative to the old coal-fired power plants. As long as rainfall obeys forecasts made about it and alterations to the environment are not inordinate *à la Aswan*, water dams are a sustainable alternative energy solution.

Hydropower, of course, is not alone in that feat. Windy, storm-lashed hills or mountain tops and coasts are good places to harness the weather, by means of turbines and wave-power machines. No doubt in countries benefiting from plenty of sunshine solar panels can deliver energy from what is supposed to be one of inexhaustible resources: sunlight (as long as the sun's energy lasts).

Not everyone is, however, convinced of this nirvana. The doubters advance technical reasons, financial constraints and a great deal of concern because claims have a nasty habit of soaring far above reality. Western governments' ambitious goals to derive most of their electricity from renewable sources by 2020 or 2030 or even 2050 are widely considered impossible. As such, they only help to prove that:

- claims are often detached from reality, and
- roadmaps are made to please the gallery rather than for any other reason.

One of renewable energy's major pitfalls, of which its enthusiasts take little notice, is that there exist limits beyond which a process cannot deliver. Hydroelectric dams, solar panels (section 4), wind turbines (section 6), and other means of alternative power production cannot continue to expand forever to fulfil the insatiable demand for electricity.

Taking the United States as an example, in 2008 wind provided 1.3 per cent of the country's total electricity generation, up from 0.4 per cent four years earlier. It is projected that in a decade it might provide 6 per cent of a bigger consumption level (and that's an *if*). By then, however, hydropower may be down to less than 5 per cent. So, where will the balance come from?

The gap which traditionally exists between wishful thinking and reasonable plans, as well as between any plans and reality, must always be kept in mind. When people say that 'alternative energy sources are slowly but surely taking a big share' of 'this' or 'that' country's energy use, they should not forget the lessons from the past – otherwise the same mistakes will be repeated. The lesson is that:

- not all plans are being fulfilled,
- roadblocks show up almost invariably, and at the most inopportune moment.

In conclusion, it is not that hydroelectric dams should not be built (although they do present environmental problems) or that wind power should not be fully exploited (even if batteries of modern windmills disfigure the landscape) or that solar energy should

5 As another example, Switzerland, too, has come up against the same limits.

not be fully used (while the state's architects object to putting panels on rooftops[6]). The message is that alternative energies will go up to a point, and no further.

There are also alternative fuels which need much more care before espousing them wholesale. Geothermal energy (section 7), for example, is a big polluter – a case where the medicine is worse than the illness. Last but not least, it is equally important to notice that alternative energies which continue to be subsidized by the taxpayer are becoming difficult to afford – particularly in the longer run.

3. Hydroelectric Power

Hydroelectricity is a renewable energy supply. The Swiss, the French and the Norwegians have done wonders in mastering hydroelectric power. In the United States the Tennessee Valley Authority (TVA) Dam was one of the best projects undertaken in the early Roosevelt years. As a source of power not only is 'white coal' non-polluting (although, as we have seen, it has environmental problems), but the TVA also helped transform a wide stretch of the US south-east from backwater to a prosperous valley.

Hydropower's history started in the late nineteenth century in England. In 1870 William Armstrong, an inventive engineer, became the first person to light his house in Northumberland with electricity produced from hydropower. He did so by building two artificial lakes and a pump house. Today, from Switzerland to western and central Europe, Russia, Chile, Canada and the United States giant dams, artificial lakes and big generators produce a great deal of electrical power.

Nowadays, most of the world's new hydropower projects are being developed in China. This includes the Three Gorges Dam, which is planned to have a total capacity of 22 gW,[7] and the Xiluodu Dam project, due to generate up to 12.6 gW from 2015. Africa could also host several hydroelectricity projects, such as:

- the Grand Inga plant on the Congo river in the Democratic Republic of Congo, and
- the controversial Merowe hydroelectric power project in Sudan, which has stirred up a mare's nest of environmental and social problems.

Some 95 per cent of Ethiopia's power and 70 per cent of Kenya's is hydroelectric. But mudslides and floods, with streams and rivers carrying off the topsoil, have turned the perceived benefits of these enterprises on their head. It is indeed surprising that so many projects fail to account for their negative consequences.

Whoever projects energy resources based on 'white coal' should remember that whether turbines are built on fast-flowing rivers or installed in dams, hydropower alters a river's course and causes disruption in the biosphere (section 2). There are also risks connected to rainfall. As droughts become more frequent around the world, hydroelectricity becomes less reliable.

With rivers alternately running as torrents, then down to a trickle or drying up completely, dams are running out of water; some are empty. Turbines have shut down. The

6 For instance, in France where the Architectes des Bâtiments de France, a state outfit, wrongly interprets its mission all the way to being contrarian to the government's own alternative energy campaign.

7 Critics say that this dam has obliterated the pristine landscape of Three Gorges.

electricity supply throughout east Africa is patchier than usual, just when governments need rich energy supplies to beef up their economies.

In mid-2009, for example, the delayed opening of a big Ethiopian dam capable of producing 300 mW has resulted in daily blackouts in Addis Ababa. The solution adopted by the government has been to set up electrical generators to supply to Ethiopia's grid, and the back-up provided has been fossil fuels.

Chile has been more ingenious in developing its white coal resources, but growing demand for energy turned the tables on government plans. In 1995 hydroelectricity supplied nearly 60 per cent of the country needs, but as the demand for power soared by the middle of this century's first decade the hydropower share fell below 40 per cent. This pushed the government, which faced problems in importing natural gas from Argentina, to reconsider and expand the contribution of white coal.

In association with Spain's Endesa, the Chilean government promoted a $2.4 billion project to build four hydroelectric dams, with a total capacity of 2.4 gW in the Aysén region, in the country's far south, where rainfall is abundant. The problem is that Aysén is a pristine area of fjords and forests.

- Environmentalists oppose the project, and
- Chile is still facing a conflict between promoting power and economic growth versus protecting the environment.[8]

While white coal does not have the CO_2 emission problems of black coal, the problems caused by building dams are most keenly felt by the elimination of as yet unspoilt areas, as well as by people whose homes are destroyed to create them.

According to publicly available information, in Sudan, the aforementioned Merowe dam, completed in 2008, required the displacement of up to 50,000 people, and also flooded areas rich in Stone Age archaeological sites. (Even at the dam's planning stage, the Sudanese government was already forcibly moving some people from the area, which led to clashes, some fatal.)

In Canada in the early 1990s an environmental coalition known as the James Bay Task Force called for a halt to the hydroelectric plans connected to the then Quebec Premier's dream of 'turning rivers into riches'. Opponents said that Hydro-Quebec's projects threatened wetlands that were the largest breeding grounds for birds on earth as well as the habitats of countless species, including beluga whales, otters and a rare type of seal.

'The James Bay wetlands are the largest feeding and breeding grounds for birds on Earth,' said Karen Lohr, head of the opposition to the hydropower plans. 'We're talking about an impact of unprecedented magnitude on the migration of birds, an impact that will be felt to the tip of South America.'[9] Following up on Lohr's argument, the price for reducing CO_2 is the elimination of wildlife's breeding grounds.

Sure enough, there are also many positive examples of hydroelectricity production, including their contribution to power supplies even in countries rich in oil and gas. More than 95 per cent of Norway's electric power comes from hydropower. Norway was also one of the first countries to adopt a carbon tax (back in 1991) in an attempt to

8 'Praying for Rain: A Dash for Gas Gives Way to a Proposed Return to Big Dams', *The Economist*, 17 May 2007.

9 *The Japan Times*, 2 April 1991.

slow global warming, as well as the first country to capture carbon dioxide and store it underground.

Among the 'pluses' of hydropower is that, unlike wind power, it can generate electricity on a grand scale, as shown by projects such as the huge Itaipu Dam power plant on the border of Brazil and Paraguay. And, unlike new technologies such as wave power, hydropower is a proven source that has been widely used for more than a century. This, however, does not alter the fact that dependence on hydroelectric power has negative aspects that are not particularly wholesome.

In addition, the cases mentioned in the preceding paragraphs are mainly from developing countries, whereas in developed countries all the dams that could conceivably be built to exploit hydropower are more or less already in place. The lack of future big hydroelectric opportunities in the West is disquieting power consumption is still reaching for the stars.

4. Solar Power

Theoretically, solar power and wind power are inexhaustible. Who would ever think that in the foreseeable future the sun will not shine and wind will not blow, or that people will be unable to capture the power of either and both. In practice, however, there are limits to the exploitation of all these forces. Moreover, they don't necessarily provide cost-competitive energy supplies for the near term – despite what their proponents say. True enough:

- giant solar farms have been built in California using the solar thermal method of turning sunlight into electricity,
- Israel has taken the lead in solar energy technology (and energy production), albeit at an individual, not central, level,
- Germany has developed large wind farms which provide input to its energy grid, and has also invested in solar energy despite its soupy weather.

But not all places on earth have California's sunshine, and not all governments are as supportive of solar power as the Israelis or of wind power like the Germans, Danes and Swedes. There is no question about the seriousness of solar technology. Installations can vary in size from limited, targeting water heating, to whole solar farms. Lower-cost solar cells have begun to be widely used, while the more expensive photovoltaic systems convert sunlight to electricity directly. All this is known and appreciated, but there are:

- technical,
- financial, and
- legal constraints when it comes to mass production of energy through solar power.

Pollution, for example, is part of the technical challenge. The raw material is semiconductors. Hence, solar energy is not 100 per cent environmentally-friendly, as claimed by its proponents. Pollution associated with solar equipment is present both at production level and with disposal of the waste – a fact rarely accounted for in discussions and decisions regarding the use of solar energy (more on this in section 5).

Another technical issue is the quest for greater efficiency. There are various competing designs for increasing solar panel performance: parabolic-dish mirrors, parabolic-through mirrors, power towers using an array of mirrors to focus the sun's rays on to an elevated platform, and fresnel systems mimicking a parabolic trough using cheaper flat mirrors.

These technologies have been largely studied for farms of solar panels, which raises the question of where to install them (see also the discussion on similar problems with wind turbine farms in section 5). America has the Nevada desert. Europe has no such landscape, and ridiculous discussions about installing them in Morocco or the Sahara only show how empty-headed some people can be.[10] This has discouraged several EU countries from going ahead with massive solar energy projects.

Moreover, both individual and large-scale energy projects need to be financed, and even before this they must obtain the appropriate licence, which means that would-be users confront a dual problem. If one is obtaining only the permission or only the appropriate financing no business is done.

- Financing is by no means unlimited and cost-effectiveness is not self-evident, and
- with regard to legal constraints, there exist conflicts of interest even in the case of governments promoting and subsiding solar energy.

Some governments find it difficult to sort out their priorities. Are the criteria regarding alternative energies primarily architectural? Or do they reflect an ingrained policy of energy conservation? France is a case in point, but it is not the only example.

Correctly, both the government in Paris and the regional governments, promote solar energy. They even finance a good part of the cost of installation, and apply a reduced value added tax to what still needs to be paid. Yes, but ... in France there is also a body that depends on the state: Architectes des Bâtiments de France (ABF). Its agents exercise dictatorial powers, having given themselves the right to veto solar panel installation, and they usually do so. The ABFs are a state within the state.

Provided that the installation of solar panels is permitted, the next question is: 'Who pays the costs?' Governments can ease the financial constraints through taxpayers' money, but they must also carefully ensure that funds are not wasted and laws are observed. At the same time, modern states are also running after too many expensive projects. California tried to be ahead of the game in alternative energy, university education and health care – with the result that its finances are now not in a brilliant state (to say the least).

Indeed, California has been boasting about America's most ambitious initiative to promote solar energy, known as One Million Solar Roofs. The state promised to pay \$2.9 billion in rebates over ten years to households and businesses that install solar panels. The federal government also contributed with a tax credit of 30 per cent of the cost of installation.[11] Subsidized solar panels are a good start, but:

- *if* the process cannot become self-sustaining,
- *then* its future is bleak, because it would not attract new investments.

---

10 Just think about terrorists taking the European Union hostage by threatening to blow up the Saharan and Moroccan solar farms.

11 'Tilting at Windmills', *The Economist*, 16 November 2006.

Germany has been more successful in dismantling the barriers to solar energy as evidenced by the fact that in mid-2008 it had already met the European Union's target that 12.5 per cent of electricity should come from renewable sources. In the process, the country's manufacturing industry specialized in renewable energy equipment – and the latter became an export item alongside cars and machine tools.[12]

An inflection point in this switch has been the Renewable Energy Sources Act, passed in 2000 , amended in 2004 and known as the EEG. This encouraged investment by cross-subsidizing renewable electricity fed into the grid. The law says that electricity produced from renewable sources must be purchased by utilities according to a generous feed-in tariff at much higher-than-market rates fixed for 20 years. Preferential tariffs, however, increase the cost of power to businesses and consumers.

5. Preferential Tariffs Feed Inflation

Fairly often, a new industry is pampered with preferential tariffs to enable it to stand on its own two feet. But *if* standing up by itself does not happen within a reasonable time and the new industry continues to require public money in order to survive, *then* subsidies become counterproductive. They promote featherbedding and lead to a negative public response.

For instance, in Germany, the government's policy has been that roof-mounted photovoltaic systems can sell power at €0.49 per kWh until 2027 (the law's limit). That's roughly 700 per cent of the German going wholesale price for photovoltaic. A fixed rate at seven times the current price of kWh:

- allows investors to reap a lucrative return on investment,
- but also significantly increases the cost of energy, as utilities that buy power at these higher rates pass the extra costs on to industries and customers through higher electricity bills.

Needless to add that the end users of electricity are not happy with the so-called *feed-in tariffs* which oblige utilities to buy all the electricity generated from renewable sources such as windmills and solar panels at a high price, for a long period. To get out of the bind, the German government has arranged that the tariff offered to new installations falls slightly each year – which is the usual bureaucratic meddling solution.

Far from being only a German phenomenon, feed-in tariffs have been adopted in many European countries and beyond, but governments have started to have second thoughts. Even Germany, which pioneered them, seems to be finding it difficult to come up with a new formula that:

- is generous enough to spur investment,
- but is not so lavish that it overburdens business and consumers.

Amid complaints that the feed-in tariff for solar panels was generating an unaffordable proliferation of solar farms, the government decided to double the rate at which the

12 However, most of Germany's electricity still comes from coal-fired and nuclear plants.

tariff decreases each year, from 5 per cent to around 10 per cent. At the other end of the equation, the glut of solar panels led to a fall of about a quarter in the price of building a solar farm – hence the subsidy is still too generous.

The economics of wind power (section 6) present similar challenges. A standard 2 mW turbine costs about €2.8 million ($4.1 million) and, under good conditions, earns about €270,000 per year selling electricity at the market rate. That revenue can double to about €500,000 or more with special state-mandated preferential tariffs paid by utilities and therefore by consumers.

But there is a downside. In a way similar to what happens with solar panels when the sun does not shine, wind farms can only be operated at two of the four wind-speed ranges. Calm air produces no power, but gales faster than 90 or 100 kilometres per hour necessitate shutting down the turbines to prevent serious damage. All this suggests that, as we will see in section 6, wind farms must be carefully managed.

Denmark is typically seen as a good example of successfully using wind power (Chapter 9). What the favourable reviews forget to state is that wind power lacks dependability as a consistent energy source. This has made the Danish electricity market more volatile and less efficient than it used to be. Critics say the country's claim that wind power accounts for 20 per cent of electricity production is quite misleading because much of what is generated is surplus to local demand – or, alternatively, presents a deficit.

- When the wind is blowing strongly, Denmark offloads excess power to Germany, Sweden and Norway at cut prices.
- When the wind stops, it has to buy power from its neighbours at higher rate, or increase output from its coal-fired power plants.

In the aftermath of all that, Danish citizens face one of the highest electricity bills in the European Union. Danish tariffs must not only account for the negative and positive premiums of selling and buying electricity, but also for preferential rates paid to thousands of local turbine owners. This should not be expected to last forever.

There is also the risk that a policy promoting premiums for renewable energies can change overnight, for reasons irrelevant to energy production. An estimated 2,350 wind turbines are installed in France. Until recently they benefited from incentives. In 2010, however, the majority of them will be taxed through a new law which taxes assets as a way of replacing the other income sources (like the one from *taxe professionalle* cancellation.[13]

The foregoing examples play into the hands of critics, who also comment that high premiums paid for renewable energies feed inflation. This becomes an important argument where renewable energy now accounts for a significant chunk of total production, as it currently does for 15 per cent of Germany's electricity.

There are also other problems – outcomes of a policy betting on more and more renewable energy. For instance, Germany's power grid cannot accept much more wind and solar power without extensive and expensive restructuring of the power transmission network – an unexpected consequence (see also Chapter 15 on smart grids).

American renewable energy projects have also experienced headwinds. Investors are fickle and go with the prevailing trend. Back in 2006 they were falling over themselves to

13 *Le Canard Enchainé*, 23 January 2010.

finance start-ups in clean energy technology. At the time, Venture Business reckoned that in 2004–2006 investment by venture capitalists and private-equity firms has quadrupled to $2 billion.[14] Then came the 2007–2011 economic crisis, and money dried up.

After the economic and banking crisis had started, New Energy Finance, a research firm, calculated that the amount of project finance devoted to cleantech projects worldwide fell by almost 25 per cent in the third quarter of 2008.[15] For their part, big American utilities have been busy slashing their investments in alternative energy. For instance:

- Florida Power & Light cut its planned investment in wind power for 2009 by 400 mW, and
- Duke Energy of North Carolina chopped $50 million off its budget for solar power.

The drying up of money in terms of subsidies and investments led several experts to predict that despite its glow, the sun will remain largely untapped as a source of electricity. The world's biggest solar farm in California's Mojave desert, where more than 400,000 mirrors cover 10.3 square kilometres, was built in the 1980s and still churns out 354 mW of electricity – enough for 900,000 homes. But since then no more large solar plants have been built.

A contrarian opinion in the United States has been that thanks to a confluence of different factors, such as federal tax credits, an increase in federal funding for renewable electricity standards and public pressure, the sun will be making a comeback. We shall see, but we should not forget that, as already mentioned, solar energy is not without ecological problems.

The easy part is that, once installed, the panels have to be hosed or brushed off several times a year. More preoccupying is the fact that they are made of silicon, and the production of silicon is not environmentally-friendly. In addition, damaged or worn-out silicon panels must be disposed of, adding to the electronic waste (e-waste) problem.

Some are now saying that it is questionable whether solar panels are significantly better for the environment than fossil fuels. Silicon-manufacturing factories are energy- and water-intensive, and the manufacture of silicon wafers uses energy from traditional fossil-fuel power generators, perpetuating the old pollution problems.

To this should be added the aforementioned problems of disposal towards the end of a panel's life, keeping in mind that these are frequently doped with toxic materials like arsenic. This sees to it that their manufacturing is not as environmentally benign as it is generally assumed.

It is generally known, for example, that China has huge environmental woes, but less appreciated is the fact that solar power is one of them. More than 20 Chinese companies produce polysilicon, the essential component of solar wafers, and the combined capacity of these plants is estimated at 80–100,000 tons, more than double the worldwide capacity outside China. The environmental problem derives from the fact that few of the facilities have installed the expensive equipment necessary to recycle the polluting by-products. Instead, some factories are simply dumping leftovers and waste, which contain high concentrations of chlorine and hydrochloric acid, in empty fields or they are stockpiling

14 *The Economist*, 18 November 2006.

15 *The Economist*, 8 November 2008.

it in drums.[16] So, when talking of alternative energies, think again about the after-effects prior to reaching a decision.

6. Wind Power[17]

According to its supporters, there is no better alternative energy source than wind power. The sails of windmills have been turning for over 4,000 years, even if it is only 60 years or so since engineers targeted the wind as a means of producing electric power and only recently that wind farms saw the light of day.

Not everything is 'plain sailing' with these wind farms and what their proponents suggest should be put under scrutiny. In May 2010 the French government said that it was going to introduce new legislation with tougher regulation for wind turbines, because of:

- failures in installing them, and
- a great deal of lost cost-effectiveness.[18]

This followed on the heels of some unexpected consequences. A major wind turbine plant installed in France in 1998, still far from being at the end of its useful life, is now a heap of iron because the direction of the wind has radically changed. In addition, its past performance suggested that this was not a wise investment in the first place.

This is not the only case in which the economies have been turned on their heads. In France 2,600 wind turbines represent only 1.5 per cent of power consumption (80 per cent comes from nuclear power) but the investments required to achieve this 1.5 per cent are out of proportion to the deliverables.

Hype has infiltrated wind turbine economics in a big way. Some articles suggest that wind turbines pay for themselves after of 15 months' usage. That's fantasy. A wind farm is really productive only for a maximum of 92 days per year. This means that:

- Every ton of CO_2 'saved' by wind power costs €266 ($346).
- First-class house insulation achieves the same saving at a cost of €2 ($2.6).

These lop-sided investment economics document the difference between theory and practice. Theoretically, the wind provides an inexhaustible source of energy,[19] and with strong winds it is like turning copper into gold.[20] A modern wind turbine with three sails can capture up to 60 per cent of the wind's motor power. That looks great, except that:

- *if* there is a wind, it often blows on average ten hours a day, which means a 60 per cent underutilization, and

16 *Business Week*, 28 March 2008.

17 See also in Chapter 9 the case study on wind-power farms on the island of Samsø, Denmark.

18 *24 Hours*, 4 May 2010. See also the Appendix to this chapter for the legal risk associated with wind turbines.

19 For a positive example, see, in Chapter 9, the case study on wind farms on the island of Samsø, Denmark.

20 As we saw in section 5, however, gale-force winds of over 90 km/h (about 25 metres per second) can damage the turbines, obliging to close them down.

- *then* the wind's on-and-off presence reduces the aforementioned 60 per cent to 36 per cent, and that's too little.

Another technical problem (always supposing that the wind blows) is wind quality, defined as *changeability*. Wind conditions not only vary greatly with location, but in any given place the wind is inconstant: it comes and goes; it is strong and weak; it takes that direction or the opposite one. In short, there are limits to the ability to get the most from investments made in:

- wind turbines, and
- the conversion of air's kinetic energy into electricity.

Worse still, there are legal issues – or, more precisely people problems. In a society like ours which is always prone to litigation, it is advisable to listen to what people say about the effects of wind energy. As reported in the French press, a whole village in Brittany, France, went to court because every citizen allegedly got persistent headaches and reportedly could not sleep at night because of a nearby forest of wind turbines.

This does not mean that wind power should not be exploited but rather that, like solar power, it should not be taken as being manna from heaven. Nor should anyone be carried away into cloud cuckoo land by his or her enthusiasm about having found, or rather having revived, a cleantech solution.

At end of 2008 wind provided only about 1 per cent of America's electricity. However, wind enthusiasts predicted that by 2020 that figure would rise to 15 per cent, although they failed to explain how this near-miracle would be achieved and how it would be financed in a sustainable way.

The pros said that the fact in May 2008 T. Boone Pickens, an oil tycoon, announced a deal with General Electric to build, at a cost of $2 billion, a 1 gW wind farm, the world's largest, proved that wind farms were taking off in a big way. This argument forgot that one swallow does not make a summer – and, indeed, two years down the line, the evidence is that it did not.[21]

Using taxpayers' money, rather than private funds, some European countries are way ahead of the United States in terms of wind power. Section 5 mentioned Germany as an example. Sweden is another, and the same is true of Denmark (Chapter 9) and Spain, but it should not be forgotten that initial investments have generally not been followed up with new important installations.

- In a well-organized society nothing happens without a reason, and
- the reason is that the economics of wind (and solar) power are not that brilliant.

Preferential tariffs feed inflation when they become the rule. This poses serious policy dilemmas that will not disappear by wishful thinking.

Under the European Commission's proposals, in its energy mix, each member state has its own legally binding target for increasing the share of renewables, such as wind and solar power. But not every EU country puts up the public money needed to promote alternative energy targets. (In several, what might have been available in public funds fell

21 'Pickens Calls off Plans for Vast Texas Wind Farm', *The Washington Post*, 8 July 2009.

straight into the big banks' bottomless pockets to save them from self-inflicted wounds and outright bankruptcy.)

In Britain, for example, alternative energy generates a mere 2 per cent of the country's energy needs, the lowest of the European Union's biggest economies. Under the EU proposals, Britain must increase the share of its renewables to 15 per cent, shifting investment to places where green energy can be produced. But all the evidence seems to suggest that Britain is looking at nuclear energy production as better investment rather than subscribing to the argument that 'when in doubt about energy policy, build more windmills'.

By contrast, that latter argument was the thrust of the energy bill approved by the US House of Representatives on 4 August 2007, even if legislators disagreed about how to reduce the fuel-thirst of American cars. Proponents of the Renewable Electricity Standard (RES) say that it will help reduce both America's dependence on imported fuel and its greenhouse-gas emissions without raising power prices. But not everyone is convinced, a fact which has shown up in the deliverables during the three years following the RES. Politicians from southern states have complained that their part of the country:

- is not blustery enough for wind power,
- or mountainous enough to provide much hydropower, and
- not sunny enough for solar power.

Southern Company, a big utility, claims that complying with an RES directive of 20 per cent of energy coming from alternatives would cost it over $26 billion by 2030. Some environmentalists have argued that there is plenty of scope for generating relatively cheap green power from farm waste in the South, or from offshore wind farms. But other protectors of the environment objected that offshore wind farms deface the seascape.

In a nutshell, this argument and counterargument rounds up some of the issues associated with wind power. Turbines cannot be installed just anywhere one fancies. Meteorologists must be employed to scout the land for best places. It is not just a question of when the wind blows, but also of how powerfully. Over the years, a difference of as little as a couple of kilometres an hour in average wind speed can have a significant effect on electrical output.

Where to place the wind farms is another challenge. Siemens' wind-power business unit reckons that the whole of Europe could be powered using offshore wind, but competition for space near the coast is stiff because this is also a tourist attraction. (Siemens' goal is difficult to achieve if only inshore sites are available.)

Nor are European environmentalists likely to sit idle if the coastline is defaced by turbines. Siting giant wind turbines within view of coastlines also causes conflicts with fishermen, shipping and the armed forces – and let's not forget the reaction of US environmentalists against wind farms defacing the pristine seascape.

Theorists say that floating turbines moored far out to sea could avoid such problems, and indeed deepwater turbines could generate much more power than in shallow water. In practice, however, this has so many negatives (including security) that while it could be practised here or there (as it is in Denmark), its widespread use is the stuff of comic books.

Moreover, no matter what the chosen solution may be, with large-scale investments meteorologists must sit in control centres making detailed forecasts (a day or two ahead) to help the energy company manage its power load. Solving power-balancing problems is a task not for the mechanical engineers who build the turbines, but for the electrical engineers who run the power networks. The latter also have to take into full account the utility's other power production facilities:

- Effective wind power depends on farms of expensive turbines; it's not like old Dutch windmills operating stand-alone.
- Nor does wind power offer the relative simplicity of solar panels installed by individuals and families to heat water during the summer months.

From an electrical engineering perspective, a country's entire grid system should be up for re-evaluation and redesign. It makes no sense to install expensive wind farms to generate electricity and operate a national grid according to nineteenth-century concepts. (A reference to this was made in section 5 in regard to Germany's wind-power production and problems posed to its power grid.)

In fact, the whole concept of power grids must be turned inside-out, from smart-grid concepts (Chapter 15) to direct and alternating current. New solutions should use direct current (DC), rather than alternating current (AC) – which was adopted as standard over a century ago, when the electrical world was different.

- DC is better suited to transporting power over long distances, and
- the less power is lost the better it is, particularly when its production is very expensive.[22]

The argument about a profound power grid renewal goes beyond DC/AC to embrace other technical issues and to employ knowledge engineering to constantly monitor the load and calibrate electricity distribution over the network. For instance, targeted consumers could be taken offline – with their prior agreement and in exchange for a lower price – if that load surges beyond a pre-set level. Provided that it is knowledge-enriched and reliable, the intelligent grid's software would be able to identify both:

- particular circuits, and
- specific appliances in a home, office or factory.

Summing up the case for wind power, its potential has been put to good use, but it is beginning to look as if its growth curve will emulate that of hydroelectricity. By late 2009 wind generated only about 1 per cent of all electricity worldwide, although it provided 20 per cent in Denmark, 10 per cent in Spain and 7 per cent in Germany. Wind power is also on the rise in the United States, which claims to have overtaken Germany to become the world's biggest wind-power generator.[23]

Wind power is attractive because it is widely available and renewable, but its technology must still be significantly upgraded. According to General Electric, when it entered the turbine business in 2002, on mechanical engineering side, the average turbine was out

22 DC cables can also be laid on the seabed where the presence of water would dissipate an AC current very quickly.

23 America's windy and lightly populated heartland states are ideal sites for wind turbines, so the country's wind-power industry seems poised to grow.

of commission 15 per cent of the time. Nowadays its downtime is less than 3 per cent, significantly reducing the cost of the energy taken out of these turbines, to about 8 cents per kWh, and it is still falling. (Coal power is still cheaper, though, at about 5 cents per kWh.)

But as already indicated, major investments have to be made, all the way from electrical engineering to plant, grid and system management. Whether we talk of solar, wind or other renewable energies with expensive farm-type solutions, they all require equally expensive grid infrastructures which in turn rest on complex and time-consuming processes of cross-country agreements in a community of states like the European Union.

7. Geothermal Power

The geothermal power we discuss in this section can be a solution for mass society, as demonstrated by power plants in Italy and New Zealand (more on this later). There is also a sort of geothermal power for heat pumps, which find their market in individual houses and small buildings. The problem with it is that:

- it requires investment beyond the reach of many families,
- it is not practical in every case, and
- long stretches of land are put out of bounds on account of the wells that are needed to benefit from the difference in water temperature between the surface and 150 to 250 metres deep.

For example, in Switzerland the authorities correctly give priority to the protection of clean water, having divided each town's territory into blue, red and green zones. Blue is the fully protected clean water catchment. In the red area, which surrounds the blue, drilling for geothermal energy is not permitted; only the green zone can be exploited.

There are sound reasons for this topological division, but it also works against the soft version of geothermal power. One of the problems with alternative energy is that the distance from source to heat engine must be short – preferably less than 12 metres, but no more than 15 metres. Longer distances make the installation costly and inefficient. A smaller house, of say 100–120m^2, could use the air–air heat pump. It works with electricity, but requires only a third or a little less than a third of what would have been needed if the house was heated by electric power alone.

The bigger brother of this heat pump is air–water. It is more expensive but theoretically (and according to the vendor I used) at least it can be used with bigger private houses and small buildings. However, my experience of having one of these types of heat pump installed was dreadful, and I feel obliged to convey this to the reader.

- The manufacturer of the heat pumps overstated by far what they could deliver, with the result that its presence became more or less redundant.
- The heating system, which included two previously-installed oil burners, was not optimized; the heat pump's engineers worked only in their corner of the system.[24]

24 The manufacturer/ vendor was not a small unimportant company, but a well-known one: Stiebel-Eltron.

- To avoid installing the necessary programmable thermostat the company which installed the heat pump manipulated and destroyed the oil burners' control panel, by using it as overall thermostat.
- Lack of optimization, reversal of priorities concerning system components and destruction of the control panel saw to it that the oil burners consumed 117 litre per day versus 56 litres per day prior to the heat pumps.
- The local electricity company, which by law had to pay a subsidy for alternative energy, found a minor technicality to pay nothing.

This was a case of the unwilling being asked by the unable to do everything wrong. My advice to would-be users of air–water heat pumps and other alternative energy, is: prior to signing a contract, oblige the vendor and installation company to deposit in a bank double the stated costs with an affidavit that the client can use that money for damages should the installation turn on its head.

It should be appreciated that the technology of this type of installation is still uncertain, and the outside temperature greatly influences its performance. The result is that these are largely individual solutions which should be thoroughly examined before implementation – with the feasibility study based on cost, effectiveness and a well-rounded environmental protection perspective. They are not solutions for a mass society's insatiable appetite for energy. Hence, the deliberate decision to focus, in this section, on geothermal power in its more classical and massive form.

The massive form of geothermal energy employs the huge reservoir of heat that lies beneath the earth's surface, theoretically making it a renewable source that does not pollute. Moreover, according to its supporters, continuing advances tend to allow engineers to use previously unexploitable, lower-temperature reservoirs that are hundreds of times more abundant than those used in the past.

All this sounds like very good news for geothermal power. I used to think along similar theoretical lines before visiting New Zealand's Wairakei geothermal power station in the late 1980s. Technically, Wairakei's statistics are impressive. It provides 3–4 per cent of New Zealand's electricity needs. Power is produced through 120 wells of which 60 are working at any time. New Zealand derives an average output of 150 mW or 200,000 horsepower (hp). But let's look at some other facts.

Wairakei Geothermal is the world's second largest power station of its type; the first was built in Ladarello, Italy (more on this later). In both of them, vast quantities of cold water are used to condense the steam – which is the raw material of geothermal power plants.

- This increases the plant's efficiency.
- But it uses vast amounts of scarce potable water (Chapter 12).

While corrosion seems to be minimized by the careful selection of resistant materials such as stainless steel and Teflon (where needed), water pollution *is* rampant. Approximately 4,000 tons of borewater *per hour* runs to waste. At present, this waste water is discharged into the Waikato River although investigations are being carried out to see if it is possible to inject a large proportion of it back into the reservoir – even though reinjection may have detrimental effects on steam production.

There is no lack of research aimed at minimizing or avoiding these effects, but the problems are complex and there is no perfect solution yet in sight. With existing technology, thermal pollution of the waterway is unavoidable. The Wairakei geothermal station is at the riverside precisely because:

- very large quantities of water are needed to condense the steam, adding thermal pollution to chemical pollution, and
- this contradicts water preservation goals – a fact which escaped the attention of the Greens and others parading in 'environmental protection rallies' and in expensive but ineffectual world conferences.

Rampant thermal pollution of a major waterway is only one of the negatives. There are also other major problems. Wairakei Geothermal *pollutes* the nearby river with arsenic, mercury and plenty of chemicals that extinguish fish life.[25] That's the bad news. Chemicals dissolved in waste water are highly detrimental to river life., The annual amounts are approximately:

- chloride – 66,000 tons
- sodium – 40,000 tons
- silica – 16,000 tons
- potassium – 5,000 tons
- calcium – 1,000tons
- boron – 1,000 tons
- arsenic – 150 tons
- mercury – 4 kg.

Nor is this the total list of pollutants resulting from geothermal energy. The temperature of the steam when it arrives at the power station is 114°C (238°F) and 148°C (198°F), for steam pressures of 1.6 bar absolute (25 psi absolute) and 4.5 bar absolute (67.5 psi absolute) respectively. This is 99.5 per cent pure steam.

- The rest is gas, 95 per cent of which is CO_2, which is inert.
- The remaining 5 per cent is H_2S and constitutes the main problem.

As if one calamity were not enough, another station, Ohaaki (110 mW) located 30 kilometres north of Wairakei is also in production, while other steam fields are under investigation (at Mokai and Ngawha). More power stations and more mercury/arsenic pollution will surely follow since it is estimated that New Zealand has reserves of 1,000–2,000 mW of geothermal energy suitable for electricity production – in short, a capacity for immense destruction.

Of course, such statistics make deceptive reading. There goes another great hope of prosperity: almost limitless electric power that the earth itself is supposedly yearning to supply. The associated pollution sees to it that, rather than a hope, geothermal power is a curse – and a major one indeed.

25 From a Wairakei Geothermal Power Station publication, February 1985.

Another country in the Pacific Rim where geothermal power might have had a bright future is the Philippines. The archipelago is part of a geology that is devastated by volcanic eruptions from time to time. In the Philippines, the Japanese financed a 112 mW geothermal power plant on Leyte in the early 1980s. Later on, they committed themselves to fund a second, but that plant was scrapped after technical problems developed at the first geothermal installation.[26]

The practical negatives these and other case studies document far outweigh the argument one often hears that 'geothermal is an inexhaustible source of energy', and that drilling through the earth's crust for just three kilometres 'would suffice to confront the energy needs of billions of people for 100,000 years'. When we contemplate new departures, it is wise:

- to examine both its present and potential 'pluses' and 'minuses', and
- learn important lessons from historical precedent, to avoid talking like a fool.

For instance, there must be good reasons why that 'inexhaustible source of energy' whose history dates back a whole century, has not taken off in a big way. The first geothermal energy plant saw the light of day in Tuscany in 1913. Interestingly enough, it is still working today, with an output about equal to a medium-size nuclear plant. Italy also has some other geothermal plants, but nothing spectacular.

What 100 years of experience prove is that geothermal processes are doable but at the same time questionable because of their huge pollution potential. This explains why, after a century of having proved its feasibility, geothermal power has not taken off even in countries with plenty of volcanic activity.[27] This is a sort of covert rejection, and it means that its future prospects are limited.

8. Hydrogen as an Energy Source

If 'renewable' geothermal energy *is* such an environmental disaster, maybe we should capitalize on hydrogen of which, again theoretically, there is plenty. Take the transport industry as a practical example. Interest in the use of hydrogen is several decades old.

'The main problem,' said Wolfgang Reitzle, then BMW's research and development chief in the early 1990s, 'is how to produce and distribute hydrogen.'[28] This is more than a matter of service stations. There is also the challenge of hydrogen production, involving challenges in chemistry and physics.

There are no secrets about hydrogen's production. It is made by electrolysis. An electric current passes through electrodes immersed in water, splitting water molecules into:

- hydrogen, and
- oxygen atoms.

---

26 *Newsweek*, 6 February 1989.

27 European countries with a volcanic background, where geothermal energy can easily be brought at earth's surface, are Sweden and Iceland. No doubt they have pondered its 'pluses' and 'minuses'.

28 *BusinessWeek*, 4 March 1991.

Hydrogen, textbooks say, is as innocent as an element can be. Its combustion ends up as water, with no carbon dioxide gas emission. But there is a hitch. The intrinsic problem in developing a hydrogen-based power economy, as opposed to one that relies on carbon compounds, is that a great deal of energy is required to produce hydrogen.

Removing hydrogen from water calls for a lot of electricity, which somebody has to provide, probably with oil or coal – thus defeating the purpose of a hydrogen engine – or with nuclear power plants. (So far, car-makers are using petrol engines that have been modified for hydrogen, though not yet optimized.[29]) And who remembers that the *Hindenburg*, the airship which crashed in a ball of fire in the United States in 1937, was filled with hydrogen?

Despite its challenges, some of which are summed up as the 'Hindenburg connection', some people consider a hydrogen-based power economy a bold vision. Others are of the opinion that, if it happens, it will move very slowly and only to a limited extent.

- For two decades, supporters of hydrogen have been arguing that it will still prompt stupendous growth for firms in the business.[30]
- This has not yet happened, even though 20 years is a long time for a new process to get off the ground.
- Critics say 'Beware', and so far they can show that there has been no real breakthrough towards a hydrogen economy.

In addition, costs matter, and hydrogen solutions are far from being cost-friendly. Today's electrolytic approaches need two to three times more money than the classical method of chemical hydrogen production, which itself presents problems. And if hydrogen is used to power cars, a transportation network will be needed.

In Germany, since 1940, kilometres of pipework channelling hydrogen have been in existence – more than 100 kilometres in the Ruhr alone. But other countries must start from scratch, and so far they don't seem inclined to go ahead with huge investments of that sort.

Motoring on hydrogen becomes a feasible proposition only *if* one ignores the need for large amounts of electricity to produce hydrogen, as well as investments, costs, production, distribution and the risks involved,. Back in 1990, Mercedes-Benz clocked 500,000 miles on prototypes, including a three-year road-test in Berlin of five cars and five vans. And, having done so, the company put the hydrogen file in the time capsule.

Interestingly enough, hydrogen-powered buses were a solution that passed the test, but for obvious reasons there has been no great enthusiasm for buses running on hydrogen. Critics pointed out that hydrogen is dangerous to handle, requires expensive new service-station gear, and at the end of the day this is another parochial solution. At best, one day hydrogen power might find a niche, but it is no answer to our society's massive and growing energy consumption problems.

---

29 Fuel cells running on hydrogen have been touted to replace the internal combustion engine.

30 *Business Week*, 4 March 1991.

9. Clean Coal and Other Alternative Energies

One of the newer energy technologies is carbon capture and storage (CCS), popularly known as 'clean coal'. Its proponents say that it has the potential to save the world from global warming, and its believers make sure that CCS features prominently in practically all the principal blueprints for reducing greenhouse gas emissions.

Part of the excitement about it comes from the fact that other options are so limited, as this book has documented. According to the International Energy Agency (IEA), the world will need over 200 power plants equipped with CCS by 2030 to limit the rise in average global temperatures to about 3°C. (Note that 3°C is, in any case, a bigger increase than many environmentalists would like to have, their target being less than 2°C.)

The carbon capture-and-storage business, however, is full of undocumented hypotheses about its contributions. These are bound to remain undocumented since there is not a single big power plant using CCS anywhere in the world. Utilities even refuse to build any, because the technology is:

- unproven,
- expensive, and
- in their judgement, a non-starter.

Most of what is heard about CCS is theory. The pros say that costs will come down with time, but they do not dare to predict by how much, or who should bear the extra cost until then. The environmentalists themselves are divided regarding the wisdom of CCS as green pressure groups worry that captured carbon will eventually leak (which is most likely):

- CO_2 capture, storage, cost, unproven hypotheses and risk of leakage are central to the problem.
- Carbon storage is expensive whatever way one looks at it, and there are good reasons why the scheme may not work at all.

An experimental plant in Denmark that uses monoethanolamine as a CO_2 captor has been running for two years, but the results are inconclusive. Alstom, the French engineering company, has built a plant in Wisconsin that uses ammonia. Experts say that it is too early to talk of deliverables, let alone express any judgement.[31]

In 2008 Vattenfall, a Swedish utility, opened the first power plant with CCS technology at Schwarze Pumpe, in Germany. (The name 'black pump' resonates with the worst characteristics of coal-fired power plants.) This is a pilot project, less than a twentieth of the size of modern coal-fired plants, and, although it is proving to be an interesting experience, the results are tentative and inconclusive.

According to expert opinions, the only serious project aimed at investigating the use of CCS principles in an actual power station has been FutureGen, based in Illinois. In this case, the project was cancelled in January 2008 because the expected cost had risen from $830 million to $1.8 billion.

---

31 Also, Exxon Mobil runs a carbon-capture facility at La Barge, Wyoming, with a network of 5,800 kilometres (3,600 miles) of pipes to carry carbon dioxide to oil and gas fields where it is welcome.

As these references document, CO_2 capture and storage is still an idea that excites only theorists. As far as the process itself is concerned, capturing CO_2 is not so difficult. Carbon dioxide reacts with a group of chemicals (the amines). At low temperatures CO_2 and amines combine. At higher temperatures they separate.

- Based on that principle, coal-fired power-station exhaust could be purged of its CO_2 by running it through an amine bath before it is vented.
- Then the amine can be warmed to release the gas where it will do no harm.

Alternatively, the coal can be reacted with water to produce a mixture of CO_2 and hydrogen in which CO_2 is more concentrated than in 'normal' conditions. All this processing, however, is expensive and, even leaving aside the risk of CO_2 escape as well as other issues, the fact remains that somebody has to pay for all that carnival – and its details.

These details are projected to cost big money, and methods on how to go about them are untested. No wonder, therefore, that governments and utility companies keep out of CCS. By contrast, CCS particularly appeals to eggheads as well as to spend-spend-spend politicians reluctant to limit the use of coal because they come from coalmining districts.[32] In theory:

- *if* dirty coal transforms itself into clean coal with CCS,
- *then* mankind's problems with energy supplies might be solved for more than 100 years.

This is, however, one of the cases where theory and practice come unstuck. There are two problems in clean coal's way: technology and cost. And that's too bad for America, Australia, Britain, China, Germany, India and other countries where coal provides a big chunk of the daily power supply.

Just one of the reasons why carbon capture and storage may prove utterly impractical is plant size. According to various studies, coal plants using CCS will have to be at least 35 per cent bigger than classical ones to generate the same net amount of power. They will also:

- consume at least a third more fuel, and
- will require the extra expense of building capture plant and injection pipelines.

In addition, if the CO_2 storage site is distant from the power plant, even more energy will be needed to move the CO_2. Estimates of the total cost vary widely, but all of them are high. A similar statement about 'scale' applies to the volumes of CO_2 to be handled.

America's electricity industry alone produces 1.5 billion tons, which would mean finding about 1,500 appropriate sites. Apart the fact that nobody really wants to have a CCS storage facility as a neighbour, no one really knows whether the country's geology will allow such mammoth constructions. Europe faces similar problems.

32 Even though everyone knows that coal is the most polluting fuel: burning it releases roughly twice as much carbon dioxide as burning natural gas.

Let me add a footnote to the foregoing paragraphs. The reader should not confuse *clean coal* with the *coal to liquids* (CTL) process. The latter targets diesel fuel from gasifying coal, but it is costly and emits lots of carbon dioxide. Both are negatives.

CCS and CTL have a common link in that *coal to liquids* proponents say that the carbon being emitted can be sequestered. We have seen what that means. Greens are dubious about CTL, pointing out that it is polluting and that the power plants use lots of water. (By extension, the Greens should also have doubts about geothermal energy.)

In the United States, the Environmental Protection Agency states that even *if* carbon emitted can be captured and stored, coal-to-liquids conversion would still emit slightly more carbon dioxide than oil. That's a big '*if*' – to end up with worse results than from oil-burning. It does not look as though CTL has a future – except that plenty of politicians are backing it, and ultimately it may be a case of the dead cat walking.

Among other alternatives, Norway has in the works an ambitious project involving an exotic new technology called *salt power*. This requires tapping both freshwater and seawater at the mouths of the country's many rivers and fjords, diverting it into tanks on either side of a semi-permeable membrane.

- Drawn across the membrane the freshwater dilutes the saltwater, and so helps to build up pressure on that side of the tank.
- The pressurized water is then used to turn a turbine to generate electricity.

The saltwater project's promoters say that it could eventually provide 10 per cent of Norway's demand for power. Up to January 2009 the project had spent ten years and NOK100 million (€12 million, $18 million) so no great sums of money have been budgeted in developing the technology. Norway also started a pilot plant on the waterfront south of Oslo, but the power it produces is far from being cost-effective, let alone competitive.

Still another idea for alternative energy is embedding piezoelectric crystals in roads. According to some, these could generate 500 kWh per 1km stretch of dual carriageway. Innowattech, an Israeli company developing this concept, says that its piezoelectric 'parasitic energy harvesting' recovers energy that would otherwise go to waste, without affecting vehicle efficiency or the road surface. It also adds that this approach:

- can be used with asphalt, concrete or composite, and
- can be installed not only in new roads, but also while resurfacing existing ones.

The way an article in *European Automotive Design* had it, Innowattech is presently working on road, railway and airport runway applications using piezoelectric generators able to recover energy from weight, motion, vibration and temperature changes. These were developed at its research facility at the Institute of Technology in Haifa.[33]

Recycling car exhaust is yet another alternative energy project. Germany's BMW made an experimental car fitted with a device to capture some of the heat lost through the exhaust and convert it into electricity. The thinking behind it is that car engines waste, in the form of heat, as much as 60 per cent of the potential energy contained in oil – and fitting a thermoelectric generator in the exhaust system can capture it.

33 *European Automotive Design*, January 2009.

The generator converts heat into electricity. The mechanism works best while driving at about 130 km/ h, which means along motorways. It is not so clear what a car running at 130 km/h will do with this electricity, nor is it clear how much the mechanism will cost and who will pay for it.

In conclusion, the critique that this chapter has made of the most prominent renewable energies does not mean they should not be tried. They should. Everything that can get us out of the current impasse is worth studying, even if it is only a partial solution. But we should always keep in mind applicability and costs.

Out-of-the-box ideas are welcome, but they should not linger forever. In 1811 the first steamboat on the Mississippi raised eyebrows, as many people thought that although it could go south to New Orleans, it would not be able to return upstream. It did. Some of the newer projects in alternative energy may also go upstream towards a solution to our society's deep power problems. However, we should not dismantle what we already have before we have properly tested a truly better solution.

Appendix: Legal Risks Associated with Wind Turbines

On 4 February 2010 the court of Montpellier (Tribunal des Grandes Instances) ordered GDF Suez, the French energy company, 35.6 per cent owned by the state, to dismantle part of its park of 21 wind turbines. In their decision the judges noted that the existence of a large installation of wind-power production next to wine-growing and other agricultural areas caused trouble well beyond the admissible limits of neighbourhood disturbance because of:

- 'the permanent noise pollution and its impact on daily life,
- 'the visually degraded landscape, and
- 'an evident depreciation of the value of the adjoining estates.'[34]

It should be noted that these 21 wind turbines did not spring up like mushrooms. Nine years earlier, on 21 July 2001, the authorities had given a permit to construct the alternative energy park to La Compagnie du Vent (which was subsequently acquired by GDF Suez). Indeed, the permit allowed the company to construct one of Europe's largest wind-power centres with a capacity of 18 mW. The judicial battle lasted from 2003 to 2010, and was sustained through expert opinions on noise levels and visual degradation ordered by the court.

The case is being appealed by GDF Suez, but the court's order for damages to be awarded to the wind turbine's neighbours is not subject to appeal. This is a case which, by all evidence, will multiply many times over. Let's face it, alternative energies are not free of negative effects and consequences; nor are they immune to legal risk.

34 *Le Canard Enchainé*, 10 March 2010.

PART II

Biofuels, The Green and Blue Revolutions, and Agricultural Policy

CHAPTER 6 Biofuels and the European Union

1. Biomass as Source of Energy

It was stated in Chapter 5 that renewable energy sources, particularly solar, wind and hydroelectric, hold some hope for meeting part of the world's energy needs – but they also present environmental problems. In addition, the amount of energy they deliver is not in quantities that are sufficient to respond to the challenge of an exponential rise in energy usage worldwide. In addition, solar and wind power are costly, requiring preferential tariffs to be sustainable.

A good way to appreciate the exponential rise in energy consumption is to take into account that today the world uses about two orders of magnitude more energy than was consumed at the start of the Industrial Revolution. Even that may be an underestimate, since precise statistics are lacking. Worse still, the amount of consumed energy keeps on rising faster than ever before, as developing nations:

- industrialize and trade,
- raise their transportation quotient, and
- give free rein to urbanization, which increases by leaps and bounds – along with accompanying pollution (see Chapter 11).

It should also be appreciated that although renewable energy sources are generally considered to be non-polluting, this does not always hold true. As the careful reader will remember, geothermal energy creates thermal pollution of waterways and destroys river life, hydroelectric power brings with it environmental challenges, including damage to agriculture (see Chapter 7), solar energy collides against the will of architectural czars, and giant turbines exploiting wind power are said to create headaches and psychological problems – apart from spoiling the landscape as they multiply.

The option we have not yet examined as a renewable source of energy is *biomass*. The idea is by no means a new one. Rather, it is back to basics. Mankind has been using different types of biomass-based energy for thousands of years: from wood in campfires, cooking and house heating to present-day conversion of agricultural crops to fuels like ethanol.

As long as high-energy plants can be sown, grown and harvested, they can serve as a source of renewable energy.

A key advantage of biofuels – also known as *agrifuels* when they are derived from agricultural produce – is that the process constitutes a true alternative to fossil fuel in transportation. Unlike hydroelectric, solar, wind and geothermal power, ethanol competes directly with oil in powering motor vehicles, which constitute the largest and fastest-growing demand for energy. Moreover, a variety of crops can serve as biofuels: sugarcane, corn, soya beans, canola and other agricultural crops, as well as waste from classical agricultural harvests.

Transport, not power production, is the domain to which most attention should be paid in reducing emissions. In mid-2007 the California Energy Commission made a well-documented study of four major polluting industries or, more precisely, end-user sectors. The score was based on millions of tons of CO_2-equivalent emissions per year (in this particular case, 2004). As shown in Figure 6.1, the results demonstrate that transport emits twice the amount of CO_2 produced by electricity generation, and agriculture emits only one-quarter of transport's CO_2.

Most importantly, for the European Union and the United States biofuels are the way of getting out of the yoke of oil-producing countries and their blackmail. This is a theme to which I will return fairly often in this book because the strategic advantage of independence in energy supplies outweighs any risks drummed up by the critics of biofuels.

The argument that food should be used for humans not for energy is a non-starter, because the energy being produced is also for *Homo sapiens*.

The concept underpinning growing energy sources by planting is no different from that which propelled the advent of settled agricultural societies some 9–10,000 years ago. It led to the division of labour and to the growth of a rural community that transformed nomadic living into settlements – and characterized social life for millennia.

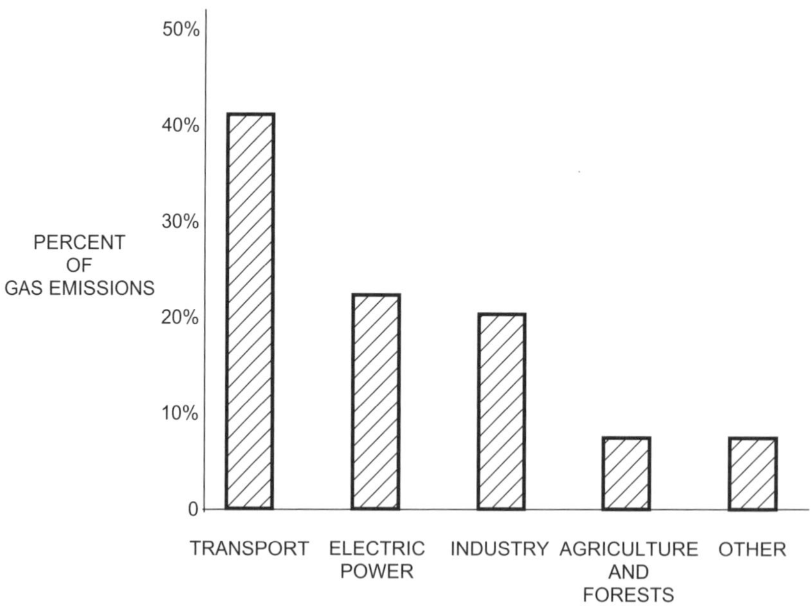

Figure 6.1 California's gas emissions by industry sector in percentages of CO_2-equivalence per year

Source: *The Economist*, 23 June 2007.

Alternative energy through biomass can give new vigour to rural life which, over the last two centuries, has fallen into decay, with significant social and economic impacts. Let's not forget also that urbanization has been promoted by energy, capital, technology and industrialization.

Capital was the enemy of the Left well before Karl Marx. More recently, energy replaced capital as the scourge of the Left because of its abundance of CO_2 and other greenhouse gases, as well as a mire of different pollutants of land, air and water (see Part Three). But there is no such thing as 'ideally clean energy'.

- Chapters 3, 4 and 5 provided evidence that there are no miracle solutions in controlling pollution, other than controlling the energy being produced.
- Birth control, the best possible solution, is anathema to the 'pro-lifes' of the Right and the Left.
- And *if* the amount of energy is drastically cut while the number of people increases,[1] *then* there will be dramatic consequences ranging from rising unemployment to the reduction of standard of living.

Vague measures empty of detail (if not altogether of content), such as the requirement to generate 20 per cent of electricity from renewable sources by 2012, are solutions specific to eggheads. They make no sense because they don't balance the 'minuses' against the 'pluses' – and fail to provide the details (where the devil chooses to reside).

Even *if* energy companies were eagerly signing up to construct wind and sun farms, which is far from being the case, there is simply not enough supply out there. As for installing solar roofs on houses, such projects are stymied by the high cost of photovoltaic cells, red tape, arbitrary refusals by local authorities and the need for continuing subsidies which hits households' energy bills, reducing their standard of living.

This means that plenty of people, from environmentalists to those who govern, are in a bind. They want more energy, but not its negative consequences. The two, however, go together, and no vague pronouncement emanating from *Fiascopenhague* conferences (Chapter 14) will change this fact. Nor is it possible to disregard the fact that energy has been behind five great surges:

- The Industrial Revolution which centred in Britain in the decades preceding 1830,
- The Age of Steam and Railways, 1830–1875,
- The Age of Steel, Electricity and Heavy Engineering, 1875–1920,
- The Age of Oil, Automobiles and Mass Production, 1910–1960,
- The current Age of Information Technology and Telecommunications, 1960–2010.

By all evidence the next age will be characterized by the end of non-renewable energy, and this will bring the end of the industrial age as we know it, if it does not provide a sustainable, renewable energy source at a relatively low cost. It is not far-fetched to think that the next 40–50 years will be called the Age of Biotech and Cleantech.

The key foundational technologies of this new age will be genetic engineering and nanotechnology. But, at the same time, their demand for energy will raise the stakes for uninterrupted supplies, bringing agriculture back to prominence with:

1 As contrasted to better conservation (Chapter 1).

- energy production first becoming equal in importance to food, and
- *then* taking over as the number one goal of expanding agricultural activity.

This in no way means that the road towards energy sufficiency through biomass is going to be strewn with roses. There are political, social and technical problems. Most critical in the biomass revolution is going to be the engineering of enzymes to break down cellulose in the creation of biofuels.

One of the greatest obstacles to massive use of biofuels (also present with other renewables) is *energy density*. The hope for a better solution lies in research. To date, scientists have approached the challenges posed by biofuels in either of the following two ways:

- Biologists have tinkered with enzymes and microbes.
- Chemists have worked with heat and metal catalysts to refine their material.

Recent R&D projects have focused on combining the two methods, to break down long carbohydrate chains to form small, useable molecules. Simultaneously, they are removing the oxygen and maintaining the energy content of the biofuel.[2]

From a biochemical perspective, a promising method is to use glucose, common in food. A new biofuel can be made directly from fructose, which is present in apples, pears, berries, melons, other fruits and, most importantly, some root vegetables. It can also be manufactured from large polymer chains found in trees, cereals and grasses.

- The resulting biofuel is energy-rich and water-repellent.
- It also has a higher boiling point than ethanol and is thus safer.

2. Self-sufficiency in Energy: Case Study on Brazil

In 2010 about 80 per cent of Brazil's electricity came from hydroelectric dams. At times of high demand or low rainfall, however, hydroelectric power has been complemented with energy supplies from thermal power plants mainly using natural gas. An estimated 50 per cent of the natural gas consumed in Brazil is imported from neighbouring Bolivia, which in 2006 nationalized the investments made by Petrobras (Brazil's state-owned oil and gas firm).

To keep ahead of game, in 2007 the Brazilian government gave the go-ahead for a third nuclear reactor at Angra dos Reis – while Ibama, the country's environmental agency, approved the construction of two new dams on the Madeira river (in the state of Rondônia). Public opinion has not been favourable to that move (for the reasons detailed in Chapter 5 in connection to hydroelectric power).

Environmentalists who declared themselves in opposition to the two new hydroelectric dams predicted that their turbines would silt up, while to make space for them, 5,000 families would be displaced (rather than the official estimate of 900).[3] It's no longer

2 Another possibility would be to use genetic modification to create plants that are more amenable to biofuel production.

3 'Scarcity in the Midst of Surplus', *The Economist*, 16 August 2007.

'Greens of the world unite', because those in Europe still favour hydroelectric power, particularly as an alternative to coal or nuclear plants.[4]

As an alternative, local environmentalists have been promoting as 'a Brazilian solution' to the coming electricity crisis – the generation of power from sugar cane, in conjunction with ethanol production. In contradiction to the British study commissioned by the Friends of the Earth (section 3), only a minority of Brazilians are sceptical of biofuels on the grounds that the technology is still fairly new and untested.

- The good news for Brazil is that it produces far more ethanol per hectare with fewer emissions than other countries like the United States.[5]
- The bad news is that, as their critics say, ethanol and similar derivatives are not necessarily so environmental-friendly (more on this in sections 3 and 4).

What the critics say, however, does not discourage Brazil's Greens, who, following a study by McKinsey, suggest that if the area under sugar cane was doubled, a rich dose of fertilizers applied and farming mechanized, Brazil's ethanol output would increase significantly – from about 20 billion litres a year today to 160 or more billion litres by 2020.[6]

To understand this argument, the reader should know that a fairly accurate, but not necessarily precise, rule is that one ton of sugar cane contains the equivalent energy of 1.2 barrels of oil, taking into account the leaves, bagasse and sugar, each of which constitute one-third of the total. Oil prices and sugar prices therefore play cat and mouse with the allocation of sugar cane to sugar or ethanol production, with a clear effect on end costs.

In 2008–2009 in Brazil from over 520 million tons of sugar-cane production, 51 per cent was devoted to sugar and 49 per cent to ethanol (to the tune of 22.5 billion litres.

Under current technology for biofuels there is a problem in balancing food and energy when they both come from the same crop. If Brazil's environmentalists carry the day, which is far from certain, experts suggest that to cover the country's growing electricity needs by 2015–2016, the share of *bioelectricity* in terms of power supply will have to soar to 17 per cent of the country's total needs. This represents an 850 per cent increase from the 2 per cent that was achieved in 2008–2009).

In Brazil people talk of a forthcoming change from an oil economy to a *glucose economy*, with sugar turned into fuels and maybe even the bio-equivalents of petrochemicals, like bioplastics. That would be a *Deus ex machina* for tropical countries, but critics of biofuel-based energy supplies say that this process will replace one set of unreliable sources with another.

We are far from being at that point, and in the meantime Brazil's leadership in biofuels has aroused the interest of investors. In 2006 Cargill, the American agribusiness group, entered into a joint venture in a sugar and ethanol mill. Its key competitor ADM, another US major, has been building a biodiesel plant in Mato Grosso state. It is also seeking

4 Bolivian officials, too, are furious about the Madeira hydroelectric projects, claiming the dams will raise water levels in their country, spreading malaria and displacing families and wildlife – which is indeed an interesting switch in long-held convictions about the unquestionable benefits of hydroelectric power.

5 The United States bets on corn-based production of biofuels.

6 'Scarcity in the Midst of Surplus', *The Economist*, 16 August 2007.

acquisitions in sugar and ethanol. All this suggests that companies are busy positioning themselves against what they perceive as energy forces of the future.

Other critics who oppose the idea of biofuels made of sugar cane on food grounds say that in the coming years ethanol might be responsible for raising food prices and for pushing cattle ranchers into the Amazon. The pros answer that sugar cane currently occupies only a fraction of the acres devoted to cattle ranching. (A wholesome part of these millions of acres has been opened up by cutting down the rainforest.)[7]

An issue waiting to hit the headlines is 'what if' sugar cane destined for biofuels becomes genetically modified, as has happened with corn. Venture capitalists at Votorantim, a conglomerate, have backed two start-ups in Campinas in São Paulo state, working on increasing energy efficiency by sequencing the DNA for sugar cane.[8]

It is neither the purpose nor the intention of this text to take sides in the battle royal for and against genetically-modified (GM) food, but it is likely that what has happened with corn will spread into other agricultural produce like sugar cane – and it will not be easy to distinguish between production for sugar and one for biofuels. Ultimately, genetically-modified foods may become inevitable anyway, as:

- billions of new mouths around the globe wait to be fed,
- classical fertilizers used in intensive agriculture pollute the environment, and
- hundreds of millions of cars need energy sources, too, so they can continue running up and down roads and motorways.

It is no wonder that in Brazil, the number of mills in operation is steadily increasing and an industry association has suggested that forthcoming investment would amount to almost $15 billion. Investments must be commensurate not only with internal sugar consumption and exports, but also with the prevailing government policy that the country's new cars must run on ethanol, or some ethanol–petrol mixture, while oil sold at the pumps is nearly a quarter ethanol.

In the European Union, too, there is a steady increase in billions of gallons of biofuels, particularly in biodiesel consumption. But unlike in Brazil, local production is not growing by equivalent numbers. This is a challenge for both EU policy-makers and farmers, as we will discuss in Chapter 8.

3. Many Problems Relate to Intensive Agriculture, and Not Just to Biofuels

Technically speaking, about 80 per cent of biofuel today made in the European Union is in the form of biodiesel, largely based on rapeseed oil blended with diesel. The rest is ethanol, made from sugar beet or grain and blended with conventional petrol. Various EU member states aim to increase the use of biofuels by transport vehicles, but Brussels issues ambitious goals – which are therefore ambiguous and do not look as though they have been set to be met.

---

7 To make hamburgers a few cents cheaper, and at great cost to the environment.

8 'Lean, Green and Not Mean', *The Economist*, 26 June 2008.

This is a pity because the European Union should be a leader in biofuels for two very good reasons: to become fairly independent of imported oil, and because this is a good outlet for the agricultural industry – surely better than the past policy of subsidies (Chapter 8). Among EU member states, over the last couple of years Germany produced the most biodiesel and Spain led in ethanol production. Poland, too, aims for a higher biofuel share year on year. There are, however, three adverse forces:

- the still uncertain effect biofuels have on the environment (section 4);
- political concerns about using food for fuel, which doesn't bother Brazil (section 2); and
- the cost of current biofuel solutions compared to oil, which creates a limit.

Regarding the first bullet, various studies on the ecological after-effects of biofuels stir up dissent. Yet, not so long ago, mid-way through the first decade of this century, the then French Minister of Finance proudly announced that within three years his country would have switched to 10 per cent of biofuel blend.[9] The minister also demonstrated a model pump for bio-ethanol.[10]

In the longer run, it is a positive development that some scientists have turned against the wider use of biofuels because it allows opposing opinions to be heard. In an article published in *Le Monde*, the French daily, on 25 September 2007, Paul Crutzen,[11] Nobel prize-winning chemist, said that biofuels are an ecological catastrophe. It is good to listen to such statements, but it should be noted that one of Crutzen's basic objections is to the intensive use of the industrial fertilizers necessary for the cultivation of biofuels.[12]

With all due respect to Crutzen, he is confusing intensive agriculture with biofuels. The cultivation of food crops requires large volumes of industrial fertilizers – and this is only extended, rather than introduced, to biofuel cultivation. There are only two ways of correcting this flaw of industrial fertilizers and pesticides which poison everything (a notion to which I subscribe):

1. birth control, or
2. genetically-modified foods.

It is time for politicians and society at large to decide what they want to do about the explosion of *Homo sapiens* on spaceship Earth. This is not an issue which can be left to drift, as per usual policy. Fertilizers and pesticides are needed to help farmers feed the astronomically rising number of mouths. Without doubt, industrial fertilizers produce so much N_2O that they have an impact on the atmosphere and on the ground. After all N_2O is their base. Shall we then stop all agriculture? As far as important decisions are concerned, it does not help to mix two different issues:

9 Governments should also require the development of a new generation of cars known as flex-fuel vehicles (FFVs) able to use blends of up to 80 per cent biofuels.

10 *Le Canard Enchaîné*, 18 October 2006.

11 Of the Max Planck Institute for Chemistry, in Germany.

12 Another opposing argument has been raised in regard to arable land. A study by Scott Wilson Group, a consultancy, commissioned by Friends of the Earth to assess the impact of Britain's Renewables Transport Fuel Obligation (RFTO) targeted land use and found that 10 per cent extra arable land would have to be found to replace food and other crops used for biofuel in Brazil, Argentina and the United States.

- emissions connected to intensive agriculture, and
- those related to the production of biofuels.

It is only normal that the latter will include the former. It is more important to find out where the core of the problem lies. Let's look once again at Figure 6.1.

At least as far as California is concerned, 40.7 per cent of total pollution comes from transportation and only 8.3 per cent from agriculture. A careful study would examine how much this 8.3 per cent would increase because of biofuels, and to what degree the 40.7 per cent will be reduced. One-sided studies cannot be called scientific.

This in no way means that everything is right with biofuels; one particular negative is the fact that they are promoted by organized lobbies. People who follow that line of thinking say that George W. Bush, who wholeheartedly espoused biofuels was not known as being ecologically-friendly. He simply promoted special interests,[13] critics say.

That can also happen with food agriculture. In the early 1990s the European Union faced a crisis of a surplus of agricultural produce, including grains, milk and butter. Since then, 15 per cent of agricultural land has been taken out of production, and its owners, particularly those who were better off, were rewarded with hefty subsidies (Chapter 8). It would have been much better if, instead of keeping the farmland idle and paying for it, it was used for biofuels.

Another argument made by American activists against biofuels is that in the two years following the passage of the Energy Policy Act of 2005 (from the beginning of 2005 to the end of 2006), US corn acreage increased by 10 per cent per year, much of it at the expense of wheat, soya beans and cotton. Trade-offs showed up along the food chain, with livestock feed prices increasing as corn prices rose. (The percentage of the annual corn crop going into ethanol in the United States rose from 6 per cent in 2000 to 20 per cent in 2006.)

However, looking at this issue more objectively, one notices that even before the US biofuels industry took off in a big way, there was increased competition for fresh supplies between rapidly growing industrial processors, livestock farmers who rely on corn and other grains as feed for animals, and less developed countries which:

- benefited handsomely from the West's agricultural surpluses, and
- were suddenly at risk of seeing the manna from heaven reduced.

'More than three billion people in the world [would be] condemned to premature death from hunger and thirst [if food is used for fuel]', Fidel Castro, of Cuban renown, charged in an article posted as the lead story of *Granma* (Cuba's newspaper) on 29 March 2007. At the same time, he argued that energy cutbacks are necessary worldwide (a theme discussed in Chapter 1) and that producing ethanol from sugar residue is fine (Cuba is a major sugar cane producer).

That's a good example of the types of contradictory statement made by people who cannot decide what they want. Not unexpectedly, other politicians sang their own song – the one that best matched *their* interests. Those maintaining that rising food and grain

13 In August 2005 US law-makers passed the Energy Policy Act, mandating the consumption of 7.5 billion gallons of biofuels a year by 2012 (compared with 3.5 billion in 2004). This stimulated a building boom of new ethanol facilities.

prices will provoke a clash among competing ends use China as an example, forgetting that China's problem is the state-promoted rapidly rising standard of living.

In conclusion, it does not make sense to criticize biofuels for problems connected to intensive agriculture. The argument of using intensive agriculture just to feed more mouths born by accident is not tenable, if for no other reason than it is a losing game. On the one hand, intensive use of industrial fertilizers – no matter for what purpose – damages the soil, air and water and, on the other, there is no reason why agriculture should be constrained to what it did in the seventh millennium BC, when it was originally invented.[14]

4. Other Headwinds Encountered by Biofuels

Sceptics say that growing crops for ethanol will employ more energy than it saves, and there will also be plenty of teething pains. Professor Tad W. Patzek, of civil and environmental engineering at the University of California, Berkeley, and Professor David Pimentel, of Cornell University, conducted an analysis of energy *input–yield ratios* of producing ethanol from corn, grass and wood biomass. Published in *Natural Resources Research*, their findings are that, compared to fuel produced:

- ethanol from corn requires 29 per cent more fossil-fuel energy,
- ethanol from grass, 45 per cent more, and
- even worse, ethanol from wood biomass 57 per cent more.[15]

(It may be interesting to note that burning wood has been the most traditional source of energy source, and the practice did not die away. During World War II in occupied Greece, where oil was a most rare commodity, buses ran on energy derived from burning wood. They pulled a specially (and locally) constructed unit on wheels which provided a 'solution' that was better than nothing. It would be ridiculous to even suggest that today's energy-intensive economy would run on burning wood, although it is done in Haiti and elsewhere.)

In their study, Pimentel and Patzek looked at the energy used in producing the crop, including pesticide and fertilizers, farm machinery, irrigation, transportation and energy necessary for distilling the ethanol. On this basis, which includes the pitfalls explained in section 3, they concluded that there is no real energy benefit in using biomass for liquid fuel. Nevertheless it is always good to listen to a contrarian opinion. This says that:

- *if* ethanol production requires large fossil-fuel energy input,
- *then* it is not contributing to environmental protection, or to the reduction of oil and natural gas imports.

14 If politicians and well-wishers really want to eliminate hunger, then they should apply themselves to instituting family planning and birth control, as Mao once did.

15 David Pimental and Tad W. Patzek, 'Ethanol Production Using Corn, Switchgrass and Wood', *Natural Resources Research*, 14, 2005, pp. 65–76.

(Note that the word 'imports' is inappropriate. Biofuels produced in the United States *do* contribute to the reduction of oil imports, no matter what may be the other arguments.)

The researchers calculated that it takes about 131,000 BTUs (British thermal units) to make one gallon of ethanol, while one gallon of ethanol has an energy value of only 77,000 BTU.[16] This amounts to a net loss of 54,000 BTU per gallon which, the study concluded, is not at all a good deal.

People who do not agree with these findings point out that one reason why biomass (which produces less than a fifth of the country's renewable energy) is energy-inefficient in the United States is because most of it comes from the liquor of paper pulp mills, which is burned to economize the heat and power of paper factories. This is not the sort of biofuel we are talking about as a solution to the energy problem. Quite to the contrary, combined with international standards, the new generation of biofuels aims to provide significant improvements in efficiency (see section 5).

For his part, Dr Howard Hayden, Professor Emeritus at the University of Connecticut and publisher of *The Energy Advocate* newsletter, states that 'to produce ethanol with as much energy as we use in transportation would require 1.1 billion acres devoted to high-yield corn production, complete with everything environmentalists hate: fertilizers, irrigation, and pesticides.'[17]

That's about 1.8 million square miles, some 51 per cent of the land area of the 50 states of the Union – too big to be accepted at first sight. Moreover, as has been already explained in section 3, there is no point in debiting industrial fertilizers and pesticides to biofuels because that is the luggage carried by modern intensive agriculture – and the best way to avoid it increasing by leaps and bounds is birth control.

Another document, a 2009 report produced by a team of scientists working in behalf of the International Council for Science (ICSU), a Paris-based federation of scientific associations, concludes that the production of biofuels has aggravated rather than bettered global warming. That study follows up on the 2007 controversial finding by Paul Crutzen (see section 3), which concluded that most analyses had underestimated the importance of nitrous oxide (N_2O) to global warming.

Taken together, the weaknesses and contradictions of the different studies we have been examining suggest that it is not enough to look at only one side of the problem. The needed radical restructuring of agriculture must be studied as a whole, not just the part of it concerning biofuels – particularly in the light of the failure of the Green and Blue Revolutions documented later in Chapter 7. A systems study should go straight to the heart of the argument that:

- the rapid increase in the earth's population makes a dual solution for food and energy both tough and urgent, and
- modern intensive agriculture is highly polluting for the environment, whether its produce is food or biofuels.

Many of the headwinds confronting biofuels originate from the fact that the use of industrial fertilizers has increased sixfold in recent decades – well before any use of canes

16 *EIR*, 12 May 2006.

17 *21st Century Science & Technology*, Spring 2006.

or grains for bio-energy. In addition, different grains have different N_2O patterns. Corn is a nitrogen-leaking plant because its shallow roots take up nitrogen from the soil: this makes it a significant contributor to global N_2O emissions while being a staple food.[18]

The broader issue, therefore, goes well beyond biofuels to the extent to which *Homo sapiens* are overexploiting and polluting the earth, which constitutes the basic resource of its future survival. This is not going to end tomorrow, and the salient problem is how to bring the unsustainable overpopulation under control. Yet, surprisingly, this is not on the agenda of ongoing environmental protection movements, to say nothing of the agenda of UN's touristic, pseudo-environmental protection gatherings.

In regard to the relative inefficiency of present-day biofuels to which Pimentel and Patzek made reference, several projects aim at developing better approaches. For instance, a process for ethanol production from *soft biomass* has been studied by Honda and the Research Institute of Innovative Technology for the Earth (RITE) in Kyoto, Japan.[19] The aim was to overcome production limitations of making ethanol from sources beyond sugar cane and corn, using:

- waste wood,
- plant residue, and
- livestock excreta – which has been, for millennia, the reason for India's sacred cows.[20]

A pragmatic argument on environmental conservation is that the real benefit from biomass would come from displacing fossil fuels not only in terms of a better emissions ratio, but also in making the European Union and the United States independent of foreign sources of oil. However, to be done on any significant scale, this will require developments in technology, investments and plenty of convincing by governments of both businesses and consumers:

- to switch to the new fuel, and
- to do so, even if ethanol has an *entry price*, since no switch is not free of cost.

Overoptimistic goals or claims should be avoided – for instance, the European Commission's demand that biofuels account for 10 per cent of transport fuel by 2020. Critics say this is a rushed schedule, but the Commission insists that it will enforce strict rules on biofuels. This faces two headwinds. One is that conversion to alternative fuels like biodiesel and ethanol in the European Union encounters political obstacles as some countries are delaying implementing EU production targets and relying on exporters like Brazil. The second obstacle is that neither the EU Commission nor the US government (or any other authority for that matter) has agreed on biofuel standards. Practically everybody does as he or she pleases. This is a major handicap which should not have happened. We will return to this argument.

18 N_2O is made by bacteria that live in soil and water. One of their raw materials is the nitrogen-rich fertilizer amply used with intensive farming.

19 *European Automotive Design*, October 2006.

20 The problem is the release of methane. In her book *La Rage et l'Orgueil*, (Plon, Paris, 2002) Oriana Fallaci makes reference to wartime Vietnam where the Hanoi regime obliged the people to urinate and defecate in different holes because the excrements were used as fertilizers and should not be mixed with urine.

5. The New Generation of Biofuels: The Need for International Standards[21]

Articles recently published on biofuels, as well as lectures about them, talk of 'a second generation'. I don't like that term because it is not only the products and processes underpinning them that change, but also the emphasis on how to use biomass as a source of much-needed energy and public attitude about them. In turn, the latter is conditioned by advances in research and engineering.

There is, for example, a new type of ethanol made not from corn or other grains, but distilled from plant waste left in fields after farmers harvest their corn, wheat or barley for food or animal feed. Theoretically, by transporting these plant stalks to *biorefineries*, farmers could rake in an extra 15 per cent in sales, which is good for them and good for society's energy needs. Practically, two questions come up:

1. How widely distributed will be the refineries to serve the stated recycling model?
2. What will happen to the calculation of cancelling out the effects of emissions caused by intensive agriculture, which is one of the basic problems?

Let's always remember that, compared to petrol, cellulosic ethanol (a term used to distinguish it from grain ethanol) when burned yields only a modest reduction in motor exhaust emissions. In addition, the wider use of cellulosic ethanol (to which the term 'second-generation biofuels' is frequently associated) is still years from commercial production in meaningful quantities, one of the reasons being that experts are starting to look seriously at:

- the cost at the pumps, and
- the real amount of CO_2 reduction that it provides.

Finding people possessing the sufficient quality and quality of skills able to analyse complex situations involving materials, costs and CO_2 after-effects is another challenge. Leaving aside the chemistry of the process, while the various arguments about new-generation ethanol still wait for validation, there is (as always) resistance on the part of vested interests in the current energy production and distribution system. This should not be discounted. With biomass,

- energy production will inevitably be much more decentralized, and
- this will tend to unsettle huge industries that have been a dominant part of the economic scene in late twentieth century.

Clearly, any massive changeover to the use of biofuels is a formidable task. On the other hand, a factor that carries weight with governments that look favourably on ethanol is that an energy system based on the biomass will be less dependent on foreign supplies and therefore less vulnerable to disruption. I subscribe to this argument which is of vital importance to the European Union and the United States.

---

21 In February 2010 in America the Environmental Protection Agency (EPA) released the renewable-fuel standard (see the Appendix to this chapter). This, however, is valid only in the United States, not globally.

Over the years, this issue has influenced several US states that have decided to take that road. In Hawaii, Minnesota and Montana all petrol must contain 10 per cent ethanol, while Washington State requires petrol and diesel to contain 2 per cent renewable fuel by volume. In the European Union, which also promotes ethanol, so far only France and Germany have really been tooling up to meet the community's goals.

However, although it is important that governments look favourably on biofuels, it is not enough on its own. A major ingredient of the new generation, *if* it is going to be successful, is specialization in applications substantiated through global standards. Closely connected to this are:

- the current almost random diversity of potential biofuels,[22] and
- the challenges this brings for optimizing the means of combustion.

Under this perspective should be seen present-day efforts aiming at a new generation of biofuels. Instead of ethanol, current research targets hydrocarbon molecules chemically much more similar to those that already power transport engines from trains and cars to airplanes. The produce is becoming known as *drop-in fuels* which can be put into existing fuel tanks and pumped through certain pipelines. According to some opinions:

- drop-in fuels can substitute for diesel and aviation fuel, and
- they also have production-side advantages.

Moreover, they satisfy the demands of end users whose overriding requirements are low cost and consistent quality. New alternative fuels will not get market leverage only by modestly reducing greenhouse gas emissions.

Design engineers in the automotive industry emphasize the urgent need for normalizing biofuels by means of well-defined international standards. This is of fundamental importance if the issues outlined above are to be successfully confronted. Biofuels bring significant implications for engine durability and treatment, as well as for fuel-handling systems designed for oil products.

While classical oil products are refined in accordance to tightly defined international standards, there is nothing like that for biofuels. For example, in the European Union biofuels can be produced from a wide variety of plant material with essentially different characteristics.[23] The success of an effort for a sustainable renewable energy source for transportation evidently depends on many players, and this includes car-makers. (As will be recalled, it was stated in Chapter 2 that the Swedish government works with Volvo and Saab to develop motor vehicles effectively using biofuels, including those derived from the country's forests. In the United States, both Ford and GM are involved in similar projects.)

Automotive engineers complain that measures taken to accommodate biofuels have forced some unwelcome compromises. This is not admissible in the longer run. The car

22 In Britain the Royal Society, Britain's national science academy, published a report that analysed the bewildering range of biofuels on the market. Because their characteristics and after-effects vary, the Society also wants the government to specify targets not for biofuel consumption, but for greenhouse-gas reduction. This was an initiative by the previous British government and it is far from clear what the present Coalition government will do.

23 Vehicle manufacturers may warranty a vehicle for use with typical French B20. But what if the engine is used with B20 manufactured from alternative feedstock, or with B30?

industry still has a lot to learn about using higher concentrations of biofuels, and the rate of advance should be accelerated *if* different targets set by governments are to be met – but this is not achievable if the current diversity continues because it makes it impossible to optimize combustion.[24] This evidently affects:

- emissions performance,
- warranty protection,
- engine performance and other important variables.

The technical problems resulting from the present lack of international standards can be formidable. It is not only extremely difficult but also unwise to develop and calibrate an engine that will run equally well from all biofuel pumps around the European Union and internationally. And it is just as tough to decide on the specification of fuel-handling components.

Technical standards should also account for ongoing research projects aiming to greatly improve ethanol's performance by modifying conventional methods for producing liquid fuels from biomass. For instance, researchers at Purdue University found that adding hydrogen (H_2) during gasification could triple biofuel output.

Chemical engineers say that adding hydrogen suppresses the formation of CO_2 and increases the efficiency of the process. This solution is known as *hybrid hydrogen-carbon process* (H_2CAR), explained in a research paper published in mid-2007 in the *Proceedings of the National Academy of Sciences*. With the H_2CAR concept:

- hydrogen would be harvested by splitting water molecules and added during the gasification step, and
- this will make the process more efficient by suppressing the formation of CO_2 and converting all of the carbon atoms to fuel.

(When conventional methods are used to convert biomass or coal to liquid fuels, 60–70 per cent of the carbon atoms in the starting materials are lost in the process as CO_2. This is due to the fact that one is using energy contained in the biomass to drive the entire process. The aim of some new-generation projects is to accomplish the complete transformation of every carbon atom in the feedstock to liquid fuel transformation by supplementing the conversion process with hydrogen from a carbon-free energy source.)

There are also other research projects like *Enerfish*, aimed to produce biodiesel from waste generated at fish-processing plants. This is run by the VTT Technical Research Centre of Finland and involves the cooperation of several Finnish and Vietnamese small and medium enterprises (SMEs). Most of the funding comes from the European Union.[25]

An American project targeting the effective use of biomass for energy purposes converts plant sugars directly into petrol and petrol-blend components, rather than ethanol. The goal is biofuels that can be used at high-blend rates in standard petrol

24 In the European Union, biofuels are principally made from rapeseed; other materials are wheat straw ethanol (the most land-grabbing), cereals, wood and sugar beet (the most economical in acreage.)

25 Entities involved in the project include the National Energy Foundation (NEF) of Britain, TÜV Rheinland of Germany, Technofi of France, Preseco Oy and Vahterus Oy of Finland, the Energy Conservation Centre of Ho Chi Minh City, RCEE Energy and Environment JSC, and AFI-Industry JSC, all three of Vietnam.

engines.[26] This product is known as *biopetrol* and it has a higher energy content than ethanol. Still another research effort aims at converting a wide variety of waste materials, from domestic rubbish to old tyres – not only agricultural products – to biofuel.

The latter process is based on a genome patented by Dr Ralph Tanner, of Oklahoma State University. It uses commercial gasifiers running at 1,093°C which incinerate the material, converting it to gas. Subsequently, proprietary micro-organisms convert the gas into ethanol by consuming the carbon monoxide and hydrogen in the gas stream.

The so-called Coskata process creates a mixture of ethanol and water. Membranes separate the two liquids, resulting in 99.7 per cent pure ethanol. Researchers claim that their system produces 100 gallons of ethanol from one ton of dry material, using less than a gallon of water.[27]

In conclusion, solving the biofuel challenge economically, in an environmentally-friendly way and in a sustainable manner requires well-defined optimal fuel compositions and established global standards. As these references document, there is plenty of concentrated research effort targeting the production of energy resources from biomass and waste. Ethanol will not replace oil anytime soon, but the fact that it has started changing a culture addicted to oil is in itself a major positive result.

6. Gas and Biofuels: The Real Challengers

The benefits expected from any new process have to be calibrated and judged against a challenger. While the general opinion is that energy from biomass is being developed as an alternative to oil, the real challenger for biofuels is not oil (for transportation) and coal (for electrical power production) but gas which serves both industries – as biofuels should do.

Gas emits about half as much carbon dioxide as coal per unit of energy produced. True enough, coal is the cheapest fuel and it is plentiful – two good reasons why it currently produces America's and China's baseload power. On 21 January 2010 it was announced that China imported 35 million tons of coal in 2009 versus 7 million tons in 2008 – a 500 per cent increase in one year.

Costs, among other reasons, see to it that coal-fired power plants run constantly to meet basic demand, with natural gas switched on when demand is higher. But when carbon taxes are added, gas could play a bigger role. A general opinion is that putting a price on CO_2 emissions will change the economics rather radically. A carbon price would mean that gas-fired plants would be switched on more often to replace coal, and biofuel power plants may well be the challengers.

Counted on the basis of *carbon equivalent* CO_2 emissions per kWh, gas is in the range of 100–220 grams per kWh, while coal stands between 200 and 350. In terms of greenhouse emissions, gas is, of course, much more polluting than nuclear energy (Chapter 4), wind, solar and hydroelectric power production (Chapter 5). As already noted, however, the latter three cannot possibly fulfil the ever rising energy requirements, and each of them has its own problems.

26 *European Automotive Design*, April 2008.

27 According to the Argonne National Laboratory, the Coskata process generates up to 7.7 times the amount of energy required while significantly reducing CO_2 emissions.

Section 5 brought to the reader's attention that biomass-based fuels benefit from intensive research and technology advancement. The same is true of gas; whole reserves are, in many cases, underestimated. For instance, the Potential Gas Committee, a semi-official body in the United States, estimated that the country's gas reserves were some 39 per cent above its earlier assessment.[28] Most of this difference came from higher estimates of gas in shale formations, which until a few years ago were expensive and difficult to reach.[29] More recently, new technology for extracting gas from shale and other dense geological formations has caused a jump in America's output of gas and seems likely to boost production elsewhere.

In their belief that these sources are here to stay, several oil majors are building up their reserves of 'unconventional' gas. Medium-size companies are, however, in trouble because in recent years leases in America have become very expensive. (In December 2009 Exxon bought XTO, and, as a recent article had it, Exxon will have no trouble absorbing XTO's debts or paying to develop its reserves. Several other independent gas producers facing similar constraints may also be snapped up in the near future.[30])

Factors affecting gas prices are not limited to exploration. At the beginning of 2010, natural gas prices remained rather weak in the European Union. As in the United States, the market lacks major storage facilities, with the result that the European gas market remains oversupplied.

Still, all over the European Union both larger and smaller companies are actively looking for natural gas trapped in shale, a resource that has transformed the gas market in the United States. Exxon Mobil is drilling in Lower Saxony, in Germany. Shell is targeting trapped gas in Sweden. ConocoPhillips has joined a small firm known as 3 Legs Resources to explore shale gas in Poland. OMV of Austria is testing geological formations near Vienna.

However, the extent of unconventional gas reserves in the European Union is still unknown, although the International Energy Agency (IEA) has estimated it at 35 trillion cubic metres. This is much less than in North America or Russia, although it is a multiple of Europe's conventional reserves,[31] and, as an ancient Greek proverb has it, it is better to have a second string to your bow.

Exploration in the case of shale gas and novel processing methods with biofuels suggest an interesting similarity between these two energy strategies. After the relatively primitive first generation of biofuels, the next generation will require very thorough technical studies and technological advances. So does the discovery and capture of the earth's unconventional natural gas supplies.

There is also another significant similarity. To put it bluntly, finding oil and gas was relatively easy for the previous generation and harder for this one, while future generations may find it impossible. More and more as yet unexplored oil reserves cannot be exploited through conventional methods because (as we saw in Chapter 3) they are located way down beneath the oceans. Something similar is valid when comparing the current generation of biofuel processes and the new-generation projects we examined in section 5.

---

28 *Business Week*, 22 June 2009.

29 According to the committee shale gas accounts for roughly 64 per cent of America's recoverable reserves.

30 *The Economist*, 19 December 2009.

31 The German Research Centre for Geosciences is working on a more detailed assessment.

Nor are the old tools able to confront the challenge. Seismology was the classical approach developed in the mid-nineteenth century to study earthquakes and it was used by the German army during World War I as a technique for locating the position of enemy artillery batteries. In the post-World War II oil boom, seismology was the most important tool employed by the oil explorers, short of actual drilling.

At the heart of this prospecting is comparing the time it takes man-made shock waves to show up at different locations on the surface. But fields that are offshore – thought to be small, deep, or both – or hidden under complex geological structures play games with seismologists. By contrast:

- The tool for unconventional gas exploration is *hydraulic fracturing*.
- This is a drilling technique in which a mix of water and chemicals is blasted underground to create fissures in gas-bearing rocks.

This approach makes it much easier and cheaper to extract gas from shale, coal and other geological formations yielding tight gas. The technology takes account of the fact that tight gas reservoirs can be thousands of times less permeable than those associated with conventional oil and gas, but with effective technology they may become prolific producers. Breakthrough technology, involving state-of-the-art multi-zone simulation is key to unlocking the economic potential of tight gas that has previously been uneconomic or impossible to develop. This is important inasmuch as experts suggest that, as far as oil and gas energy resources are concerned in the coming years, the big four will be:

- tight gas,
- shale gas,
- tight oil (Chapter 3), and
- coal-bed methane (CBM).

In contrast to these 'big four' and their unprecedented requirements in capital and technology, the extraction of energy from biomass is far more straightforward and comprehensible. At the bottom line, new-generation biomass processes (section 5) and unconventional drilling on land or offshore are the only sure way to meet the world's insatiable appetite for energy while paying attention to environmental prerequisites.

Limiting ourselves to only one option – gas or biofuels – should be avoided because, in the end, the only possibility may be coal. Therefore, the discovery of large reserves of gas in shale formations is no signal to downplay biofuels. In addition, unconventional gas reserves cost more and have much less certain prospects than classical gas deposits. This gives rise to the following questions:

- How much can be produced economically?
- How quickly can effective production take place?
- What will be the final cost to businesses and consumers?

Moreover, nobody can at this moment guarantee that extracting gas from shale sources will not emulate developing oil from oil sands which has proved to be considerably more carbon-intensive than extracting conventional crude oil. At a time when energy security is high on every country's agenda, it is hard to resist the wisdom of advancing on two fronts rather than putting all the eggs in one basket.

7. Wholesale Dependence on the Status Quo: The Wrong Bet

Today, all countries depend on oil as energy source, but some do so much more than others. Therefore, they are most exposed to the possibility of oil being used as a weapon by its producers. If they want to, the big oil-producing and exporting countries can strangle the world's economy by controlling the delivery of oil. No wonder, therefore, that France, Sweden and Brazil have sought to liberate themselves from the oil producers' yoke:

- through nuclear energy,
- through a mix of strategies (Chapter 2), and
- by opting for biofuels.

The United States, Japan, Germany, Italy, Switzerland and the Benelux countries are particularly vulnerable to oil producers' blackmail because their level of industrialization requires huge amounts of energy. Japan and the continental European countries do not have any oil resources worth talking about, Britain's reserves are running out, and it seems to me that biofuels are the best bet for coming up from under in terms of energy supplies.

As for the United States, for the first time in its history, it depends on imported crude for more than 50 per cent of its oil needs. This did not happen for lack of trying to find a home-based solution, capitalizing on the fact that the methodology, equipment and skills to drill for oil and natural gas have been vastly improved.[32] Existing evidence suggests that in continental America natural gas and oil is running out.

In a bid to open up more areas for fossil-fuel development, the Bush administration unveiled a proposal for a National Energy Strategy and sought to encourage the growth of the domestic oil industry (in part by offering tax breaks and deregulation). But the results have not been that positive. Anecdotal evidence suggests that the average cost to produce one barrel of oil in America is about double that of a barrel's cost in Saudi Arabia and other Gulf nations, where huge deposits lie relatively near the surface.

Critics say that the Bush administration's plan for new oil findings was a version of status quo – and therefore ambiguous and inefficient. It allowed drilling in the Arctic National Wildlife Refuge, but also directed the Interior Department to permit drilling only if it 'will result in no significant adverse effect on fish or wildlife, their habitat and the environment' – a sugar-coating to please ecologists. This inconsistency violated the principle that:

- you have to make tough decisions, because you cannot have it both ways, and
- if you try to hit two birds with one well-placed stone, you will get neither.

32 For instance, the companies that provide the geological and exploration service have introduced supercomputers to spot drilling sites and sophisticated equipment that can even drill sideways.

This is no endorsement of drilling in the Arctic. It is only written to emphasize that a good commander-in-chief would have decided either to drill or not to drill – without 'ifs'. George W. Bush has not been alone in that effort of 'double hit, double miss'. Other heads of state and prime ministers have followed the same path, and Barack Obama is no exception. His forte is 'in' and 'out' of promises made for the control of CO_2 emissions. After all he has won a Nobel Prize for peace of mind![33]

But all this does not add up to a policy. Law-makers in Washington are being asked the wrong question. The US energy industry has been so certain that Obama's ill-studied unilateral cut in CO_2 and other gases would fail that, months prior to the mid-September 2009 vote, the American Petroleum Institute staged a press conference to predict that the American Clean Energy and Security Act, known as the Waxman-Markey Bill, would fail to pass through the Senate.[34]

The act's key feature was a cap-and-trade mandate to impose increasing limits, or caps, on US industrial emissions of CO_2 and other greenhouse gases. The public was against it, fearing that it would lead to job losses and increased fuel costs. Few voices were raised to say that cap-and-trade CO_2 is not a policy but a mockery (Chapter 10).

Even people in the Obama administration have backed off because the issue is controversial given the backdrop of a depressed economy. Beyond that lies the fear that such legislation would make energy more expensive than elsewhere, and it won't take long before that begins to have a debilitating effect on American exports and the economy.

Critics say that cap-and-trade is a device aimed at perpetuating the status quo. A better approach for the United States would be to follow the road Britain took: switching from coal to gas for energy production which would cut associated CO_2 emissions by about half. By all evidence, the United States has so much gas that it might last for 90 years (section 6), and there is talk of converting the LNG import terminals into export terminals. That's good news for the American economy and for the control of emissions.

- It provides breathing space until the new-generation biofuels are ready, including the urgently needed production and quality standards.
- It makes it possible to avoid friction with Congress, since converting to gas and adopting a solid biofuels strategy is a solution that looks forward, not backward.
- This approach also provides a blueprint for the European Union's energy policy which is presently in a shambles (Chapter 2). Instead of building all these pipelines (and unwisely subsidizing some of them, particularly the most questionable like Nabucco), the EU Executive should have looked at an operational and financial plan based on two legs:
  - short-term, on the conversion from coal to gas, and
  - medium- to longer-term, capitalizing on standardized, advanced–design, next-generation biofuels.

Today's measure, taken out of the fear of adverse reactions by environmentalists and other Greens, is nothing other than extending the status quo – while holding it together with scotch tape. The European Union also needs transborder mergers of energy

33 'War is Peace' wrote one of the banners against the unjustified Nobel Prize quoting Orwell's 'New Talk'.

34 This was the bill the environmentalists had been counting on: it passed the House of Representatives by a slim margin in June 2009, giving the wrong message that carbon legislation was imminent.

companies to create big entities able to put private, rather than public, capital to work to make the dual strategy successful.[35]

The message provided by the references and case studies in this chapter is that today the search for energy is a world apart from days of the oil prospectors, who roamed the earth in search of conventional and easily accessible petroleum deposits. Wildcatters and engineers of large oil companies still travel, but now they are looking thousands of feet under the surface of the earth, beneath oceans, ice caps and mountains. Technology helps the new explorers map those layers of rock holding enough energy resources to be worth investment. Satellite images, airborne radar and real-time pattern recognition (using supercomputers) help geologists map the earth's surface in their effort to detect traces of gaseous hydrocarbons. Seismic surveys carried out on land and water record differences in how rocks reflect shock waves. They give the geologic picture needed to judge the probability of success.

- Aerial surveys measure magnetic fields.
- Magnetotelluric studies measure magnetic and electrical fields.
- Detected variations may signal an interesting rock layer.

Likewise, coordinating the excellent research on biofuels done by independent projects in American universities and providing for quality standardization through a serious international conference – rather than events like the Copenhagen circus of December 2009 (see Chapter 14) – will put technology at the steering wheel of an effort to:

- kick the oil habit, and
- assure self-sufficiency in energy supplies for the United States and the European Union.

Reality, of course, is never quite as tidy as theory, but what I am writing is not just theory because there are successful longer-term examples from France, Sweden, Brazil to Britain (for conversion from coal to gas). What does not exist today, in the European Union and in the United States, is an imaginative policy for energy independence – and this is unforgivable.

Appendix: The Renewal Fuels Standard by EPA

As the largest biofuels producers, the United States and Brazil account for 85 per cent of the growth in that industry. It is projected that ethanol production in America will reach 978,000 barrels per day (b/d) by 2015 from 684,000 b/d in 2009. The use of biofuels has been promoted by the Environmental Protection Agency (EPA) and ongoing commitments to the US Renewable Fuels Standard (RFS).

EPA paints a rather ambitious picture of biofuels' future and wants the amount used as transport fuel to go from 49 billion litres (13 billion gallons) in 2010 to nearly three

35 A policy greatly handicapped by rampant economic nationalism in the European Union. See D.N. Chorafas, *The Business of Europe is Politics*, Gower, Farnham, 2010.

times that amount in a dozen years (by 2022). By all evidence, the largest part of that major increase will come from various advanced biofuels, rather than ethanol made from corn.

The biofuels industry itself is in a process of concentration. Despite the failure of a number of biofuels firms in late 2008 and early 2009, corn-ethanol production has continued to grow. This highlights the efficiency of the various protections it enjoys, but standardization also helps. According to some opinions, the RFS's role in ethanol production and consumption reaching 10 per cent of total US fuel demand by 2014 will be significant.

Such projections did not escape the attention of big Western oil firms, like Exxon Mobil and Royal Dutch Shell. While seemingly confining themselves to trading biofuels and blending them, they are investing heavily into second-generation biofuels made not from sugar and corn, but from straw, algae or agricultural waste. But they also hedge their bets, as revealed by the $12 billion joint venture unveiled on February 2010 by Shell and Cosan, a major Brazilian producer of ethanol.

CHAPTER 7

The Green and Blue Revolutions that Have Never Been

1. *Agflation* and the Daily Bread

In late 2006 and early 2007 rapidly rising prices worldwide for fruits, eggs, grains and other crops, called *agflation* (a new label), began to drive up the cost of breakfast, lunch and dinner. Year on year, the price of orange juice rose by 25 per cent, eggs by 20 per cent, but milk by only 5 per cent.

Underpinning most of these price rises was a sharp in increase in the prices of corn and wheat. This caused some people to ask: 'Did the world run short of these crops? If so, why didn't farmers grow more, weather permitting?' That is exactly what they were doing. The International Grains Council, an industry lobby, estimated that global production of grains was increasing, rather than the other way around, but demand for grains was growing even faster.

Critics said that the reason has been the use of grains to make biofuels, like ethanol. This was only partly true, because the demand for meat increased sharply and a great deal of grain was being used to feed livestock. The increase in direct human consumption, because of decades of soaring population growth, was also a major factor. The argument about biofuels and livestock conveniently forgot that:

- domesticated animals are raised and eaten by *Homo sapiens*, and
- cars, locomotives, ships, airplanes and other big users of energy are also driven by humans and for humans – not by tigers, even if ExxonMobil has a tiger in its logo!

The Greens have been quick to point out that over seven years the corn used to make ethanol consumed a fifth of America's corn crop. But at the root of the demand for both energy and food we find exactly the same factor: the uncontrolled growth of *Homo sapiens*. Their demand weighs on commodity prices.

Leaving aside the fact that it is the Greens and other eggheads who campaigned for decades against nuclear energy (Chapter 4) and pushed for alternative sources of supply, it takes a very narrow mind to think that agriculture should remain as it was in 8000–7000 BC –more or less when grains became domesticated. Agriculture should evolve, like everything else, in nearly 100 centuries.

- Thousands of years ago, the mission of agriculture was to feed people.
- Today it is both to feed people *and* provide them with energy supplies.

Nor is it true that the link between fuel and food prices necessarily increases the burden on the 'poor', if for no other reason than many of the poor today drive a car (a rundown and polluting car perhaps, but a motor vehicle nevertheless). It is *as if* nobody in developing countries has ever considered that Western standards of living, is for them, unaffordable and unsustainable.

Even if it is wasteful of resources, this Western living standard is precisely what most (if not all) developing countries want to imitate, but cannot afford. As for using part of global agricultural production for biofuels, it is proper to keep in mind that oil is widely used as constituent of many fertilizers. Turning grains into alternative energy squares the circle:

- oil to fertilizer,
- fertilizer to grains,
- grains to energy, substituting for oil.

It is normal for developing countries to aspire to a better standard of living. But they should work for it. We all aspire for something 'more', but can we afford it? In addition, this higher standard of living is precisely the reason why birth control is so important and so urgent.

- At any given point in time available wealth is what it is.
- *If* there are more mouths to feed, *then* there will be less for everyone.

The big-hearted donor who would steadily transfer his wealth to a recipient who is mismanaging his life is found only in fantasy. Any serious wealth management policy should also take care of adversities which have the nasty habit of happening so often. *Stem rust* which has been spreading in the world's great bread baskets, like Punjab, is a case in point. Wheat rust is a fungal infection that:

- attacks the plant's stem, and
- forms lethal red pustules.

Its presence is nothing new. It has plagued crops for centuries. The Romans had it under control, but periodic epidemics in the early twentieth century destroyed almost a fifth of America's wheat harvest. Its reappearance in several parts of the globe puts in doubt the Green Revolution's claims that it has been definitely controlled, while its recent rapid spread makes quick containment all but impossible. Worse still, wheat rust may become endemic again.

Nature has its own way of taking revenge on its spoilers. The Greens and their social scientists should understand that the twentieth century's megatrend in population explosion has put under stress the earth's:

- food, and
- energy supplies.

This has not changed in the twenty-first century; if anything, trend has accelerated A spiralling world population of 6.8 billion, projected to reach 9 billion in a couple of generations, creates a horde of mouths to feed and cars to drive. The growing affluence of the developing countries also has the effect of accelerating dietary shifts towards meat products, which in turn means that all this extra livestock consumes part of the grain produced by farmers.

Another issue, discussed earlier in Chapter 1, is that unstoppable urbanization increases the pressures on arable land. This leads to the use of mountains of fertilizers as well as water shortages. Put these trends together and you get the picture of what can be expected in terms of shortages.

A crucial factor that many people, as well as governments, choose to ignore is the growing dual impact of burgeoning populations and the consequences of growing individual wealth in emerging economies. Back in 1985, no more than a quarter-century ago, the Chinese consumer ate 20 kilos of meat per year on average, but current projections indicate he will consume over 55 kg in 2011 – almost a 300 per cent increase.

- It takes eight kg of grain to produce one kg of beef, and
- quite clearly this tripling of meat consumption in a country of 1.4 billion people, pushes up the demand for grain.

Nor should one forget the status symbols. Theoretically, the booming Chinese car industry produces small vehicles. In practice, the best-selling General Motors car in China is the Chinese-manufactured Buick. That has a direct impact on oil and food markets; it takes a big quantity to fill up a Buick's fuel tank –whether with oil or ethanol.

This is no criticism of the evolving Chinese taste for big cars, because people in the developing countries do have the right to aspire to higher living standards – including status symbols. What is unforgivable is their policies of uncontrolled population explosion which nowadays is worse in Africa and south-east Asia, but is also beginning to take off in China in violation of Mao's wise policies of limiting reproduction to one child per woman in order to reduce the future population.

- Increasing wealth in China, India, Brazil and other countries is stoking demand for meat, boosting the demand for cereals to feed the animals, and is adding to global energy consumption by home appliances, TVs, PCs and motor vehicles.
- In absolute terms, the use of grains is linked to the growth of the world's population, whether it is for bread, meat, oil or ethanol, since all of these products are tied to economic growth.

Probably, though by no means definitely, this change in global consumption, because it has been slow and incremental, has escaped the attention of governments and their social scientists. However, they are paid to be watchful, and agflation is one of the key factors they should watch.

While ethanol may account for some of the rise in the prices of crops, corn is also fed to animals which are now being reared in greater numbers. Beyond that, the overall decline in stockpiles of all cereals is in direct relation to the number of mouths to feed as well as to the governments' wrong agricultural policies, which tend to be stuck in 7000 BC.

Not only is the European Community's Common Agricultural Policy (CAP, Chapter 8) a mess, but so are the policies followed by the WTO and the international community.

Between them, they demonstrate that issues concerning the production and distribution of agricultural produce are being managed by bureaucrats sitting in ivory towers. Agriculture is notoriously a boom-and-bust industry, and goals about the future should be:

- realistic,
- flexible, and
- realizable.

However, the reality is that the so-called Green and Blue Revolutions (see sections 2 and 3 respectively) did not materialize, and the wrong sorts of subsidies and guidelines have distorted the producers' policies. Whether based in Brussels, at the WTO or anywhere else,[1] these self-appointed groups of saviours are nothing other than a bunch of bureaucrats setting rules on the trading of foodstuffs and trying to micromanage a situation with many more unknowns than they can count, let alone comprehend.

2. The Green Revolution? 'Tell Them I Lied'

Earl Long was first elected governor of Louisiana in 1939, in a campaign noted for its wild promises and a liberal stance towards money and alcohol. On assuming office, however, Long immediately reneged on his promises and raised alcohol and other taxes. An aide plaintively asked him: 'What should I tell the people?' Anecdotal evidence suggests that, without missing a beat, Long answered 'Tell them I lied.'

If true, the quotation is directly applicable to all these demagogues (or false visionaries) who, in the years after World War II, talked about the world being on the verge of wiping out hunger because of the approaching *Green Revolution*. Though significant improvements have been made in agriculture:

- they were massively based on fertilizers and pesticides, and
- their deliverables were swallowed up long ago by the monster of the population explosion.

Even the promises made by biotechnology bear no semblance to what we have got, not because of failure to try but because demand outstripped the most optimistic projections of supply. That's the fate of the Green Revolution in a nutshell.

A growing number of people now suggest that contrary to what opponents to genetically-modified (GM) food are shouting in the streets and in the media condemning molecular engineering, its deliverables may be the best antidote to the overkill of the environment by pesticides and fertilizers whose residues pollute the earth and seep into streams, lakes, rivers and seas. There are billions of tons of pesticides lying around the ecosystem – and poisoning it.

According to certain opinions, *if* Monsanto and other makers of genetically-modified seed, such as DuPont and Syngenta, are able to win over the opinion of a sceptical public (and of policy-makers), *then* they will be able to determine how acceptable their innovations will be over the coming years. 'European farmers have been denied the

1 D.N. Chorafas, *Globalization's Limits. Conflicting National Interests in Trade and Finance*, Gower, Farnham, 2009.

right to choose,' say proponents of GM food, who view it as the better alternative to the massive use of pesticides and fertilizers.

As if to disprove these claims of discrimination against GM, on 2 March 2010 the European Commission approved the cultivation of genetically-modified potatoes, while Greens, environmentalists and some European ministers slammed the so-called 'Frankenfoods'.[2] 'This is a bad day for European citizens and the environment,' Friends of the Earth said of the green light given for the Amflora potato to be developed by German chemical giant BASF.[3] That's one-sided because:

- It's no secret that after two decades or so of gains, global food production has slowed.
- More recently, on-and-off, it is falling below consumption, with the shortfall covered by stockpiled food supplies.

Stocks are run down as everyone tries to live off last year's stores. Drought-damaged harvests evidently make a bad situation worse, leading to steep drops in world grain stocks. Grain production has slowed in several populous countries, including China, India, Indonesia and Mexico, due to soil erosion and scarcity of water.

For instance, in the late 1980s it was estimated that American farmers were losing some six tons of soil for every ton of grain they produced. A survey that came to my attention at the time measured the loss of topsoil, from US croplands at an annual level of 3.1 billion tons – or 13 per cent the world's total losses.

Far from being an exception, this is part of a pattern observed the world over, and a harbinger of coming major adversities in food. As soil conditions deteriorate, during the last two decades the inherent productivity of roughly a third of the world's agricultural land has been falling, although increased use of chemical fertilizers often masks this deterioration for a while.

- Worldwide, farmers are losing an estimated 24 billion tons of soil each year, and
- this is roughly the amount of topsoil covering all wheatland of the Australian continent.

Just as critical are the falling water tables, due to greater use of irrigation, galloping industrial usage of water supplies and, as always, the vastly increased population. In fact, a good deal of the industrial expansion of developing countries – India and China included – is achieved by overpumping, which significantly lowers the water tables.

In China, which has over one-fifth of the world's irrigated land, there is a growing scarcity of water and the water table in the area surrounding Beijing and Tianjin is dropping nearly one metre per year. As a result of overpumping, several states are suffering from falling water tables, and wells that are going dry. Thousands of villages now rely on watertankers for their drinking water.

2 The EU Commission also allowed three GM maize products to be placed on the European market, although they are not grown in Europe.

3 *Costa Headline Sat News*, 3 March 2010. Modified vegetables and cereals have long been a matter of fierce debate in Europe and the Commission stressed that the Amflora would only be for industrial use – including animal feed, which ultimately is recycled into people's food.

In Russia, irrigation diversion (including intense cotton-growing in neighbouring Uzbekistan and Turkmenistan) have sharply reduced the flow into the Aral Sea, which today is nearly a tenth of its original size and shrinking further. In other countries, extensive building of new homes cements the earth and shrinks cropland, particularly used for grain.

Even in countries like China, where economic reforms boosted annual grain production by half between 1976 and 1984, greater yields run out of steam. A great deal of this overall loss of momentum is due to overpumping of groundwater for irrigation, as well as from bringing land unfit for agriculture into production.

- Between 1950 and 1981 worldwide, the area under cultivation expanded by 21 per cent.
- Since then, however, it has fallen because of the abandonment of severely eroded land and general land shortages, including the diversion of good cropland to non-farm uses.

Yet, despite this mirage of the 'Green Revolution', the world population has continued its galloping expansion unabated, while leadership in family planning and birth control has been distinguished by its total absence. Rather than being embarrassed by agricultural failure, in many developing countries governments have been looking the other way, while the pressure of mounting population pressures has pushed farmers on to land that slopes too steeply and is insufficiently fertile to sustain meaningful agricultural production.

In Africa, formerly better-off countries like Kenya and Zimbabwe stagger under burdens that threaten to overwhelm them. Farmers in both nations used to produce enough food to feed their citizens, but Zimbabwe is bleeding from the forcible government-sponsored acquisition of white-owned farms which has halved its agricultural production and has led to mass migration because of hunger; and Kenya has sunk into riots and tribal civil war.

The United States and the European Union are no exception to this general trend. The US Department of Agriculture has on several occasions admitted dangerous drops in water table levels. Anecdotal evidence has it that 14 million of the 52 million acres of American irrigated cropland was being watered by depleting water tables, with the drop ranging from 15 centimetres (6 inches) to 1.22 metres (4 feet) per year.

Like the population explosion, this massive destruction of the earth's resources is an unmitigated global disaster. Fed by lack of foresight and lack of policies able to effectively protect current and future agricultural land, the vicious circle is intensifying. As world demand for food climbs:

- soil erosion increases,
- water tables sink, and
- this is leading to food shortages and unfulfilled promises about the Green Revolution.

Miracles do not just happen. Instead, a long list of measures must be urgently taken, based on tough decisions. Beyond the unquestionable requirement for immediate and effective control over population growth, future improvements in world food supplies (let alone the chimera of eradicating hunger) depend on:

- stopping rampant urbanization,
- reversing land degradation,
- putting the brakes on the exploitation of water resources, and
- making a fundamental choice between the intensive use of pesticides and fertilizers (with known catastrophic consequences) and genetically-modified grains (with an unknown aftermath).

Ultimately, achieving a balance between food and people may depend more on the success of family planners than that of farmers. Yet, both are not only necessary but also crucial parts of a solution – if there is one. Account should also be taken of the fact that there are no miracle technologies on the horizon that are likely to lead to quantum leaps in food output. Evidence is provided from:

- overreactions to the GM hybridization of corn, and
- overestimating the increase in fertilizer use, said to be 18 times higher in 2010 as compared to 1950.

The near tripling of irrigated areas in the hope of spreading the Green Revolution during these six decades has brought disaster to the environment and contributed to land erosion, as well as dwindling water supplies. It has reached the point where any devastated grain crop in North America or the European Union is sobering news for the 100-plus countries which depend on grain imports from these breadbaskets. But neither Western farmers nor Western taxpayers can continue to provide subsidized grain either at preferential prices or as foreign aid forever.. African and other nations need to take a close look at their food shortage problems and their causes, and find a way of reducing their dependency on Western aid.

3. A Second Grand Illusion: The Blue Revolution

An integral part of efficiency is the ability to learn from past failures, and human history has plenty of examples from which to derive lessons. One of the key stories worth examining is the evolution of the fishing industry which has a high content of efforts, successes, failures and dashed hopes.

In the background of most failures has been the overkilling of fish stocks. Some people in positions of responsibility, who care about the survival of the fishing industry, have recognized that protective measures are needed to sustain the current yield. However, this is largely a political issue, and, when it comes to negotiating catch quotas, politicians:

- are much more concerned about pressure groups and jobs at stake in their domestic fishing industry, and
- much less interested in discussing the implications and measures associated with preserving fish populations for future catches.

Statistics on how fish species regenerate after a long pause in hunting are indeed impressive. A silver lining of World War II, when commercial fishing completely ceased in the North Sea, was that, left to their own devices, fish populations recovered in terms of numbers. But as the fishing industry built up its capacity again, the overkill began. Catches became bigger and bigger under the fallacious impression that the so-called *Blue Revolution* would last forever. The world's fish harvest rose:

- from 49 million tons in 1965,
- to over 110 million tons by the end of the 1990s.

At the turn of the century, however, worrying trends emerged. Even if the 110 million tons was based on biased statistics, because much of the increase came from farmed fish, the overkill had had devastating effects on wild species. In addition, it was revealed that, among the catches of wild fish, fewer and fewer were high-value species, like cod and haddock and more and more were of less desirable kinds.

Examples of the manipulation of marine species to feed the rapidly multiplying *Homo sapiens* are legion. One of those on the road to extinction is the bluefin tuna. The hypocrisy behind its imminent extermination comes from the fact that the bluefin's survival was 'managed' by an intergovernmental body, the International Commission for the Conservation of Atlantic Tunas (ICCAT). But the job it did was so mediocre that it was dubbed the International Conspiracy to Catch All Tunas (ICAT-plus).

- In 2007 the scientific advice regarding the survival of the species was to catch *at most* 15,000 tons of tuna in one year.
- ICCAT imposed a limit of 30,000 tons.
- But, because nobody cares, the actual catch was 61,000 tons.[4]

As if ICCAT was not enough of a disaster, there is also the Convention on Trade in Endangered Species of Wild Fauna and Flora (CITES). This has become a battleground between the destroyers of the world's fauna and flora and those trying to protect it. Conflicts of interest are the order of the day, and this is a major problem because CITES is one of the few places where global environmental laws are made. It has the mandate to:

- ban trade in a species,
- limit a trade, so that it has to be carried out using permits,
- or, find ways to do nothing – hence condoning extinction.

The third bullet explains why bluefin tuna, of which 80 per cent is consumed in Japan, was refused a listing in a vote in March 2010, even though practically all scientists agreed that it qualifies for protection. Arguments given by Japan, Norway and Iceland for rejecting the proposal was that there is already an organization that exists to manage the bluefin: the ICCAT – whose mandate ranges between that of an International Commission for the Conservation of Atlantic Tunas, and the aforementioned International Conspiracy to Catch All Tuna.[5]

---

4 'Sleeping with the Fishes', *The Economist*, 20 October 2008.

5 'Fishy Business', *The Economist*, 25 March, 2010.

By passing the ball back and forth to the 'other regulatory party', CITES and ICCAT, the two international supposedly regulatory bodies, have a veto on life and death of marine species. And they condone extermination. Nor is the bluefin tuna the only marine animal condemned by *Homo sapiens*.

According to statistics published in 1998 by the United Nation's Food and Agriculture Organization (FAO), about 60 of the world's various commercial fish stocks were being harvested near or beyond sustainable levels. This is not surprising. It is more difficult to understand why, since then, nothing has been done to effectively stop the slaughter.

As might have been expected, the sharp decline in fish stocks has caused strife between nations aiming to protect their fishing 'rights'. Canada and Spain had a serious row about Greenland halibut in 1995 (in the so-called 'Halibut War'), and the Malaysian navy allegedly killed Thai fishermen suspected of illegal fishing in Malaysian waters.

The original skirmishes expanded into a battle for fishing grounds, both at international conferences and at sea. When, in 1972, Iceland increased its fishing zone from 12 to 50 nautical miles (having previously extended it from four to 12 nautical miles in 1958) and then extended it to 200 nautical miles three years later, its long-running dispute with Britain became known as the 'Cod Wars'.

More recently, less developed countries have excelled in the overfishing of fish stocks, and have been exporting growing quantities to developed countries. This led to the projection that if overfishing continues, it could hurt poor countries because their people rely more heavily on fish for protein than citizens of wealthier countries. The mirage of the Blue Revolution has also been fed by the huge subsidies governments paid, and continue to pay, to keep their fishing industries afloat. These subsidies explain:

- why the world's industrial fishing fleet grew twice as fast as fish catches during the 1980s and 1990s, and
- why, despite overcapacity in the industry, few fishing boats are being retired worldwide.

Even if most subsidies are given out by governments in developed countries, they are having an increasing effect on fishing in developing nations which are matching the thrust to get the most food out of the seas. Worse still, they do so, paying scant regard to the fact that when fish populations dwindle, their recovery is in no way assured. In the aftermath, catches continue to decline, and once-rich fishing grounds yield little fish. At the same time, newly discovered supplies of fish can no longer offset what is lost through overfishing.

4. Getting Marine Life Ready for the Table

Rivers, lakes, seas and oceans are being progressively depleted of their marine life, and nature can no longer offer us the riches to which we have been accustomed. The existence of this rich endowment was the central theme of the Blue Revolution, which is becoming a fading memory.

Why not replace the depleted wild species by raising stock in fish farms? When the search began for alternative sources of food in the 1970s, it seemed logical to set up marine aquacultures where fish could be farmed on a large scale. This process, however, did not help humankind overcome hunger because:

- pretty soon unexpected consequences occurred, and
- the limits that are associated with any process became apparent.

Like the land-based agricultural economy (section 2), a marine agricultural economy involves many challenges, a good many of them situational. One of the problems of maritime cultures, which have been widely tried, is that most seafish do not reproduce very well in captivity. Herrings, for instance, alter their feeding habits during their different stages of development. The intricate food chains in the seas and oceans cannot be emulated in a breeding tank, at least not at an economically viable cost.

As one problem followed another, the widely heralded Blue Revolution became grounded. Where sea farming is practised, thousands of fish are enclosed in pens, where some of their food and their faeces sink to the muddy sea bottom, creating a fecund environment for algae.

The west coast of Japan provides a good case study because the Japanese were enthusiastic about efficient sea farming, yet encountered major problems that were not envisaged at the outset. Another lesson comes from a case study I did on salmon farming in the pristine waters of Stewart Island, New Zealand (more on this later).

New Zealand seriously considered at sea farming, for both local consumption and export. In the 1970s oyster and mussel farms were established experimentally in Paterson Inlet, but they did not prosper. Yet, if the risk of poisoning because of polluted waters is discounted, mussel farming seems to make sense because they are relatively undemanding creatures: they do not require much space and feed on plankton. By contrast,

- most fish used for breeding on fish farms are meat-eaters, and
- the amount of animal protein fed to these fish is many times the quantity they yield.

That sort of negative food balance might only be economic in the case of high-value upmarket delicacies such as oysters and lobsters. But this would not be much help to the poor and starving populations of the globe. Nor are such solutions environmentally-friendly

Talking of the New Zealand experience, in 1981 fish farming again became a topic when British Petroleum announced its intention to use know-how from the Scottish lochs to establish a salmon farm. Big Glory Bay on Stewart Island was the chosen site. By 1983 large pens had been systematically installed in the sea. Tankloads of salmon fingerlings were brought from hatcheries on the Canterbury and Otago rivers on South Island.

- The Scottish experience was helpful.
- However, local fish farmers described their enterprise as 'learn as you go', and for about five years they were happy with it.

During these first five years everything was going well, although business and profits were affected by the exchange rate of the NZ dollar. But exchange rates change over time and, as the value of the NZ dollar plummeted, business got better. Then, on 3 January

1989, catastrophe struck. The first visible signal was discolouration of the seawater. The fish farmers checked closely and found that the fish in one of the salmon pens were not looking good, although the fish in the other pens looked healthy. However, within 24 hours every pen was in trouble.

The cause of the 1989 fish poisoning at Big Glory Bay was an algae: Heterosigma. It developed in a similar way to that which struck fish farming at the Norwegian fjords in 1988. Still, even today nobody can say for sure what killed the salmons in the pens. Two hypotheses have been put forward and, as far as I understand, it was difficult to decide which one was dominant:

- toxins from dinoflagelates,
- suffocation through the algae blocking up the gills.

As the fish farmers helplessly observed their stock die, they saw that small fish were surviving better than big fish, probably because they need less oxygen and are more active when the algae collect around them.[6] This catastrophe cost local fishermen 150 tons of pen-raised salmon.

Ironically, algae is not altogether unwanted or unwarranted. As Ron Tindall, (the then) director of the Stewart Island station of the Department of Conservation, commented during our discussion: 'Algae are an ecosystem method of wrapping up chemicals and bring them out of the water. Nature tries to adjust itself – and by adjusting causes lots of other problems.'

Nutrients (including fish food and fish faeces) was one of the reasons for the fast multiplication of the algae. The prevailing weather, warm throughout the winter, seems to have also contributed, as did the lack of strong currents. Severe accidents tend to have many contributory factors.

Accidents like that of Stewart Island are frequent in all parts of the world where there is an active salmon industry. In 2009 and 2010 it was the turn of the Chilean industry to be hit because of an outbreak of a fish disease. Chile's output of Atlantic salmon has been hammered by a virus that causes infectious salmon anaemia which kills the fish by attacking their red blood cells. Because of it, Chile, which was the second biggest producer of salmon, saw its output plunge more than 75 per cent in two years.[7]

Environmental groups have also accused Chile's salmon industry of overcrowding its cages and using too many chemicals. As the case of New Zealand's sea farming shows, to be healthy the raised fish would have to be farmed at a much lower density than currently practised. But this is only one side of the problem.

- The whole picture of intensive agriculture and fish farms at sea has to be revisited, in conjunction with birth control of *Homo sapiens*, and

6 The theory about lack of oxygen has been discarded, as at Stewart Island fish farm examinations showed plenty of oxygen being available to the dying fish. An aftermath from poisonous algae seems more likely, but it is not proven beyond doubt.

7 'Disease Sends Salmon Prices Leaping', *Financial Times*, 17 February 2010.

- most particularly, the question of nutrients used for all sorts of bred animals – from cows[8] to fish – has to be examined most carefully with health and quality, *not* quantity, as a guide.

The role of nutrients deserves attention. Other problems aside, they are typically very high-protein and hence rich in nitrogen and phosphates. This is exactly what makes algae bloom. Stewart Island was a pristine, unpolluted environment, located far away from big cities. Nor did it suffer from the effects of pesticides, as Stewart Island has no intense agriculture. The Blue Revolution has had unexpected consequences, and its potential as a solution to the food problem was simply a mirage.

5. Lessons from the Failure of the Blue Revolution

An important lesson learned from the Stewart Island experience is that when the pens in which fish are raised become crowded, the fish become stressed and the quality of their meat rapidly deteriorates.

Yet, it does not take a genius to appreciate that the post-World War II efforts, collectively described as an attempt to create 'plenty of food for all' through a Green Revolution and a Blue Revolution have collectively failed, even if their proponents (shamelessly) say that 'they reckon to have saved at least 1 billion lives'(!!!). These failures taught some lessons, and it would be a crime against humanity to ignore them.

- While the status quo prevails, governments don't even scratch the surface of what is meant by sustained development.
- Overpopulation means misery for all, and it stands in stark antithesis to the stated goal of quality of life.
- Far from providing increased productivity, the extensive use of chemicals is poisonous to nature and to all animals, including the human species.
- Nature does not accept having its hands twisted by *Homo sapiens* or any other species, and it pays back the abuse it suffers.

As long as man made modest use of the food available in the seas, marine fauna was not endangered. With 'improved' fishing, transport and processing techniques, natural supplies were exploited with increasing ruthlessness – and this was just as true for land-based agriculture.

The lesson learned from the Norwegian fjords to Stewart Island is that when algae blooms do appear, they are able to spread death in cunningly effective ways all the way up the food chain. Recent research on unwanted consequences has led scientists to the belief that when some types of phytoplankton touch the gill filaments of fish, they release a toxin. They suspect that there may be a link between algae bloom and the deaths of some marine animals, and the more daring express the opinion that this may find its way to humans through interconnections and interdependencies. Technology provides very little by way of support in controlling algae blooms. Tests are not risk-free and, in a

8 Just keep in mind the devastation British farmers suffered some years ago because of mad cow and foot-and-mouth diseases.

number of cases, scientific investigations prove to be among the worst polluters. Here is a real-life example.

At the time of the Stewart Island catastrophe discussed in section 4, the Water Quality Centre run by New Zealand's Department of Scientific and Industrial Research, wanted to discharge up to 25 grams of the pesticide chlordane into a harbour to assess the chemical's effect and rate of dispersal. It therefore applied to the Auckland Regional Water Board for a permit to carry out the experiment,[9] but Manukau residents, as well as environmentalists, lodged strong objections and the 'experimental' pollution was avoided.

No blunder, no matter its size and consequences, seems to constrain bureaucrats from doing something silly – sometimes in the name of science. If fish farming is a subject wrapped in great uncertainty, what about engineering a revival of the riches of seas and oceans? Some people put it this way: if fish farms are not the great success story they were supposed to be, why not concentrate on policies which allow wild fish to regenerate their stock?

This essentially means turning the argument around 180°. As we saw in section 3, fish farming has been particularly promoted because of the significant decrease in catches because of overfishing. No international agreement has been able to reverse that trend. Consider ICCAT, which sets catch limits for tuna and other large migratory fish (section 3). Time and again, however, its efforts have failed to do what is necessary to give the prized bluefin tuna in the eastern Atlantic and Mediterranean a real chance to survive. At a meeting in Brazil in November 2009 the Commission approved an annual quota of 13,500 metric tons for 2010.[10]

- This was below the then prevailing quota of 22,000 tons.
- But it is not the complete moratorium recommended by the Commission's own scientists.

Over and above that, the illegal catch sometimes equals the authorized quota. Scientists say that overharvesting has caused a 72 per cent decline over 50 years among adult bluefin in the eastern Atlantic and the Mediterranean, the fish's primary spawning grounds. The ICCAT scientists are right when they say that fishing for bluefin should be shut down completely until the fish have reached sustainable levels, but, like birth control, this is taboo with the politicians.

Yet, there exist some very interesting precedents. During World War I and World War II fishermen were forced to stop fishing. When they resumed their activities, they caught more and bigger fish than they had previously, but only for a few years. As the fishing spree accelerated, formerly abundant stocks of flounder were so overfished that the nations bordering on the North Atlantic and mid-Pacific oceans were obliged to sign an agreement to protect them.

The same story has repeated itself during every forced interruption of overfishing and exterminating marine life. But the message is not getting through, and cases where protecting a fish species takes priority are rare.

9 *New Zealand Herald*, 10 February 1989.

10 'Are Lower Catch Limits Enough to save the Bluefin Tuna from Extinction?', *Scientific American*, 16 November 2009.

In Peru, a fishing nation, the euphoria of the 1960s gave way to deep despondency. In a few years the Peruvian fishing industry had developed from being an insignificant sector of the economy to become number one worldwide. Here is a brief review of what happened in the post-World War II years.

- In 1964 this industry produced 10 million tons of fish – 40 times more than in 1955.
- By 1970 this had risen to 12 million tons – a fifth of world production.
- Then suddenly the miracle was over. Catches began to decline and by 1978 the figure had dropped to a mere 3.4 million tons.

A similar story of extinction took place in 2009 with the bluefin tuna, discussed above. Despite ICCAT's quotas and supposedly watchful eye, the numbers of bluefin tuna – fished in the Atlantic and the Mediterranean for thousands of years – declined by three-quarters.

Cod, too, has suffered from its popularity among *Homo sapiens*. Fifteen years ago the stock crashed to under 10 per cent of the level considered sustainable for fishing. In response, the US government stopped issuing new fishing permits and imposed restrictions:

- limiting fishing days,
- capping daily hauls,
- regulating gear, and
- closing areas to fishing boats.

These restrictions have proved to be insufficient. Despite the new rules the Gulf of Maine cod population is still at barely half the level required for long-term survival, and the population at Georges Bank, 60 miles (100 km) offshore, is just over 10 per cent above the level at which the fish could avoid extinction.

The formerly rich stock of salmon in British Columbia is heading the same way. All evidence suggests that the causes of the decline are overfishing and the destruction of spawning habitats. Scientists and environmentalists also blame unauthorized fishing on Canada's indigenous peoples, the so-called First Nation.

Critics say that another reason has been diseases spread from open-pen salmon farms. Many migrating fish pass two-dozen salmon farms planted in their way in the Broughton Archipelago, in the name of the Blue Revolution. Scientists have found that sea lice and other parasites from fish farms can infect the passing stock, reducing their chances of survival. This threat has not been properly investigated by the federal government's fisheries department because, according to some, it suffers from a conflict of interest since the government both:

- promotes salmon farming, and
- is charged with protecting wild salmon.

What all this documents is that the sustained development of fish for food should not be taken for granted. We all suffer from a conflict of interest while destroying the habitat of other animals and denying them a place on earth and in the sea.

With the rise of socialist ideas in the late nineteenth century and the twentieth century's 'social scientists' who joined the 'pro-life' movement, as well as the so-called Greens and other eggheads, politicians and ordinary people came to the belief that the population of other species would continue to rise indefinitely, in order to serve the rapidly expanding species named *Homo sapiens*. Now we can see that ultimately nature makes sure that humans pay the bill for the extinction of other species.

6. City versus Farm in Less Developed Countries

With the population explosion unabated, between half a billion and a billion people are seriously undernourished worldwide, and many countries are seeing their per capita daily calorie supply steadily decrease rather than improve. Malthus was right.[11] Food production in African countries has fallen behind population growth, and the negative margin is becoming increasingly easy to detect.

There is plenty of evidence that Africa is a Garden of Eden in decay, and it is sad to see a total lack of government initiatives in birth control (or, for that matter, in the control of AIDS). If necessity is the mother of invention, foreign aid is the mother of inertia – and Africa gets plenty of foreign aid.

Anecdotal evidence suggests that, because of population explosion, even India, which for decades successfully struggled to increase production, has a lower food-grain output per capita now than in 1900 under the British Raj. Problems in other countries are especially grave, precisely for the same reasons. Most wounds are self-made.

- Haiti must cope with virtually total deforestation.
- Egypt and Mexico must put up with encroaching desert and burgeoning cities which contribute nothing to agriculture but have huge populations that need to be fed.

There is a Pandora's box of wrong policies behind these happenings. One of the evils is the tendency of developing countries' governments to favour city-based projects, generally supporting urban residents at the expense of farmers. A rational policy requires the opposite: more spending in the rural sector, both to increase food production and to stem the migration to urban slums.

This opinion, which is publicly expressed by people who are able to see further than the typical politician is anathema to the tunnel-visioned authorities that direct the fortunes of the less developed world. All the while that very short-term considerations overrun both common sense and the longer-range perspective:

- high birth rates ensure that the migration to the cities increases, and
- this situation is compounded by governments that are leaving the countryside as a virtual desert in terms of essential infrastructure.

11 Malthus built a beautiful model with three variables: an old maid, cats and mice. The old maid raised cats and their multiplication decimated the mouse population. Many cats died. The mouse population then exploded. The cats had more food to eat and they multiplied. The old maid raised even more cats and the cycle repeated itself.

For most of the past 30 years, investments in agriculture have declined relentlessly. Most developing countries are investing only around 5 per cent of public revenues in farming. This small and wrong-way investment is a carbon-copy of what happens in the West, where the share of investment in agriculture has also been falling.

Prior to this reversal in priorities by less developed countries, staple crop yields were rising by 3–6 per cent annually. Now in some countries they are rising by only 1–2 per cent a year, and in others they are flat or declining – while their populations continue to increase.

Worse still, agriculture is not even a sideshow for the governments of less developed countries even if farming is far and away the single most important economic activity they have. Statistics document the effects of this double whammy:

- more births, and
- less agriculture.

According to the UN's Food and Agriculture Organization (FAO) the number of malnourished people in the world rose to over 1 billion in 2009, up from 915 million in 2008 – an increase of nearly 10 per cent. A quarter of a century after the famine in Ethiopia, which dramatized the failings in its food system, famine is again stalking the Horn of Africa. In Somalia the government and its citizens are much more interested in fighting it out in an endless civil war than in farming the land for food.[12]

Typically, the point of a social explosion is crossed when the government budget can no longer afford food subsidies and increasing volumes of food imports, while donors get tired of helping people who cannot help themselves. At this point, the increasingly expensive and ineffectual welfare programmes:

- crack under their own weight, and
- take the nation down the drain along with them.

Instead of opting for the rational solution of shifting economic benefits and incentives towards agriculture, supporting the people remaining in the farms to increase local food production, most less developed countries do the opposite. Why? Because when domestic benefits shift, powerful interest groups (those wielding political power) stand to lose and they do everything they can to stop the process of a return to agriculture.

Through this city-versus-countryside conflict the economies of less developed countries reach a situation of massive unevenness. Farm villages are still emerging from the Middle Ages, while the cities cannot house their sprawling numbers. The result is a disruptive pattern of counterdevelopment, with the local society creating an ever-widening gap between:

- wants, and
- means.

Urban protests that seem to be sweeping a large area of the planet give governments, particularly in less developed countries, a reason not to initiate reforms towards a more

12 They fill the gap by pirating ships and asking for ransoms.

fundamental agricultural infrastructure. In the poorest of these countries, where food security is the central (and most elusive) goal, outside agriculture, directing even more economic incentives at the cities is the worst possible way of being in charge.

In the aftermath, several states in Africa and Latin America have allowed the effects of increasing poverty due to their population explosion to be passed on from one generation to the next. Unemployed slum-dwellers, their uncontrolled procreation and their governments' procrastination contribute more and more to this vicious cycle of poverty.

One man who thought that such wrong-way policies and associated government inefficiency is really criminal is Robert McNamara. He said as much when in the mid-1960s he headed the World Bank. In his years of tenure he tried to change the prevailing patterns, but his well-intended efforts ended with a revolt on his hands.

What this revolt was really about was not the damning of inefficiency; it was the evolving change in control. Dr McNamara's sound managerial viewpoint was quite different from that of politicians in less developed countries, accustomed to handouts. Yet, it was not they but the president of the World Bank who had the necessary education, experience and dedication to see that:

- the Blue and Green Revolutions had passed by like a mirage,
- greater efficiency was one of the first essentials for eliminating poverty, and
- mismanagement at government level ended by distorting the necessary focus and associated concentration of skills on which a good solution depended.

Mismanagement, of course, is not exclusively found in less developed countries. It is also a sickness that afflicts Western societies. For instance, in the wake of the two oil crises of the 1970s, as the big global banks tried to recycle the glut of petrodollars, politicians discovered an easy way out of budgetary difficulties by what seemed to be a simple process of signing up for huge loans.

In the middle of a bonanza, the banks failed to observe the basic rule of calculating the more distant credit risk such transactions involved. This was particularly true with regard to lending to developing countries. Walter Wriston, former CEO of Citibank once memorably said: 'Countries don't go bankrupt.' But they do.

Given the energy crisis and food shortage problems, but also the megalomania of local politicians, by the late 1970s bank lending was eclipsing aid-giving in large areas of the world. Spending beyond one's means at all levels of society – household, local government and central government – created *debt inflation* which followed the population explosion's pathway.

As if getting indebted for imports of energy, food and other goodies was not enough, the proceeds of many of these loans were directed to the cities and put in non-productive businesses like building prestige projects. It should have been evident to the rulers that if income-producing investments were not made, their country would not be able to pay the interest on the loan, much less return the capital. But officialdom did not care. Instead, all hope of avoiding bankruptcy turned towards international philanthropy.

7. The Destructive Forces of Unstoppable Foreign Aid and Beggars' Policies

'Only the strong can be productive and only the productive can be free,' John Foster Dulles, the former US Secretary of State and international lawyer, once suggested. By this definition, many of the nations that got their independence after World War II but continued to be hostages to their uncontrolled population explosion are not really free. The reasons are home-made: irrational policies, low productivity and ever-mounting international debt are keeping them in chains. They have become hostages to fortune – characterized not only by dwindling food supplies, but also by a curious unwillingness to:

- take get a grip on themselves, and
- do what it takes to change their destiny.

In the first years after independence, during the 1950s and 1960s, the economic plight of many less developed nations did not look as dismal as it does today. Those who really tried to come up from under – India being a good example – progressed more than was realized at the time, but not every newly independent country made the effort. During this period the newly independent countries that tried hard to join the developed nations increased their GDP at a rate of 4.8 per cent a year. This is impressive enough, and in fact better than Britain and Germany achieved during their nineteenth-century industrial boom.

By contrast, other countries (the majority) stagnated and, six decades down the line, they are much worse off than they were under the foreign Raj. Pakistan, India's neighbour and former half-brother, as well as Zimbabwe in Africa, are examples that rolled all the way to the abyss.

For these countries 'aid', corruption and self-complacency proved to be the destructive forces. For several decades, foreign aid was thought to be the way of making a new country sociologically, educationally and even technologically advanced. We now know that this claim was overblown and, besides, in a large number of cases, the money that poured into these countries during the fat aid years was misused. As it became increasingly difficult to make financial ends meet, it became evident that:

- aid had been massively misdirected towards pockets, rather than towards the economy,[13]
- crucial agricultural development milestones were being relegated to the backburner;
- too much was being spent on industry, without the proper consumption patterns to support it,
- too little was being invested in infrastructure (schools, hospitals, transport system, dependable warehousing), and
- quality of life was seen by third-rate local politicians as a non-issue, while birth rates run wild.

13 A sure fate of funds directed to less developed countries 'to fight pollution of the environment', in the wake of the Copenhagen circus.

Tired of nonchalant inefficiency, leftist ideologies, a skyrocketing number of mouths to feed and bottomless holes in the pockets of the politicos, the richer countries started to spend proportionately less of their income on aid. This was justified by a climate heavy with distrust and disillusionment. In the aftermath, *manna* (food, energy or other goods) was no longer falling from heaven.

There is an old Greek saying that a fish rots from the head down. Many of the less developed countries' failures in food and energy lie squarely on the shoulders of the leadership, and nothing has been done over the years to correct this flaw. Confronted with limited resources, most political leaders (and not only in developing countries) are either cowards or unaware of the advantages of zero population growth. Yet, the advantages are compelling:

- reduced children mortality,
- high levels of literacy,
- a stable workforce,
- more qualified men and women,
- properly planned economic and sociological evolution,
- lower crime rates,
- less environmental pollution, and
- less pressure on physical and financial resources,

Alert political leaders, particularly of nations living in climates or deserts with minimal crops, would have appreciated these significant advantages. Instead, with their debt exploding, such countries became the global equivalent of the slums in big cities, and this created schizophrenic scenarios. On the one hand, there was no alternative to some sort of aid if these countries were eventually to fend for themselves. But on the other hand, the donors became disenchanted from lack of progress. As a result, in their search for fuel and food many developing nations have been destroying the environment on which they depend for their subsistence.

The vicious cycle of overpopulation and poverty is sure to lead to still greater poverty in the years ahead. Evidence is provided by the uprooted trees and bushes which have caused the desert to encroach, mile upon mile, on to once-habitable land. The floods sweep down mountain slopes with ever-greater fury. As wholesale destruction of the environment becomes a policy:

- frustration mounts,
- while aid continues to be reduced.

Hunger riots and protests have occurred in Bolivia, Burkina Faso, Cameroon, Egypt, El Salvador, Guatemala, Haiti, Honduras, Indonesia, Ivory Coast, Jordan, Kenya, Liberia, Mauritania, Mozambique, Nicaragua, Nigeria, Uzbekistan, Yemen and other nations. Some experts suggest that approximately 2 billion people will be soon suffering from malnutrition – roughly the equivalent of the population explosion.

In reality, the crisis is a declaration of the bankruptcy of the political system, in each of these countries and in a global sense, including the drive to 'Westernization' – that is, abandoning local and tribal cultures, without offering an affordable and sustainable alternative under local conditions. Uncontrollable globalization is not an answer; it's a curse.[14]

One of the ironies of this energy crisis has been that it has brought public attention to the fact that the current supply chain for food and energy rests on a very weak foundation. When drawing up its short- and long-term plans, every government has to recognize that food and energy shortages are here to stay. And they have to appreciate that hopes for a *Deus ex machina* through the Green and Blue Revolutions are ill-placed.

14 D.N. Chorafas, *Globalization's Limits. Conflicting National Interests in Trade and Finance*, Gower, Farnham, 2009.

CHAPTER 8 The European Union's Common Agricultural Policy and its Travails

1. Globalization Cannot be Allowed to Destroy Western Farming

Agriculture was humanity's first organized enterprise. The domestication of crops and livestock was instrumental in the development of communities and the rise of the arts – which provided humankind the stepping stone towards civilization. Trade came later, capitalizing on the fact that settled agricultural communities wanted to exchange their wares among themselves and with producers of other goods.

From the beginning, a good part of commodities being traded were agricultural, followed by artefacts. The notion that agricultural self-sufficiency more or less preceded trade speaks volumes about the position each of them has vis-à-vis the other. Agricultural self-sufficiency is not a matter of pride but of substance – and of subsistence. Whether one realizes it or not, agriculture is central to:

- geopolitical independence, and
- self-sufficiency *if*, as it happens from time to time, the worse comes to the worst.

Table 8.1 shows that in Western countries today agriculture represents a small fraction of the gross domestic product (GDP). Even so, this is a very important sector of the economy – one that cannot be given away no matter what the reasons might be. Charles de Gaulle once said: 'A country that cannot feed itself is not a great country.' What nation would ever think of:

- outsourcing *all* of its agricultural produce, and
- then waiting on other people's goodwill and efficiency to feed it with the essentials?

Table 8.1 suggests that, in the United States, services and manufacturing account for just over 99 per cent of economic activity and agriculture less than 1 per cent. What this table does not say is that the country simply cannot do without this 0.9 per cent. In the European Union, services and industry account for 97.8 per cent of GDP with agriculture providing a trifle in percentage terms.[1] And then what?

---

1 The 0.9 per cent in the United States and 2.2 per cent in Europe are fairly equivalent because in America agriculture is industrialized, whereas in Europe there is a patchwork of farms dotting the landscape.

Table 8.1 Contribution of sectors of the economy to national/regional GDP

	Agriculture (%)	Manufacturing (%)	Services (%)
United States	0.9†	19.7	79.4
European Union	2.2	28.3	69.5
Brazil	10.1	38.6	51.3
India	23.6	28.4	48.0

Note: † Fifty years ago, this stood at slightly over 2 per cent, roughly where European agriculture finds itself now as a percentage of GDP.

Source: Statistics based on a business meeting in which was discussed the importance of different sectors of the economy

Statistics are an important element in decision-making, but dry statistics never reflect all of the facts. The people who produce the 99 per cent and 97.8 per cent of GDP need to eat. The bigger the sum of the secondary and tertiary sectors, the more important becomes the contribution of the primary sector of the economy. Taking the European Union as an example:

- Are the people producing nearly 98 per cent of GDP supposed to relegate their food to farmers in underdeveloped countries, with low health standards, situated halfway round the globe?
- Who is supposed to control the quality of what these farmers' produce? That is, their overuse of antibiotics and the effects of poor sanitary conditions, pesticides and antibiotics on consumers' health?

Two-fifths of the world's population has nowhere to defecate except open ground. This is not unusual. As a result, the food of billions of people is grown on ground that contains their and their neighbour's faeces. Who the hell in the developed world would care to eat them? Food produced in less developed countries and exported to the West is also contaminated by the flies that lay their eggs in human waste. Like it or not, many – too many – people live in filth. They often die because of it and, with globalization, they export it to other countries and their consumers who are unaware of prevailing unhygienic conditions.

Several of these points are made by Rose George in her book *The Big Necessity*.[2] She knows what she is talking about because she has met activists who travel in rural India, provoking villagers to see with their own eyes the vile heaps deposited close to their homes. She also visited Chinese peasants who light their homes and cook their food with biogas generated from their own and their pigs' fermenting excreta.

Not everybody is aware of these facts (some evidence of which I myself have seen in rural India), which explains why many people in other countries buy and eat food produced under abysmal sanitary conditions. Some people say never mind how the food

2 Rose George, *The Big Necessity. The Unmentionable World of Human Waste and Why It Matters*, Metropolitan Books, New York, 2008.

is being produced, vote with your supermarket trolley instead – and shoppers go to the supermarket several times a month. However:

- public health is much more important than price discounts, and
- attention should be paid not only to the parlous sanitary conditions, but also to the damage done to the environment by pesticides and antibiotics which find their way into the food chain.

Another factor to be considered is the effect of environmental pollution caused by long food transportation routes. Since globalization 75 per cent of the world's food is allegedly not distributed and consumed locally. The longer the food supply chain, the greater the environmental damage done by transport.[3]

As for the argument about consumers propping up farmers because of Western governments' subsidies, it rests on the wrong premise that consumers pay more than the 'world price' for their food.[4] This is nonsense because there is no world price for food items, for the simple reason that the quality of food and its effects on health have a far greater influence on 'world price' than the quality of shoes and apparel.

There are also local conditions that lead to deliberate decisions to support farmers. The Swiss, Norwegian and Icelandic governments are often cited as big subsidizers of their farmers. But agricultural workers' wages left aside, weather conditions in these countries are diametrically different from those in Brazil and Thailand. Moreover, in each of these European countries, the cost of every product or service is higher, not just of food. In addition, Swiss, Norwegians and Icelanders are fully entitled to:

- care about the supply chain of their food, and
- pay the market price for it in *their* country.

Supporting local agriculture does not mean that every country should be fully self-sufficient in food. But nor should it be fully dependent on other countries to feed its people. Between these extremes lies a wealth of possibilities, with preference given to short supply lines. It is silly to stick to the extremes and fail to consider alternatives.

Without doubt during the last three decades there has been a shift towards what has become known as *global outsourcing* of food procurement. However, as with so many other types of outsourcing, its deliverables have left much to be desired. Examples of what has happened so far with agricultural products, raised under alien 'regulations', have been entirely negative.

In addition, we should always remember to account for security factors which become prominent under conditions of stress. What will happen if there is a disruption in an overstretched supply chain? In case of war? It is most surprising, and disquieting, that these questions aren't raised in the various WTO trade rounds, the failed Doha Round being the latest and most unsuccessful of them all.[5]

---

3 And the less is shoppers' ability to know about the tons of pesticides, fertilizers and antibiotics used to produce that food.

4 Besides that, consumers are being asked to subsidize alternative energy (Chapter 7). Why shouldn't they subsidize their clean, healthy food?

5 D.N. Chorafas, *Globalization's Limits. Conflicting National Interests in Trade and Finance*, Gower, Farnham, 2009.

The risks are present and real. When we talk about food, and not about shoes, we must be three times more careful. We may walk barefoot, if need be, but we cannot live without reliable and healthy food supplies whose origin is known, safe and dependable.

- Down to basics, food is a national defence issue. Interruptions in its supply can be used as a formidable weapon.
- The consumer is confronted with *health hazards*. Bird flu, for example, originated in South-east Asia among poultry being reared for food.
- Food production must observe prudent standards. Extensively used antibiotics in food production damage the human immune system.

Experts point out that protection for farmers is not a twentieth- or twenty-first century novelty. It has almost always been present, and it has become part of the fabric of society – particularly when self-reliance is equated with national security. This is precisely the case today. Not only as food producers but also as agents of the economy, farmers help in maintaining the rural infrastructure, preserving the supply chain and assisting in providing recreational opportunities.

2. A Bridge Too Far for Globalized Agriculture

Three million Chinese grow apples for export. Some of these are exported whole – like Chinese apples to France. But the majority are made into apple juice and sold all over the world. There are no standards for what this juice contains, neither are there any controls for pesticides finding their way into the juice – or for general sanitary conditions. Many people doubt that China exercises stringent health controls, and there is no evidence that countries importing Chinese apple juice test for health standards.

On the other hand there is plenty of evidence of pollution in Chinese water supplies. The pollution of Tai lake, China's third-biggest freshwater body, provides an example. In 2007 the lake was choked by toxic algae fed by the phosphates from the human and industrial waste that had been poured into both the lake and its tributaries. For more than a week, a stinking growth of the toxic algae disrupted the water supply of 2 million people living on the lake's shores. It was one of China's biggest environmental scandals:

- In the city of Wuxi, residents queued to buy bottled water, and
- the Yangtze River was diverted to flush the algae out.[6]

As far as water pollution is concerned, in India matters are no better. According to an article in *The Economist*[7] a short while after the Chinese debacle, the waters of the world-famous Ganges River was diverted for industry, agriculture and dozens of upstream cities. Ganges is now a cloudy brown soup of excrement and industrial effluent, and people are contracting typhoid, polio, jaundice and other water-borne ailments.

By official standards, water containing more than 500 faecal coliform bacteria per 100 ml is considered unsafe for bathing, and at some points along the river the water contains

6 'A Lot to be Angry About', *The Economist*, 3 May 2008.

7 'Up to their Necks in It', *The Economist*, 17 July 2008.

60,000 bacteria per 100 ml, yet people are still bathing in, and continuing to use, the Ganges water. Think of using that water for agriculture and then exporting the products to other countries and displaying it on supermarket racks to the unaware consumers.

The World Trade Organization (WTO) should have looked into these major health hazards associated with producing countries, but it has not. Its bureaucrats inhabit a sort of parallel universe in which countries negotiate not on the quality of what they produce or on the dependability of the supply chain, but just on tariffs and subsidies.

Although many countries have queries about food quality, these are taboo for the WTO. Yet, globalization and world trade greatly depend on one critical factor: the quality and reliability of supplies. Lack of attention to these issues makes the global agricultural system vulnerable. An analogy can be made to the boom-and-bust of structured finance, which saw banks distribute risk to unaware investors through specialist vehicles such as conduits.

- In the boom years, they did not worry about the creditworthiness of borrowers whose subprime mortgages were securitized and sold.
- But when the subprime bust came, homeowners lost their houses, investors got burned and banks went bust.[8]

When they reach global scale, agriculture, manufacturing and finance share a common problem which could be termed the *fragility of interdependence*. The greater the interdependence of the elements within the system the wider the effects of disruption because of failure in one part of it – all the way to systemic risk and eventual collapse.

The price and quality of agricultural produce don't always correlate, but behind much lower prices often hide health hazards. A paper presented to a UN conference in 2004 showed that American rice cost $331 per ton to produce, compared with $70 for Thailand and $79 for Vietnam. But to the question of who is going to control any health hazard associated with Thai-produced rice:

- the answer is 'Nobody!', and
- similarly, nobody is currently controlling overstuffing foodstock with antibiotics as in the case of Thai-raised shrimps.

European countries have been stupid enough to accept foreign agricultural produce without detailed standards and steady application of verification procedures. They impose ridiculous requirements on their own farmers, but they exercise no controls at all on the produce of less developed countries farmers – which is totally irrational.

Kari Redse Haaskjold, a Norwegian dairy farmer, cares for 20 dairy cows on a farm where rain and snow force her to keep the livestock indoors for at least half the year. Besides that, Oslo's new animal welfare law forces her to provide her cattle with 'more' *covered* space to roam.[9] When it comes to imported milk and meat, however, the Norwegian government does not even pay lip-service to 'cow welfare' in producing countries.

---

8 D.N. Chorafas, *Financial Boom and Gloom. The Credit and Banking Crisis of 2007–2009 and Beyond*, Palgrave Macmillan, Basingstoke, 2009; and D.N. Chorafas, *Capitalism Without Capital*, Palgrave Macmillan, Basingstoke, 2009.

9 *Wall Street Journal*, 13 December 2005.

European farmers cannot survive when governments, on the one hand, take supposedly 'humanistic' initiatives at cow level and, on the other hand, negotiate lower tariffs that open wide the gates for produce from farmers on very low wages – and even lower standards of care for the health of their animals and of consumers.

In a globalized economy, policies like the Oslo initiative are worse than nothing, because they hurt the country's own farmers while allowing farmers in less developed countries to do as they please. And because global health and farming standards are today's, not tomorrow's, issue, wise countries make sure that, world trade or no world trade, they drive for self-sufficiency in foodstuffs.

China provides an example on this policy. Its production and consumption account for more than one-fifth of the world's total grain supply. But China is self-sufficient in cereal (mainly wheat, corn and rice) and is only a net importer of soya bean. China's self-sufficiency in cereal has helped grain prices remain stable. Moreover:

- grain production has picked up in recent years, and
- grain trades are small relative to domestic demand.

As Michel Barnier stated in a letter to *The Economist* when he was the French minister of agriculture (he is now EU commissioner in charge of internal markets):

> *The future of agricultural policy is not about 'assistance', it is about rendering services to society, be it in terms of food security, the environment, rural areas. So ... several agricultural models can coexist. It is viable economics to cater to and invest in high-end products, which are inevitably linked to the diversity of regional identities and produce.*[10]

To protect local agriculture, Western governments provide financial support to farmers, and they use tariffs to assure that there is no negative effect from the unfair dumping of foodstuffs (many of which have been raised under doubtful hygienic conditions). *If* a deal is cut on throwing away tariffs, *then* cheaper but lower-quality and less dependable agricultural products (from grains to beef, milk and cheese) would flood Western countries:

- wiping out local farmers,
- putting the health of all citizens at risk, and
- increasing the Western nations' already loaded health-care budgets.

Why should the United States and the European Union accept wiping out *their* farmers? It is indeed tragic to see that not only the bureaucrats in Brussels and the WTO but also the eggheads in the various EU member nations are making the farmers' lives more difficult instead of providing them with training and hope for a better future.

10 *The Economist*, 11 March 2008.

3. A Common Agricultural Policy?

In his March 2008 letter to *The Economist*,[11] Michel Barnier also wrote about the need to overhaul the European Union's *Common Agricultural Policy* (CAP) – to position it against future strategic challenges: growth, rural development and the basic requirement of producing more and better food in Europe.

The CAP was established in 1962. Since the outset it has been characterized by ill-structured measures lacking details and the ever necessary study of after-effects. The net result was that the European Economic Community (EEC, as it was then known) set up a complex mechanism of a publicly managed agricultural market, full of price regulations. This was supposed to hit two goals at the same time:

- protecting the interests of producers, who need a stable income, and
- those of consumers, who want reasonable prices.

Many of the CAP's basic principles, announced at the time of its institution, can be summarized in a few paragraphs. The EEC defined a *single market* among the original six sovereign states for selected agricultural products. It abolished tariffs among them and harmonized prices for these specific products.

The underlying rule has been 'community preference'. Member states committed themselves to satisfying their domestic needs with supplies from other member states, unless goods were unavailable. Imports and exports with nations outside the common market were regulated by common trade tariffs.

A new entity, the European Agricultural Guidance and Guarantee Fund (EAGGF), was given the mission to watch over a parity price for different products. *If* a given product could not find a buyer on the market, *then* the EAGGF would automatically buy up the surplus, using its intervention funds. This was supposed to guarantee stable prices (as in today's China) and a decent income for farmers. By contrast:

- *if* prices rose too high,
- *then* the European Union could sell its inventory and drive prices down.

The unwanted consequence has been that parity prices created a massive incentive to expand production. Theoretically, these parity prices were adjusted permanently by common-market organizations (CMOs) run by the EEC. In practice, the suppositions underlying these future adjustments were not guaranteed by the facts.

The chosen approach was similar to that of the European Coal and Steel Community, with each CMO employing a permanent dialogue among member states to steer a distinct agricultural sector: cereals, pork, poultry, dairy products, fruits, vegetables and wine. National interests, however, came into the picture in a big way and disturbed a mechanism which might have had a chance.

The situation deteriorated as the EEC expanded. A rapid increase in membership opened the community's doors wide, and the demolition squad started its good work. In 2005 Tony Blair (of Iraq, Afghanistan and British farmer wars fame) threw a fit against the CAP, and this intensified already existing EU rivalries.

---

11 Ibid.

Blair had his fans. Applauding his ravings, the *Guardian* wrote:

> *The single thing rich countries could do that would most help developing ones would be to dismantle subsidies for agriculture. Such countries would allow poor countries to compete fairly in areas they are good at, while releasing well over $380 billion a year, currently wasted on subsidies, for the west to spend on other things.*[12]

Just like that, with no consideration given to important issues like:

- food quality,
- health standards, and
- the future of European farmers and farming as whole.

This does not mean that, without Blair's intervention, the CAP would have been an unmitigated success. One of the first charges against it was overproduction which created 'butter mountains' and 'milk lakes'. This is what happens when bureaucrats without agricultural and marketing experience try to micromanage a complex process.

More serious, though less discussed by politicians and the media, was the underproduction of other goods that the EEC kept importing from the rest of the world. In addition, by mobilizing the GATT (later transformed into the World Trade Organization), the Blair government embarked on a campaign to kill the CAP, which was accused of being:

- too expensive,
- anti-environmentalist,
- privileging just a few landowners instead of small farmers,
- too elitist, killing the poor in Africa, and more.

Nobody ever explained how the gap was 'killing the poor in Africa', but just the mention of it was too much for the leftist politicians of the European Union, so they turned against their own people, the farmers. It has also been an argument for 'more globalized trade' in the name of free markets, no matter what might be the aftermath.[13]

The greatest failure of the European Union's Common Agricultural Policy has been the lack of new goals, which would give agriculture a reason to be dynamic and productive, rather than featherbedded through subsidies. Nothing is more demoralizing than staying too long on the defensive.

- Biofuels are a new *raison d'être* (Chapter 6), which will give EU farmers a future and perspective.
- Making the European Union independent of oil imports, and therefore self-standing as an industrial power, will eliminate the current crisis of confidence in energy supplies.

12 *Guardian*, 15 November 2005.

13 D.N. Chorafas, *Wealth Management: Private Banking, Investment Decisions and Structured Financial Products*, Butterworth-Heinemann, London and Boston, MA, 2006.

The problem is that instead of providing leadership, the European Commission is trying to find a way of concealing its agricultural policies. Undeterred by its past failures to balance the scales, it specializes in a watershed of proposals to restructure farm spending.

As may be expected, these poorly studied, uncertain and debatable restructuring plans fall far short of a liberal farm policy (favoured by Britain), while they sharpen a debate in France regarding support for agriculture. By all evidence, the most controversial and inconsequential plan has been the 2003 CAP reform (discussed in section 5). Besides the many enemies it made in the European Union's agricultural community, it also entailed:

- distorted effects, and
- divergence between EU prices and the international prices of a number of agricultural products.

This does not mean that other revisions were more successful. For instance, the European Union's 2008 proposals, which covered the period to 2013, were broadly aimed at moving the Common Agricultural Policy further away from solutions like quotas, which keep prices artificially high, and away from payments to farmers linked to the amount they produce. The then agriculture commissioner, Mariann Fischer Boel, said that the new measures were all about freeing farmers to meet growing demand and respond quickly to what the market is telling them. In reality, they were old measures in new clothing.

They were also poor compromises. These proposals were followed by months of wrangling over the final shape of agricultural policy between Britain and France. In this case, though this is not always so, the French were right because their policy promoted the European Union's food independence.

Correctly, the French thesis[14] also called for ambitious health standards both inside and outside the European Union. Imports could be curbed if foreign farmers don't meet EU standards not just on health, but also on animal welfare, environmental protection and labour laws (whose non-existence in developing countries allows them to sharply cut prices).

It is plain nonsense that the European Union obliges its farmers to uphold high standards in food production (which is fair enough) but opens the backdoor to admit substandard or unhealthy food products. The latter find their way to consumers and raise the social security's health bills. It is high time for the European Union to:

- sort out its farm priorities,
- put its agricultural production system in order, and
- study, through simulation, the outcomes of its different decisions and their unwanted consequences, before finalizing them.

The milk crisis which erupted in 2009 provides an example. Since the new agricultural directives of 2008, EU milk quotas have been on a slow path towards abolition, which led in 2009 to plunging prices. 'Europe is deaf to farmers' cries' screamed the front pages of

14 The French paper was released in November 2008.

newspapers in Belgium after hundreds of tractors rolled into central Brussels on 22 July 2009.

The European Milk Board, a farmers' lobby, wanted EU dairy production curbed by 5 per cent immediately, either through quotas or by paying farmers to limit production voluntarily. This, however, is another market-meddling approach promoted by the fact that European farmers are suffering.

- From 2007 to 2009 prices for raw milk fell by half in some countries,[15] and
- they have been tumbling below the cost of production on many farms, leading to the farms' extinction.

Nor did European consumers benefit from this fall in prices. Instead, they kept on paying pre-2007 prices for their bottles of milk. Solutions are made more complex by the fact that milk production within the European Union is uneven. Denmark, the Netherlands and Italy want to produce more milk, but in France milk production is well below the level established by quotas.

'There is a deficit of 600,000 cows in France to reach the quota level,' says Henri Brichard, president of the French Milk Producers Foundation. 'This is a deficit of 600,000 tons of milk out of 24.5 million tons allocated to France.'[16] This same phenomenon of underproduction of milk also exists in other EU countries and also characterizes two big milk exporters: Australia and New Zealand.

Critics say that the reasons for the big and growing milk deficit in the European Union can be traced to the silly decisions taken by the Brussels bureaucracy when the CAP was restructured. The financial aid provided to the agricultural sector is no longer linked to production. A farmer can reduce his milk delivery and increase his cereal produce without any loss of financial support, and as it takes up to three years to turn a calf into a milk cow, the situation in the next few years can only deteriorate.

4. From Chaos to Uncertainty: The European Union's Agricultural Goals

Sustainable agriculture with produce which is healthy[17] and cost-effective has never been a clear goal of the European Union's agricultural policy, or for that matter the policy of any state in east, west, north or south. Instead, governments have a tendency to capitulate to the drive of the day – from protectionism to 'free-trade' dictatorships.

Both glut and penury are failures in agricultural policies. Gluts cannot be unloaded on to the world's markets except at rock-bottom prices, like the sales of warehoused Dutch butter as animal feed to the Soviet Union in the Cold War years. Arbitrary corrections lead nowhere. In 1968 Sicco Mansholt, the then Dutch EU agricultural commissioner, created the so-called 'Mansholt Plan' which:

15 In addition, the 2007–2009 deep recession hurt global sales of cheese, and cheese-making consumes nearly half of all EU milk.

16 *Le Canard Enchainé*, 15 August 2007.

17 The term 'bio' is deliberately avoided because it has become terribly corrupted since it became a marketing buzzword.

- took 5 million hectares of useful farmland out of production, and
- through lavish subsidies 'convinced' 5 million EU farmers to abandon agriculture.

In 1984, 16 years later, milk quotas were established, aimed at lowering production, and the principle of 'reducing spending' was adopted. Another two years down the line, in 1986 – coming out of the 1986 GATT Punta del Este Uruguay Round – a group of 19 agro-industrial emerging countries – including Brazil, Argentina and Indonesia, but also developed countries like Canada and Australia – met in Cairns, Australia. They denounced the CAP and claimed their right to export cheap food into the European Union. The term has two meanings: low-cost and low-quality.

Instead of protecting the European farmers, the second-raters in Brussels continued their silly business. The farmland of the European Union and its people were victimized by a policy of mandatory *set-aside* totalling 3.8 million hectares. Subsequently, the Uruguay Round pressured the European Union to open up to the world market and to decouple subsidies from production.

In 1992 Ray MacSharry the then Irish EU agricultural commissioner, imposed a lowering of the guaranteed prices of agricultural products supposedly 'compensated' by direct financial aid to producers. The guaranteed price of cereals was reduced by 35 per cent, and beef by 15 per cent. For European farmers this was both a psychological and a financial killer, since it obliged them to:

- accept the decoupling of subsidies from production, and
- live on permanent aid – subsidies now being converted into handouts – instead of the pride and pleasure derived from their work.

Worse was still to come. To the destruction of European agriculture created by the MacSharry 'reforms' was added the warfare of the GATT Blairhouse agreements. The European Union unwisely accepted that it would produce *less than 30 per cent* of its own oilseed and high-protein food needs. Consequently, Europe remained totally dependent on the goodwill of giant food cartels (see section 6).

Not to be left behind in the process of agricultural devastation, in 1995 the newly created World Trade Organization (WTO) imposed the Agreement on Agriculture (AOA). This was a curious WTO code of conduct aiming to increase runaway free trade and cartel power, by classifying subsidies into three 'boxes':

- a green box of authorized subsidies that 'did not distort' international trade,
- an orange box, which permitted some interim but decreasing domestic support, and
- a blue box theoretically targeting the 'dismantling of protectionism' through the reduction of export subsidies of developed nations by at least 35 per cent between 1995 and 2000.

All this corresponded to over 20 per cent of volume, to the detriment of EU farmers. *As if* this was not enough, it was followed by the so-called 'Agenda 2000' which set out to limit spending on the CAP, diverting funds into various environmental schemes.[18]

18 The CAP represented 81 per cent of the European Union's budget in 1985; 65 per cent in 1995; 44 per cent in 2005; and is planned to represent less than 37 per cent in 2013.

In the meantime, various EU governments did their own good work, whose net outcome has been to make the farmer's life miserable. The signal was given by the Blair government, and the excuse was an outbreak of cattle disease known as foot-and-mouth in 2001. The result was the devastation of a whole class of British farmers. Unlike BSE (the so-called 'mad cow' disease), which is due to a tiny little virus in hibernation[19] and is transmitted to humans, foot-and-mouth has been around for hundreds of years and typically spreads virulently among other animals. The killing of thousands of cattle with foot-and-mouth was very expensive for farmers – British farmers in particular. The decision of the Blair government was largely political and economic, the latter in the sense that the large reduction in livestock herds resulted in an upward surge of meat prices for farmers – other than those whose fortune was decimated, of course.[20]

These extreme cases aside, European farmers were hurt as the European Union adopted the decoupling of subsidies from production demanded by MacSharry and the WTO. French opposition led to partial decoupling and sector-by-sector, even case-by-case, arrangements. Nevertheless, European farmers were reduced to being mere gardeners of their landscape. This has been a huge:

- social,
- economic, and
- political mistake.

Critics have plenty of good reasons when they say that EU policies – agricultural policy being one of them – undoubtedly hit the wall because they are guided by short-termism. Milk production and its quotas is one example of unintended consequences. Fishing presents another case. On 22 April 2009 the European Commission published a consultative green paper on the common fisheries policy (CFP), suggesting needed changes because an estimated 88 per cent of the European Union's stocks are overfished (see also Chapter 7 on the failure of the Blue Revolution).

As a result of radically reduced fish stocks, most of the EU fishing fleets are barely profitable or lose money despite direct aid and subsidies like tax-free fuel. Ironically, in several member countries the cost to public budgets of subsidizing fishermen exceeds the total value of the catches – which means a taxpayer who eats local fish pays for it twice.

Another basic problem where EU member nations fight it out is overcapacity. Many fishing fleets are two-thirds too large. A solution of sorts has been already applied in New Zealand and Iceland, where professional fishermen hold individual fishing rights that can be sold or leased to others. The EU Commission suggests that such rights should be tradable between national fleets, creating an EU-wide market. But:

- This has created friction among countries with big fishing fleets.
- The core problem of overfishing is not being addressed in a direct way.
- In fact, it is not being effectively addressed anywhere in the world, since the Blue Revolution revealed itself as a mirage.

19 A submicroscopic particle like the tobacco virus, which has the nasty habit of regenerating. Modern medicine and pharmacology have no way of neutralizing it. It was transmitted to cows because man converted them to carnivores by feeding them with food made from the carcasses of other animals.

20 Continental EU wouldn't buy British beef for a very long time after the outbreak was eliminated and a long time after cattle movement was resumed and animals slaughtered for food again in the UK.

The same sound of an uncertain trumpet is heard in connection to matters concerning the health standards of agricultural goods. France and Germany demand that foreign competitors must apply EU rules on hygiene or face high import tariffs (although health standards that are inspected would evidently be the better alternative). Opponents, like Britain, claim that if food were genuinely unfit for consumption, it should be banned, not taxed.[21] But who is to say whether it is fit or unfit?

Moreover, environmentalists and many others want to see legislation that leads to the most hazardous chemicals being withdrawn from the market. In the opinion of Wolfgang Reinert, an official at the European Commission's directorate on Health and Consumers, new EU rules must embrace the philosophy that something should be done to ensure that a producer can sell his product only after he has proved it can be used safely.[22]

Reinert is right. However, a firm EU policy on health standards for every item entering into the food chain should not be left for 'some time in the future'. It should be studied *now* and implemented.

The enemy of this policy is not only inertia, but also embedded interests wrapped up in a 'free trade' label by multinational organizations which play games with people's health. In this, they are helped by the supranationals. The WTO is a dictatorship whose bureaucrats decide what is permitted and what is not in farming, and the European Union bends over backwards to please them.

- The more the global bureaucrats issue *diktats*, the more world food depends on a small number of global suppliers, while trade distortions become commonplace.
- The more the population explosion continues unabated, the more common become the stresses on the earth's resources.

In conclusion, embedded interests, an exploding world population, the struggle to raise standards of living and the battle for energy resources have led to a huge *mis-entente* on matters as basic as agriculture. 'World agriculture has entered a new, unsustainable and politically risky period,' said Joachim von Braun, head of the International Food Policy Research Institute (IFPRI) in Washington, DC.[23]

5. Flaws in the EU Common Agricultural Policy

The unfortunate one-sided initiatives taken by politicians and bureaucrats, who have lost contact with reality and with the interests of their electorate, can have disastrous consequences on European agriculture. The various ill-conceived 'liberalizations' take no notice of the fact that from the nineteenth century's Corn Laws in England to the Treaty of Rome, legislation singled out farmers for special *help* rather than for punishment – because agriculture has always been a concern in Europe.

21 An example is a ban on imports of American poultry washed in chlorine, to remove bugs picked up during intensive rearing. The Americans can do what they want, said one of the EU's agricultural ministers; Europeans have different standards.

22 *The Economist*, 5 July 2008

23 *The Economist*, 19 April 2008.

Since the 1950s the Common Agricultural Policy has proved to be the most costly of the European Union's initiatives because it has been full of flaws. Even if its aims were laudable, in that it was designed to ensure that the populations of the European Union's six founding states would never again go hungry, special interests and near-sightedness dotted it with lots of holes and loose ends.

The goal of providing the citizens of the original Common Market with plenty of agricultural produce has been reached in a flimsy manner. As so often happens with different government initiatives, poor management saw to it that the CAP, more than any other aspect of EU legislation, became the *bête noire* of Eurosceptics and opponents of the European Union. The CAP's proponents say that its goal is to:

- increase agricultural productivity,
- guarantee farmers' incomes,
- stabilize agricultural markets,
- assure a food supply for everybody in the European Union, and
- guarantee that food reaches European citizens at reasonable prices.

These goals have not been studied in unison, as a system. The main link among them is that the CAP is anchored in Article 33 of the Treaty of Rome, establishing the European Community, which also contains policies and measures to improve rural development and the environment – in one go.

In addition, Article 39 of the Treaty of Rome promised a policy assuring security of supply; Article 40 offered a choice of ways in which these goals might be achieved; and Article 42 says that specific provisions of agricultural policy take precedence over normal competition rules. This has been turned on its head with the WTO agreements signed by the European Union.

At the start, the CAP supported the European Economic Community's farmers' income through import levies, export subsidies and guaranteed prices for products that would otherwise have remained at a much lower world market price level. But by the 1980s these measures had led to high budgetary costs and an oversupply of agricultural products because the whole enterprise had become like a runaway train, with no driver in charge.

As a result, it became inevitable that the European Union would introduce various measures to limit expenditure and production, impose restrictions on vineyards, establish ceilings on national aid for various products and come up with the idea of 'setting aside' agricultural land. These decisions were taken:

- piecemeal, without a master plan covering the longer range,
- in the absence of careful study and simulation of the consequences, and
- under the heavy burden of horse-trading, as every EU government wanted to gain the most for *its own* farmers.

These three bullets explain why today many experts consider the CAP as an unreasonably complex assembly of instruments and regulations only partially covering a wide range of mostly interconnected issues. These include price supports, trade controls, prevailing regulations, labelling standards, conservation policies, imports, exports and more.

Together, half-baked rules, superimposed clauses and a variety of regulations boil down to budgetary costs and subsidies that don't satisfy their recipients as proved by the fact that protests about falling farm incomes came from the CAP's beneficiaries. Therefore, for technical and political reasons from time to time the CAP underwent a major face-lift which:

- added to its complexity, and
- reduced its transparency to both taxpayers and beneficiaries.

This was the case with the 2003 extensive CAP reform whose main goal was to meet the requirements of international trade liberalization agreements – and only secondarily to alleviate EU budgetary pressures and make EU farming more competitive. That reform lowered guaranteed prices substantially and introduced a *single farm payment* to replace many of the previous payments to farmers that were tied to production. It also led to abuses and to distinctly unamusing results like subsidizing the rich (see section 5).

The pros say that the CAP's 2003 version had a positive outcome, and anyway it was inevitable. They add that no matter the resulting complexity, the establishment of an integrated common market for agriculture was pivotal in the formation of the European Union (which is an overstatement).

Critics answer that the 2003 and other CAP revisions were nothing more than patches. Some add that the whole idea has been lop-sided. In the post-World War II period when the memory of food shortages was relatively fresh, the productivist emphasis of the EEC's founding fathers was understandable. Nowadays, it doesn't make much sense. In fact, fundamentally, neither argument – pro or con – makes much sense.

An EU agricultural policy must consider all players, avoid favouritism, and be fairly simple as well as transparent. Moreover, it should support the weakest part of the farming community and it must be cost-effective. As we have seen so far, and will see further on, CAP honours neither of these prerequisites – a fact made much worse by the European Union's apparently uncontrollable and unstoppable enlargement.

- Direct payments to farmers in new EU member states, which are unrelated to output, have been rising continually for no purpose.
- However, the budgeted share of expenditure on the agricultural sector cannot be redimensioned because poorer and poorer countries continue to join the European Union, and many of them have a huge amount of corruption.

Taken together, enlargement and corruption assure diminished support for western European farmers. The 2003 Luxembourg reform of the CAP further disfavoured western European farmers by decoupling direct payments from production, with subsidies becoming a sort of single farm payment. This decision accentuated the wrong-way risk, Agricultural subsidies today go:

- to the wealthy in the European Union (see section 6), and
- to the poor countries of the steady and unwise EU enlargement in south-eastern Europe.

Needless to say, all this leads to a major budgetary squeeze. The European Commission proposes that, in any one year, payment appropriations, which have to be met from member states' contributions, must not exceed a certain ceiling linked to their GDP. This simply ignores the fact that:

- CAP expenditures have the nasty habit of steadily rising, and
- they rose even faster with the accession of Romania and Bulgaria, both poor agricultural states.

It is beginning to be belatedly recognized that endless increases in CAP subsidies are part of the European Union's *enlargement fatigue*. They also have two major risks associated with them. The first is that the agricultural objectives of the Treaty of Rome are in tatters as agricultural produce is relocated towards eastern Europe. The second big risk is that the booming CAP budget cannot be met. Margaret Thatcher got her money back in Britain's agricultural policy financing. And Germany is no longer the European Union's locomotive as it used to be, while overdependence on its financing is unfair because Germany is a net importer of agricultural produce.

The irony of expanding CAP expenditures because of irrational but unstoppable EU enlargement should not be overlooked. It comes at a time when an economy which had no difficulty in financing agricultural subsidies has its own problems to deal with and can no longer afford to pay for other people's pleasures. Germany is the largest contributor to the EU budget and, in its way, this violates the principle of 'just return' that the different recipient countries want to see applied to *their* money.

6. Controlling the Global Food Trade's Health Hazards

It has been a deliberate choice to include in this chapter the themes of mass farming and intensive food marketing, because they correlate among themselves and with agriculture. Nowadays, intensive marketing is done through two media: the Internet (which gains in popularity) and what in the 1950s was called 'the cyclops' or 'one-eyed beast'.

'Your favourite TV dinner' is a different type of health hazard to those described in previous sections. But it leads all the way to *obesity*, and it can be just as destructive. Obesity is also promoted by fast-food chains, which have spread over the globe, and some of the wares marketed by big global organizations.

The Big Food Industry is a development of the last four decades. It earns hundreds of billions of dollars, pounds and euros per year in major and minor countries, its income being in excess of the government budgets for education and health services. Because Big Food is steamrollering people's consumption habits:

- its methods have sometimes come under scrutiny, and
- reports by the health insurance outfits of many states pinpoint the added cost due to obesity.

Obesity rapidly spreads around the globe, and not only among more affluent nations. According to certain opinions, life expectancy in Europe will soon start to decrease for the first time in 500 years. Obesity is promoted by several factors imprinting themselves on the public mind, including intensive advertising of food products and false information on labels such as:

- cereals 'essential to the bone structure',
- eggs that are 'anti-cholesterol', and
- a variety of foods falsely advertised as being 'fat-free' (though with plenty of sugar) and so on.

Quite clearly, the message drummed into consumers' eyes and ears by fast-food chains and big food companies' products should be sincere. Today, it is not. Sausages? You don't have to be afraid of getting fat. Biscuits? They don't contribute to obesity. Beer? You can drink it like mineral water. Salt? It has been limited – and other lies.

The effects of intensive publicity on public consumption of food has led some critics to ask for such products not to be advertised two hours before dinner. But, having chopped off cigarette advertising and its revenues, governments are reluctant to take a step that:

- may ruin the TV stations by depriving them of income, and
- runs contrary to the interest of powerful food lobbies who often dictate government decisions.

According to some estimates, about €2 billion is invested annually by the Big Food Industry in France alone, and 75 per cent of that sum goes to television. This is money which serves to promote obesity, and the irony is that, among other metrics, the obesity of EU citizens has become a curious sort of measure of 'marketing success'.

Global procurement of food products, including countries way behind in health standards, is another major concern. In a way emulating the securitization of subprimes by big banks – which led to the severe economic and banking crisis of 2007–2011 – all sorts of questionable food products are packaged and sold. (A better handling would have consigned them to the rubbish bin.)

Then come epidemics like bird flu (H5N1 virus) of 2006 and the H1N1 virus of 2009. What else can be expected when the movement of eggs and poultry from one country to another continues taking place without controls, and the indiscriminate use of poultry manure as agricultural fertilizer is still in the news. While the H5N1 epidemic was raging, the UN Food and Agriculture Organization (FAO) estimated that Nigeria imported some 1.2 million day-old chicks annually.[24] And several experts suggested at the time that many of these chicks were arriving in Nigeria from countries with domestic H5N1 infections, like China and Turkey.

As with mad cow disease, which was promoted by inappropriate feedstuff, experts attributed a much of the trouble to what the mass-produced chicks ate. They believe that the central role in regard to avian flu and other epidemics has been played by the industry of raising all sorts of birds for commercial consumption without properly investigating the potential consequences. Yet,

- in a mass society there are serious reasons for doing analytical studies, and

24 *The Economist*, 23 February 2006.

- without research that attacks the origins of an epidemic there are no valid protective measures.

To avoid the usual fire-brigade approach, a serious study should factor in the likely effect of virus transmission in conurbations of with 5, 10, 15 and 20 millions of people – as well as of virus origination in factory farms, and in subsequent processing and mass marketing. Chapters 6 and 7 also highlighted the malicious effect of chemicals, pesticides and fertilizers which settle into, and become permanent features of, the food chain.

There is no escape from the fact that the controls necessary in a mass society are totally different from those that prevailed in previous generations. Many agricultural scientists argue that the thousands of tons of chemicals used today have widespread and alarming short-, medium- and longer-term consequences for food products. Decisions to curtail their use are not kids' stuff:

- On the one hand, cutting down on the use of poisonous substances would reduce food production roughly by a quarter.
- On the other hand, the health threat from the intensive use of pesticides, insecticides and other chemicals is certainly most serious.

Nor is the prevention of future epidemics in livestock being paid the attention it deserves. In particular, little or nothing is being done in connection to the spread of serious animal illnesses through global trade. The whole bird flu business started in south-east Asia, in heavily industrialized bird farms which raise food for export under conditions that are anything but hygienic.

Although migratory birds were first thought to be the carriers of the flu virus, it was subsequently established that some of the migratory circuits of wild birds are bird-flu-free, while the commercial circuits have had more than their share of the virus. These commercial circuits:

- handle both birds and eggs,
- produce and deal in animal food, and
- according to experts, trade in contaminated animal foods.

In fact, a country that never had a human bird flu case is Laos, because the government closed the borders to industrially raised chicken from Thailand. By contrast, nobody stopped wild birds from flying over Laos. In addition, while small numbers of birds raised by families developed certain mild forms of the flu, they never developed into global threat because of the small numbers of chickens at each farm and their distance from other farms.[25] The global threat came from the large-scale, intensive raising of chickens, ducks, turkeys and other birds for export in food farms of less developed countries.

The industrial food given to chicken and other birds worldwide is partly composed of animal faeces, feathers of dead birds and other (un)appetizing ingredients. And, as the *Canard Enchainé* aptly noted in an article, '(it is curious that) nobody asks why the moment the bird flu reached France it hit 11,000 turkeys of an industrial bird reproduction

25 'Bird Flu Crisis: Small Farms are the Solution Not the Problem', *Grain*, July 2006, at: http://www.grain.org/seedling/?id=437

facility.'[26] This led to 43 countries boycotting French poultry. But it also raised serious doubts about the wisdom of continuing with the WTO's global trade negotiations without first *studying* and *solving* health-related problems in massive agricultural enterprises.[27]

In conclusion, the effectiveness of health standards and of quality control from production and packaging to the marketing and advertising of food products should be significantly increased. It is safe to assume that the world's hunger for protein is not going to abate. The more massive the food trade becomes, the greater must be the controls and the more focused the health standards.

7. Wrong-way Risk with the European Union's Agricultural Subsidies

Theoretically, but only theoretically, one of the basic objectives of the European Union's agricultural policy is to increase competitiveness in the agricultural sector as well as provide for growth and employment. In practice, while statements like that make nice reading, they contradict other objectives and are therefore unrealistic.

According to the Brussels bureaucracy, competitiveness is to be increased through education and research promotion, as well as by means of unspecified 'trans-European networks'. To get this magic going, it is projected that funds will be significantly increased – except that the coffers of member nations are empty. Worse still, EU agricultural funds are being transferred from the fields to the cities.

The story not told by cold statistics is that during the twenty-first century the considerable increase in the funds for the European Union's structural policy was mainly due to the integration of the ten new member states from eastern Europe and the Mediterranean followed by the (unwarranted) accession of Bulgaria and Romania – which meant an accentuation of the economic heterogeneity of the member states.

- This has been a structural change for the worse, and the agricultural budget for the 'old' EU member states shrank.
- The farmers continued to receive money, but only half of what was allocated to the agricultural fund.

According to the EU Commission, the aim of this switch in structural policy was to assist regions with below-average convergence. It does not take a genius to understand that new countries that are below the average of convergence should *not* be admitted to the European Union. By rushing into the various 'enlargements' the second-raters running the European Union's fortunes have created new problems before solving the old ones. There is also the danger that, as with all statutory promotion measures:

- undesirable incentives are being created and
- mismanagement is turning the supposed 'more efficient' structure upside down.

---

26 *Le Canard Enchainé*, 15 March 2006,.

27 Happily for everybody the Doha Round is by now only a dark spot in history. See also D.N. Chorafas, *Globalization's Limits. Conflicting National Interests in Trade and Finance*, Gower, Farnham, 2009.

This simply means that the mistakes made with the CAP over several decades are being repeated and magnified. The EU bureaucrats who imagined all that as a way of justifying *their* wages did not even care to submit their regional agricultural policy to critical examination, which should have benefited from:

- simulation,
- dissenting opinion, and
- a detailed look at unexpected consequences.

Oligarchic decisions, which have not been carefully studied but have been made in a rush, are in the background of most of what comes into the envelope of *structural policy*. Nothing has been learned from the fact that the CAP was unsuccessful both in reducing rural poverty and putting EU agriculture on its feet – and for good reason. In an application of Pareto's Law,[28] an estimated 80 per cent of farm support has been going to:

- the richest 20 per cent of farmers, and
- landowners whose only contribution to farming is that they keep their real estate out of agricultural produce altogether.

Prince Albert of Monaco has been the only EU landowner honest enough to admit to this fact. In 2005 he received €300,000 as an annual subsidy for the 700 hectares he owns in the Aisne (north-eastern France). By contrast, in the south-east of the country a big landowner who received €866,290 for 1,733 hectares and another one in Aquitaine who got €811,755 for 1,500 hectares kept quiet.

What this reveals is that the EU Executive uses taxpayers' money not to support the farmers, but to promote inequality. In Britain, in 2005, Prince Charles received the equivalent of €435,000 in agricultural subsidies and the Queen even more.[29] This, of course, has never been the objective of an agricultural policy.

- A well thought-out plan should help the struggling farmers.
- It should not further enrich large landowners.

That silly outcome came from an ill-studied measure aimed at removing the worst trade-distorting aspects of the CAP. The message this and similar examples convey is that effectiveness in the use of agricultural funds has gone out of the window. Money simply goes where politicians want it to go.

It should need no explaination that the EU's Common Agricultural Policy does not come free of cost. Unfortunately, however, every time the CAP is massaged, it comes out in worse shape than before the change. No wonder that richer countries are increasingly reluctant to fund the various unwarranted bonuses.

28 In the late nineteenth century Vilfredo Pareto, a Swiss-Italian economist and mathematician, established that typically a small part of one factor correlates with a big part of another one. It later became the 80–20 rule which was somewhat inaccurately dubbed Pareto's law.

29 *Le Canard Enchainé*, 9 November 2005.

Britain, for instance, has been fighting to save its annual rebate – a legacy of Margaret Thatcher's time. Here is a brief reminder of the facts and events. Back in 1979, when it became clear that, because of its small farming sector, Britain got few CAP receipts while being one of the bigger net fund contributors, Thatcher *asked for her money back*. Several bad-tempered summits later, she got it. Since then, the European Commission has calculated the gap between:

- what Britain pays in, and
- what it gets out every year.

Having done this, it makes a rebate of 66 per cent of that gap, paid back to Britain a year in arrears. (This has done nothing to weed out the significant difference of opinion on how the CAP's goodies should be shared.)

Jacques Chirac, as president of France, and Tony Blair, as prime minister of Britain, frequently clashed before and during EU 'summits'. Most notorious was the mid-June 2005 summit in which Chirac demanded that Britain give up its EU budget rebate, Blair insisted that the CAP should be renegotiated. As president of the European Commission, José Manuel Barroso issued a gloomy warning, saying that:

- a failure to come to an agreement would be disastrous, and
- it might mean that the European Union would sink into a permanent crisis and paralysis – *as if* that had not happened already.[30]

This has not been an isolated case. Financing the CAP has given rise to widely different net contributions and receipts, thereby producing winners and losers. Of the 15 member countries, prior to the 2004 enlargement:

- the Netherlands, Germany, Sweden, Britain (even with the rebate) and, to a lesser extent, Austria, paid out more than they received,
- Denmark, France, Italy, Finland just about broke even,
- wealthy Spain, but also Belgium, Greece, Ireland, Portugal and filthy-rich Luxembourg, got more than they paid out.

It should also be appreciated that the European Union is by no means alone in providing agricultural subsidies. Other countries with high producer support for agricultural goods are Canada, Iceland, Japan, Mexico, Norway, South Korea, Switzerland and Turkey. In Turkey agricultural subsidies have persistently increased over time, and they will be paid by west European taxpayers *if* Turkey joins the European Union.

All this inevitably raises the question 'What aid policy should the European Union follow?' followed by the query: 'Where will the money come from?' Neither Brussels nor anybody else has come up with the right answers. As for the 'summits', generally (though wrongly), it is decided to decide nothing.

30 This 16–17 June 2005 event was bound to be fraught, because it followed the French and Dutch referendums that decisively rejected the EU constitution.

PART III

Business Consequences of A Deteriorating World Environment

CHAPTER 9

Strategic Inflection Points, Business Risks and Opportunities

1. Business Consequences of Major Inflection Points

A *strategic inflection point* is a time in the life of a society, an economy, a company or a person when the fundamentals of its existence, and therefore of its future, are about to change. An inflection point can be a threat, but it also provides the opportunity to break out of the status quo and thrust into a higher level of achievement. Precisely because they represent exogenous change, inflection points involve risks, particularly when they are not properly attended.

This is the case with the poorly focused surge of interest in the control of carbon dioxide (CO_2) emissions[1] (which is the theme of Chapter 10), environmental pollution, overexploitation of the earth's resources and other related problems. As is always the case, however, these threats and projects that are designed to counter them also create the background for developing business opportunities.

This dual play between severe problems creating a strategic inflection point and major opportunities (visible or latent) is no twenty-first-century phenomenon. It has existed since the beginning of time, one of the earliest examples being the switch of human societies:

- from hunting to agriculture, and
- from nomadic life to settlements.

Historians say that the life of the first farmers was characterized by more anxiety and greater health problems than that of hunters. Not only was there anxiety about the new environment, but also the people applying themselves to agriculture suffered more skeletal wear and tear from hard work in the fields, lost the protection they had established for millennia as cave-dwellers, and caught diseases from domesticated animals. Societal life, too, underwent a twist. Hunter-gatherer society depended on sharing each other's hunting, which made it sort of egalitarian. By contrast, farmers – particularly successful farmers – hired the labour of others, fought (or bought) the control of the water resources (Chapter 12) and promoted generations of conquerors.

Friedrich Engels equated agriculture with a loss of political innocence. It would be more precise to say that relative stability, as well as economic and political might made

1 CO_2 emissions are theme of Chapter 10.

possible through agriculture and the domestication of animals (not only cows, ducks and pigs, but also other men as slaves), gave rise to three two important changes – after-effects of this inflection point:

- the need for energy sources,
- the concept of territoriality, and
- conquest to enlarge one's domain.

One of the reasons why the hunter-gatherer society was healthier was the fact that the weak who could not hunt died. (Ancient Sparta espoused this policy with its custom of leaving newborn babies unprotected out on the mountainside overnight.) The agricultural society put up with illnesses, of which there were plenty because of close proximity to others, measles transmitted from cattle, plague from rats and more.

Another strategic inflection point in the history of *Homo sapiens* was the early to mid-nineteenth-century Industrial Revolution. Rural peasants swapped their hovels for textile and steel mills or for laying down the rails for trains. Numerous chronicles describe an urban misery with poverty, pollution and, again, illness albeit from a different origin. Make no mistake, however:

- The agricultural society not only radically altered the people's lives, but also provided the resources necessary for a quantum jump in civilization – by making endeavours from learning to the arts affordable.
- The industrial society built on the agricultural society's assets, including those of business and trade, and therefore it, too, must be seen in the context of a progressive change over the conditions preceding it.

But any change is accompanied by a downside, which is rarely visible at the start. Usually, but not always, downsides show up much later as a result of failures, a series of excesses, an unexpected mass effect, or all of them together. That's the sense of unexpected consequences.

Business activities are generally thought to create upsides, but downsides are their alter ego. Throughout the industrial age we have been creating ecological crises for ourselves and our habitat. In fact, that's not really 'news' because it has been taking place for tens of thousands of years, but it has now reached a new inflection point characterized by the unprecedented impact of man on his environment.

Overpopulation (Chapter 13) and overexploitation of the earth's resources have raised the likelihood that climate changes will become irreversible. Yet there can be no factual answer to this because scientists have an imperfect understanding of the interaction between the land, forests, lakes, air, rivers, oceans, clouds and weather. Under normal conditions, the oceans give out and absorb far more carbon dioxide than is created by human activity.

- But man-made changes see to it that we no longer live under 'normal' conditions, and
- there are also contradictory hypotheses in regard to the effect of ongoing changes in the sun's energy.

In the nineteenth century, Hermann von Helmholtz formulated what has become the standard cosmological theory based on *entropy*.[2] His theory of *heat death* (from entropy) stated that the universe is gradually running down and will eventually reach the point of maximum entropy where all available energy will have been expended and no more activity will occur – a state essentially corresponding to eternal rest.

On the basis of the hypothesis that man-made climate change is indeed on its way, governments are ready to invest significant capital in an effort to stop (but not necessarily reverse) the ongoing degradation. This is neither easy nor self-evident, even if (theoretically) nearly everybody agrees that one of the major goals should be to reduce fossil-fuel air pollutants and curb CO_2 emissions.

As will be recalled from the preceding chapters, superficially the clash between fossil fuel and renewable sources of energy looks like a miniature war. In reality, it is much deeper than that because it is associated with a strategic inflection point with far-reaching physical and social consequences. Solutions to carbon dioxide and other environmental challenges designed around low environmental damage – therefore low emissions – require four basic global changes:

- finding new energy sources,
- developing environmentally-friendly technology,
- making major investments in not-so-well-known processes, and
- having a leadership able to inspire confidence that a U-turn is achievable.

All four bullets involve risks and opportunities, including business opportunities. But, for the time being, global leadership is lacking. What exists is national heads of state typically fighting each other about who can avoid having to cut CO_2 emissions. Issues of industrial exports and employment are behind this policy of escaping from commitments.

Regarding capital, while seed money may be put up by governments, sustainable major investments for environmentally-friendly energy sources can only be made by the private sector. If these prove to be profitable, they will mushroom. Today, business leaders are split over this issue, partly because visibility is poor and largely because they can perceive that governments are far from being decided, let alone committed, on what they want to do.

One issue on which both those 'for' and 'those' against a U-turn agree is that no private party can pay the hundreds of billions of dollars, pounds or euros guesstimated as necessary to radically alter the current CO_2-producing structure. Most business leaders look at it as being a government responsibility, but governments running a huge deficit in the aftermath of the 2007–2011 deep economic crisis have no funds available to allocate to such projects – and this encapsulates in one sentence the economic challenge associated to the most recent inflection point.

2 In physics, the degree of randomness in a certain situation can be measured by the *entropy* associated to it. The lack of patterning, disorder or disorganization of a system is known as its entropy. The statistical measure of the negative of entropy (negentropy) is the same as that for information. (It has been a deliberate choice not to include in this footnote the Second Law of Thermodynamics.) See D.N. Chorafas, *Risk Pricing*, Harriman House, London, 2010.

2. Shareholder Activism for Environmental Protection

Not long ago, in one of the leading US corporations a group of shareholders requested that the board of directors report on how the company ensures that it is accountable for its environmental impacts. This report had to address not just one but all of the product lines which constitute its business, in all of the regions and communities where it operates. The report, shareholders said, should contain the following data:

- how the corporation makes available information regarding its emissions and their environmental impact,
- how it integrates accountability for environmental issues into its code of conduct and ongoing business practices, and
- to what extent the corperation's activities have negative health effects on individuals living in different communities, particularly those of the economically poor.

The board recommended that shareholders should vote against this proposal because 'the company is anyway committed to operating in an environmentally responsible manner in every place it does business'. The board noted that the company's environmental policy clearly states that it will comply with all applicable laws and regulations, as well as apply responsible standards where laws do not exist.

However, environmentally active shareholders were not happy with the position taken by the board. They responded that even though business plans make reference to key environmental drivers, set targets in focus areas and identify projects or actions to achieve those targets, there is no guarantee that these plans are effectively carried out. Nor is there anything wrong in reporting about them to shareholders.

These were, however, minor shareholders lacking the voting power to see their petition through. Only one of the major shareholders sided with them, but the majority rejected the environmental motion on the grounds that 'the board knows best'.

In another company, shareholders requested that the board of directors adopt quantitative goals, based on current technologies, for reducing total greenhouse gas emissions from the firm's products and operations. They also asked management to report to shareholders on its plans to achieve environmental goals at reasonable cost (which was a vague statement and a contradiction).

In this case, too, the board recommended that shareholders should vote against this proposal because the corporation's views and actions on climate change were widely available – for example, in executive speeches. Anyway, the chairman said, the company would continue to seek opportunities to improve efficiency and reduce emissions, while maintaining leadership in promoting shareholder value.

The rejection of the motion proposed by a group of shareholders further noted that in pursuing its business objectives on behalf of those who invested in the firm's equity and in meeting society's aspirations for a better future, the company believed the primary opportunity lay in improving energy efficiency (which, in that case, was true).

What these two real-life cases help document is that people may have diametrically opposed opinion about environmental control, yet each party has reasons to believe that it is right. In addition, while the costs of complying with some of the regulations are measurable up to a point (though still uncertain) there is a much trickier problem:

- estimating the benefits of an improvement in the environment, and
- comparing the after-effect of such an improvement with the costs of achieving it.

Environmental protection cannot be a blank cheque because, if it is, abuse will be rampant. It is only normal for both people and companies to want to know the size of the bill. In the final analysis, consumers will pay all expenses. The snag is that most environmental-costing techniques are simple guesswork, even if they are widely used. Indeed:

- there are many cost–benefit approaches, all of which invariably produce different outcomes, and
- none is readily able to resolve conflicts between politicians and their constituencies over the right level of environmental regulation.

Worse still, the various cost–benefit analysts don't ask the right questions, which is a prerequisite to firm answers. Added to this is widespread misinformation in matters under discussion, which also shows up in shareholder activism in favour of the environment and in board responses.

Newspapers, television and radio are interested in sensationalism not in public education. For their part, politicians of all colours have amply demonstrated incompetence as analysts, let alone educators, on environmental issues – repeating like parrots their easily predictable party lines. As for the well-meaning and enthusiastic protesters against environmental pollution, by and large they are thinking with their legs rather than with their heads. Critics say that:

- none of these groups has made a serious effort to assess numerically the risks associated with environmental pollution, and
- no effort at all against the number one enemy of the environment: population explosion which is taboo (Chapter 13).

Reduced to treating rather disconnected issues, by keeping out of discussion the most important subject of birth control, environmentalists proceed on the basis of half-truths, biased statistics, emotive phrases and sometimes ignorance. Yet, simply because the environment is now inseparable from the fabric of our society, we must have a reliable system for analysis and education, and this should become a lifelong effort. Some of the essential questions to answer in a factual and documented manner are:

- What price the ozone layer?
- What price the poisoning of the earth? Of water?
- What price the constraints on economic activity, including increased unemployment?

Answers to these questions are indispensable as a starting-point for any intelligent analysis leading to serious decisions with a long-lasting impact. Shareholder activism is welcome and necessary, but on its own it will fizzle out. Business leadership for environmental protection and quality of life is only one of the pillars on which a U-turn from the destructive practices of the past and the present must be based. The other three are:

- political leadership,
- religious leadership, and
- a decision by the ordinary citizen to abandon the current culture of environmental abuse and spoilage of resources.

General lines and vague pronouncements mean nothing. What is important is *detail*. Decisions based on *detail* are fundamentally different from 'be good and do good' pronouncements. And the approach must be holistic. When a government decides to build a motorway through a nice valley, it is inevitably removing a few environmental rights from future generations. Sometimes social service and environmental protection are incompatible with one another. Therefore, one of them must give up its position.

3. Global Warming Liability Risks

Section 2 posed the questions: What price the ozone layer? What price constraints on economic activity? Superficially, these seem to be miles apart. Close examination, however, reveals that they correlate. They also share the presence of certain risks which have been consistently underemphasized, even though they are real. Liability risk is an example.

In the background of rapidly rising liability risks, which are highly relevant to all industries, lies the fact that the occurrence of extreme weather events such as drought, flooding, heatwaves and windstorms has increased in recent years. These often result in loss of business and, for the insurance companies, financial losses.

Whether or not they are due to environmental pollution, global weather disasters have increased not only in the average but, most particularly at the long leg of the liability risk distribution, as shown in Figure 9.1. In turn, the changing pattern of liability risk because of worsening weather conditions is altering the assumptions of exposure by insurance underwriters and reinsurers.

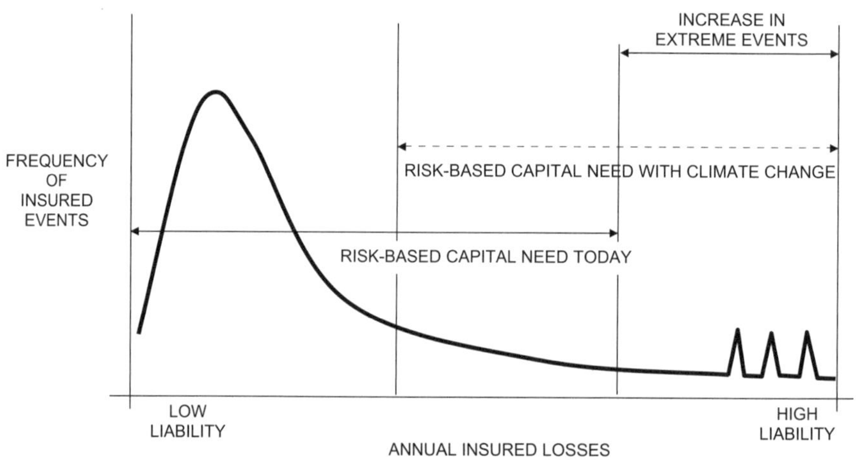

Figure 9.1 The impact of climate change on loss distribution for the insurance industry

Companies offering protection for weather-related damage are now facing new challenges, many of them connected to the higher occurrence of extreme events. To protect themselves against a higher level of risk, insurance companies are increasing their premiums – and this impacts on other companies and society at large.

- Also in the past, weather-related losses have caused insurers to go bankrupt.
- But, today, formerly extreme events occur more frequently, and failure to compensate through premiums can lead to withdrawal of insurance coverage.

The leg of the risk distribution in Figure 9.1 is characterized by greater uncertainty about frequency and intensity of weather-related losses. This complicates risk mitigation actions, increasing the capital needed to cover extreme events. Not without reason, insurers say that they need more capital to bridge the gap between losses in an average year and significantly higher losses in an extreme year.

Not long ago, a reassessment of insurance risks concluded that climatic change has become a crucial factor in exposure, having evolved significantly during the last two decades. Over this period, the Intergovernmental Panel on Climate Change, comprising 2,000 scientists from 40 countries, has issued three assessments that I know of (1991, 1996, 2001), documenting the consequences of global warming and its risks. The National Academy of Sciences and American Meteorological Society have also issued statements supporting these assessments.

An obvious risk is greater property damage from floods, stronger storms and hurricanes (like typhoons and cyclones) primarily in coastal areas. Global warming models predict intense events.[3] Catastrophe modelling gives insurers a picture of aggregate exposures which can be used to recalibrate assured risks and associated premiums. They also assist in:

- focusing the collection and sharing of insurance data,
- conducting climatic assessments in conjunction with financial-sector assessments,
- training actuaries in new risk events and exposure synergies, and
- improving regulation and supervision of the insurance industry to ascertain its viability.

All these notions are important because climate change is expected to lead to a shift in the distribution of insured losses towards higher values, with a much greater risk at the tail of the graph in Figure 9.1 and a bifurcation between expected and unexpected losses. Expected losses are projected to increase by a smaller amount than unexpected losses associated to factors vulnerable to weather-related events.

- All this is part of the inflection point discussed in section 1.
- The consequences will be adverse effects on life and health insurance, as well as on asset management.

At the same time, a volatile and changing business climate will encourage the creation of new insurance products, like swaps and options related to drought and flooding.

3 Whether because of CO_2 or of sunspots.

According to this school of thought, while weather-related events are exposing some insurers to extreme risks, they also provide opportunities for:

- product development,
- new potential income sources, and
- diversification, which all help to reduce concentrations of exposure.

A combination of more stringent underwriting, increased premiums and deductibles may help to provide insurance coverage profitably, if pricing is done in association with worst-case scenarios.[4] A critical part of worst case is litigation impacting on insured parties and possibly the insurance firm itself,[5] as well as first-class identification, monitoring and management of weather- and environment-related insurance risk.

Gaps in the definition of risk and in legislation must also be filled. For instance, greenhouse gases like CO_2 (Chapter 11) do not appear to be a pollutant as defined in pollution exclusion policies. This erodes insurers' defences against litigation resulting from global warming. In addition, litigation may also trigger directors and officers (D&O) liability policies which have an exclusion for actions arising out of a pollution event.

According to Dan R. Anderson, Schultz Professor of Risk Management and Insurance, School of Business, University of Wisconsin-Madison:

> *... corporations and their insurers operating in the United States and Australia, as non-signers of the Kyoto Protocol, will have the greatest exposure to global warming litigation. Even if exclusions are put into general and D&O liability policies, they will offer no protection for any existing and older occurrence based policies ... [Therefore] insurance companies' best defense would be to bring pressure on their policyholders, and even governments, for reducing greenhouse gas emissions.*[6]

Obviously, insurers can reduce their financial vulnerability to extreme weather-related losses by means of risk transfer, through weather derivatives. Those exchange-traded are usually linked to widely followed measures like temperature, wind and rainfall, while over–the-counter (OTC) bilateral deals are tailor-made for specific risks. Tradable indices are another financial protection product. An example is the global warming index, a tradable benchmark for global investments in weather derivatives.

By contrast, catastrophe bonds transfer a specified set of risks, like natural disaster, from the insurer to the capital markets. The bond structure ensures that interest and/or the principal are lost when a predefined catastrophic event occurs. Relatively newer as a financial instrument are weather-related collateralized debt obligations (CDOs). CDOs, however, have been one of the major contributions to the 2007–2011 descent into the financial abyss.[7]

---

4 D.N. Chorafas, *Stress Testing for Risk Control Under Basel II*, Elsevier, London and Boston, MA, 2007.

5 Hurricane Katrina victims have initiated litigation against the oil companies for property damages caused by hurricanes resulting from global warming. *Daily Telegraph*, 4 March 2010.

6 D.R. Anderson, 'Will Global Warming Liability Risks Exceed Property Risks?', *Progress*, No. 44, The Geneva Association, December 2006.

7 D.N. Chorafas, *Financial Boom and Gloom. The Credit and Banking Crisis of 2007–2009 and Beyond*, Palgrave Macmillan, Basingstoke, 2009.

The examples included in this section concerning a new generation of risks confronting the insurance industry are by no means exhaustive. At present, nobody has the whole list of horrors lying in wait for us. According to some experts, if the status quo retains the upper ground, by 2030 the world will be facing a perfect storm of food, energy and water shortages caused by climate change, and it can be expected that the insurance industry will develop a panoply of instruments offering coverage – at a price.

In fact, not just the insurance industry but the whole of humankind is in peril as it attempts, with limited will, to gain the upper hand in a war against man-made environmental catastrophes that threaten to destroy the current 'way of life'. Society at large is gripped by anxiety, but there are some private initiatives that are successful and can serve as guide posts. One of them is the wind-power solution at the Danish island of Samsø, discussed in section 4. Another is the biofuels initiative which could provide Europe with independence from Middle East oil. The principles have been discussed in Chapters 6 and 8. Sections 6 and 8 of this chapter bring recent developments to the reader's attention.

4. Wind Power in the Danish Island of Samsø: A Case Study

For reasons which will be evident in the second half of this case study on renewable energy, according to a United Nations classification, Denmark and Norway are the least corrupt countries in the world. This is important because corruption turns any environment project on its head, as the reader will have the opportunity to appreciate later on.

Good news first: the solution the citizens of the Danish island of Samsø[8] found to become independent energy-wise from fossil fuels can also be a model for many other rural areas, though not for cities. The island has some 4,300 residents distributed in 22 villages. Their produce is mainly agricultural. They also have a small cannery.

In terms of energy supplies, the farming community in the island of Samsø has been classically dependent on fuel from the mainland. Today, it generates its electricity from 21 wind turbines on land and offshore, and it exports more electricity than it imports oil.

The windmill project came into focus in 1997 when the Danish Energy Ministry announced a contest. One of the country's regions was to be selected to test how effective renewable energy could be in a real-life environment. Part of the deal was that the winning region would subsequently achieve a zero carbon footprint, without continuing assistance or subsidies from the government.

Correctly, the ministry thought that self-sustainability would make the results more readily transferable to other regions of the country, or even abroad, and that the project's costs could be better contained. In counterpart, there were price incentives. As in other EU countries, the law requires Danish electric utilities to buy wind energy at prices significantly higher than the cost of other energy processes. This hidden subsidy, which is common currency with renewable energy worldwide:

- can turn a wind-turbine project into significant moneymaker, and

8 About 16 kilometres off the Jutland peninsula.

- the profit potential didn't escape the attention of a local bank which invested €2.5 million ($3.7 million) in the project.

It is important to note that the bank and other investors did not throw money at the problem. Personal initiative preceded the bank's investment, even before the Energy Ministry's call for a region to become model of energy efficiency. The original promoter of the renewable energy idea at Samsø was Søren Hermansen, a visionary and former farmer, who is now managing an energy academy in Ballen.[9]

Thanks to personal initiative and effort, rather than lavish subsidies, the wind power of Samsø and that of other projects now produce about 20 per cent of the electricity used in Denmark. This is more than any other nation. The government provided credit guarantees to manufacturers for turbine construction, rather than direct money. This proved to be a wise strategy, but the real key to success was individuals like Hermansen who:

- put themselves to work, and
- harnessed general public support.

Other Danish communities, too, were provided with incentives to build wind farms. Today, an estimated 75 per cent of Denmark's 5,200 turbines are owned by cooperatives, with more than 150,000 families and institutions as members. Many of these communities are dedicated body and soul to alternative energy, which is not a generally encountered attitude.

True enough, the fact that the government obliged power companies to purchase wind power at fixed rates guaranteed investors a dependable stream of income. This made the members of the cooperatives more prone to come forward with investments. It also put a tap on possible corruption created by too much cash upfront.

The fact that the cooperatives were assured a good and steady income from alternative energy sources induced them to be creative beyond wind power. Being inventive, Samsø's residents also produce energy from straw and from milk. The latter is a novel idea.

The opportunity for exploiting the 'milk channel' of green energy comes from the fact that the body temperature of a cow is 38.5°C, while the milk has to be cooled down to 3°C. Farmers installed a heat exchanger near the milk tank which used the heat released while cooling the milk to heat the farmer's house.

This reinvention of the sacred cow Danish-style is a concept with interesting potential for agricultural settlements, and its applicability should be examined not only in the European Union, but also in all milk-producing countries. It must be kept in mind, however, that it is useful only in the context of individual farmer initiatives.

Energy solutions good for rural areas are by no means applicable to urban living. I was told that not so long ago the Egyptian ambassador visited Samsø and, after touring the facilities, he said that the number of people resident there could fit into three apartment buildings in Cairo!

---

9 Ballen is an old Danish fishing village. Visitors from around the world come to it to examine the equipment and infrastructure the Danes have developed.

5. A Counterexample: 'Gone With the Wind'

Samsø provides an excellent example of harnessing wind power, but other projects have been consigned to the rubbish bin. In theory, in the absence of individual initiative, the government can finance large-scale investment in alternative energy, but, in practice, it often ends up throwing money at the problem.

The European Union provides several examples. More than €6 billion from EU structural and agricultural subsidies have been allocated to renewable energy over a 13-year period (ending in 2013), and much of it has been wasted. According to dependable sources:

- it is impossible to quantify the level of fraud in public spending on wind energy, and
- while one of the stated reasons is that investigations are scattered across different countries, the real reason is greed and corruption.

The European Wind Energy Association, which represents 600 manufacturers and members in 60 countries, including some outside Europe – and is obviously a proponent of wind-power subsidies – denies that there is any corruption. This is seriously contested by others who point out that in 2008 wind power attracted more than €11 billion in investments and that there have been widespread scams.

Police investigators have been busy collecting evidence. In 2009 in France five Corsican nationalists were jailed and fined for skimming €1.54 million in European subsidies for wind farms. In Italy, 15 people were arrested in a case code-named 'Gone With the Wind' – a Ponzi scheme aimed at reaping as much as €30 million in EU aid.[10]

In 2009 Oreste Vigorito, president of the National Italian Wind Energy Association, was arrested in connection with the 'Gone With the Wind' scandal and accused of participating in an illegal scheme to collect European subsidies. Vigorito's lawyer said that he is innocent of any wrongdoing, but the prosecutors think otherwise.[11] The fact is that:

- easy money can be made in lavishly financed alternative energy projects, and
- this not only attracts the unscrupulous, but also stimulates the appetite of others who see an opportunity for enriching themselves.

Still in Italy (in western Sicily), a separate investigation code-named 'Operation Aeolus' resulted in the arrest of several people who face trial. Prosecutors alleged that organized crime is using kickbacks and bribery to make money from alternative energy, giving a council member $75,000 and a Mercedes to vote in favour of a wind park.

In Spain, in 2008, premium prices paid to wind producers above market rates reached an estimated €1.2 billion; ultimately this cost has been borne by consumers. According to information in the public domain, in 2009 Spanish prosecutors were investigating deals connected to wind turbines scattered among the more than 700 wind parks that produce 16,740 mW of energy (mainly in Leòn, Galicia and north-eastern Castile).

10 'Wind Energy is Attracting Investors, and Corruption, *The New York Times*, 15 December 2009.

11 *International Herald Tribune*, 14 December 2009.

In Galicia, the former head of industry and energy is facing charges of influence trafficking for granting approvals for seven wind parks developed by his brother-in-law. In the Canary Islands, which feature no less than 44 wind parks, two investigations led to arrests of public officials and developers on charges of influence trafficking and bribery.

Not only public money, but also big companies have been targeted. The Danish Vestas Wind Systems, a leading manufacturer of wind turbines, revealed in 2009 that it was the victim of a €12 million fraud scheme.[12] (Denmark's wind turbine industry is led by Vestas Wind Systems, which has about a third of the global market.)

One of the reasons for Ponzi games, influence trafficking and other scams is that, as far as alternative energy is concerned, there is practically no industry regulation. Critics add that government controls are patchy at best, and they are often reduced to zero by corrupt politicians and rogue entrepreneurs trying to literally create money out of EU subsidies. According to one opinion, this is a gold rush, and everyone wants a wind park to exploit.

In addition, as will be recalled from Chapter 5 with reference to biofuels, there is a conspicuous lack of standards in alternative energy. Confronted by total confusion about creating standards, the authorities have also failed in establishing a transparent system for public information about:

- land leases,
- performance goals,
- cost-effectiveness targets, and
- connections between wind developers and municipalities.

The lack of technical and implementation standards, appropriate regulation and close supervision of how funds are used is creating immense opportunities for fraud. This is highly regrettable, as well as counterproductive to the interests of an alternative energy policy.

A consensus opinion which emerged from my meetings is that there has been too much of a free-wheeling approach, which has harmed the wind energy industry. Also, although there is a need to ensure that the large amount of capital available through a variety of subsidies is always properly used, this has led to plenty of ambiguities that discourage private initiatives.

The message to be retained from all these examples is that not only is there a clear absence of EU, American or any other policy, but also a system that is not properly supervised – all the way from wind equipment standards and its sourcing to financing, management and operational efficiency. Governments, both in the European Union and the United States are on the wrong track and this will hurt their industries – particularly as China is planning to conquer the market in wind power in the same way that it is doing with solar power.

Last but not least, the absence of well-established standards and of regulatory authorities raises questions all the way to the survivability of wind farms, even with preferential tariffs which cannot last forever. For instance, in Germany a wind farm's feed-up tariff is nearly €0.84 per mW versus a market price of €0.40 to €0.70. This gap

12 The company said that three top Spanish employees, who are under investigation by the authorities in Barcelona, issued payments for non-existent services to companies under their control, shifting the money in separate businesses.

has been half-closed by the government reducing the subsidies, an initiative which has turned investors' expectations of getting their money back from alternative energy on their head. If subsidies are taken away, wind farms' cash flow will generally be negative with the result that they may become non-entities.

6. The Future of the Biofuels Industry in the European Union

The problems confronting the wind industry and other alternative energy projects are by no means new. Since ancient times and the allegory of Prometheus and Epimetheus, the production and use of power has been a moving target. Energy systems are constantly evolving, and no approach can ever claim to be 'the final' or 'unquestionably the best'.

- No matter whether it is based on fossil fuels or other ingredients, power production today is just one cog in the wheel of a much larger process.
- One thing we can be sure of is that the growth in energy usage will continue and change will occur – although we are less sure of being in charge of that change.

This holds true all the way from avoiding the types of scam discussed in section 5 to knowing our position relative to that of our competitors. After all, energy production is a huge industrial sector with a global impact, as well as a significant effect on employment. In Chapters 6 and 8, biofuels were promoted as alternative energy sources with the dual goal of:

- making the European Union independent of imported oil, and
- providing an answer to the present strategic inflection point, able to promote agricultural and other business opportunities.

To become a reality, however, this requires a significant amount of strategic planning, which has not been present in the European Union and the United States in terms of other alternative energies. For instance, in 2008 China became the largest manufacturer of solar panels in the world. This says a great deal because, by that time, renewables, including solar energy, was attracting $150 billion in new investments.

As far as biofuels are concerned, the main producers today are the United States and Brazil, with the European Union lying behind. This is inexcusable and shows a lack of foresight as well as confidence by the various EU governments and the European Commission.

Statistics tell an interesting story. Measured in billions of gallons, ethanol production in the European Union is small fry compared to that of the United States and Brazil. Particularly bad is the nine-year trend leading to the 2010–2012 forecasts.

- From barely 2 billion gallons per year in 2001, the United States reached 13 billion gallons in 2009 – more than a 650 per cent increase in eight years.
- Over that period, Brazil went from 3 billion to 8 billion gallons – an almost 270 per cent increase.
- Correspondingly, the European Union moved only from 1.3 billion to 2.7 billion gallons – a small fraction of Brazil's and the United States' biofuels output.

Comparatively speaking, the figures are better for biodiesel, where the European Union is ahead of both the United States and Brazil (the latter is a minor player). Far-sighted management at the EU Executive would have made sure that European production in both ethanol and biodiesel moved way ahead, capitalizing on the agricultural potential of the old continent.

One of the reasons for the slack is the leftists' campaign that grains should not be used to produce energy but should feed hungry mouths. Contrary to what the eggheads say, however, the European Union has no obligation whatsoever to feed, in a charitable gesture, countries that are unable to feed themselves. (Protracted aid donations teach people not to work but to wait for handouts, instilling apathy and making their plight worse.)

Some people suggest it is only natural that Brazil and the United States are moving ahead in ethanol because they can easily find spare land and increase the efficiency of farming. That argument is wrong. If both Brazil and the United States have plenty of spare land, so does the European Union since the CAP pays landowners to take their land out of production through the set aside scheme. And there is no reason whatsoever why the European Union cannot, or should not, increase the efficiency of farming.

Capitalizing on the current strategic inflection point, rather than paying farmers to keep land out of production (Chapter 8), the European Union should focus on helping them with premiums and advice on converting to energy-related products – everything in excess of what the internal market requires in foodstuff. Only when internal energy needs have been fully satisfied, should there be room for agricultural exports.

The EU countries don't have Brazil's major problem of establishing who has the right to what piece of land. One of the key challenges facing Brazil is that land tenure becomes less and less certain as you approach the equator. In the Amazon this ambiguity leads to galloping deforestation, as farmers cut down trees to claim *de facto* ownership of land and the government looks the other way.[13]

In the United States, biofuel distilleries account for about one-fifth of the corn crop. The product mix of EU farmers for biofuels (including biodiesel) may be different, but there is no reason why this 20 per cent cannot be a first guideline, taking the current US policy and its results as realistic goals.

Another lesson from America is that, over the longer term, the Energy Independence and Security Act of 2007 sets aggressive new targets for renewable fuels usage. The original objective was 7.5 billion gallons of renewable fuels usage by 2012. However, the new target for all renewables has been increased to 36 billion gallons by 2022. The Act has also:

- increased the 2008 mandate to 9 billion gallons of corn-based ethanol, and
- called for additional ethanol usage up to the 15 billion gallon level by 2015.[14]

13 In other parts of the country the disputes take the form of farm invasions by the Landless Movement (MST), which tends to go for farmland that is not too far from cities. A loosely organized movement in a big country, the MST appears in lots of different guises, its actions being largely motivated by *Lebensraum* because of the population explosion.

14 The 15 billion gallons will represent about 10 per cent of total petrol consumption in the United States, as that is the optimal blend for most American-built engines.

The United States has also set new targets for advanced biofuels such as cellulosic ethanol and biodiesel. It goes without saying that the European Union should do no less.

A lesson must also be learned from the fact that, over the last few years, America's transportation and agricultural systems have been working in unison. More and more energy inputs for the former come from the latter, which up to a point counterbalances the fact that the United States, like the European Union, made the mistake of switching from primary reliance on railroads to greater dependence on cars, trucks and airplanes.

- Currently, cars and trucks account for the most passenger and freight traffic respectively.
- Little attention has been paid to the reality that these are less efficient than other modes of transportation, which have been cut back.

The urge to grow biofuels rather than import massive volumes of oil from producer countries in no way means that all problems have been solved in terms of their manufacturing and usage. However, the European Union and the United States have the necessary laboratory power to tackle these problems, as well as to significantly improve the efficiency of engines using biomass rather than coal and oil.

7. Greater Efficiency in Energy Production is Long Overdue

Leaving CO_2 and other pollution worries aside for a moment, but keeping in mind that all over the globe dependence on massive utilization of energy is increasing by leaps and bounds, the question evidently comes up: 'Why doesn't the power-production industry drive hard for greater energy efficiency with biofuels?' Here are some answers.

In the United States there exist two organizations, the National Action Plan for Energy Efficiency and the Alliance to Save Energy, but the results they have so far produced are far from stellar. Critics say that the reason can be found in lobbies and in the existence of powerful disincentives, while incentives are not properly tuned.

The nature of disincentives is explained by borrowing a leaf from Duke Energy's[15] 2007 *Summary Annual Report* which states that many utilities don't invest in efficiency programmes, 'because the current regulatory framework is biased against investments in energy efficiency in favor of putting steel in the ground'. Despite the energy crisis, bureaucrats did not 'change that regulatory paradigm so that earnings from energy efficiency are on a par with earnings from investments in new power plants'.

This is evidently negligence on the part of the government, and it is counterproductive because energy efficiency is crucial not in one, but two, major fronts: first, in the struggle for affordable new energy sources and, second, in the battle to combat climate change by reducing CO_2 emissions. Incentives for significantly improving energy efficiency should account for the fact that new forms of cleaner energy are not yet convincing enough as alternatives to oil and coal in the sense of:

15 One of the major American energy utilities.

- massive energy production at the time when required by end users,[16] and
- cost-effectiveness which will attract private investments massive enough to make the difference between alternative energies and coals and oil.

Energy efficiency should be a short- and long-term objective, and this demands that attention to it focuses on both fossil fuels and alternative sources – not only the latter. In terms of economies in energy consumption (which is a sort of efficiency at the user end), the key lies in public relations campaigns to confront the fact that, with few exceptions, people don't particularly care about energy *supply* except when it is disrupted. But today virtually everybody cares about energy *services*. These range from the basic needs like cooking, heating and lighting to the 'musts' of modern society – from all sorts of appliances and industrial processes to wide mobility, computers, televisions and mobile communications. But conservation is far from being a major concern (Chapter 1).

The status quo is promoted not only by bureaucrats but also by the consensus view which rests on the false belief that there is no problem with a continuation of present trends which are considered to be a sort of comfortable entitlement. The irony is that:

- with increasing frequency many people say they worry about CO_2 and greenhouse effects,
- but nobody wants to pay the price for putting them under lock and key. Everyone wants to 'get back to nature' – but by car or airplane!

When in my conferences I ask if the participants think that it is time to change present habits, invariably the answer is 'Yes'. But replies to the next question, 'How?' show that everybody who wants to change would also like to see it happen at somebody else's expense– not their own. Nearly everybody thinks that way.

This basically negative response is evidence that the system that enables society to produce energy does not lend itself readily to a flexible, quick changeover. Because the engines designed for petrol and diesel can be used with biofuels with only minor adaptation, and the present system of service station remains *as is*, the changeover will be smooth.

By contrast, this is not true with other alternatives – for instance, in transportation with electric cars (Chapter 6). When the models of usage significantly change, quick switches can lead to chaos. Change also has to be planned most carefully in power production, never forgetting that the current infrastructure of power plants and energy distribution networks may last another 20 to 40 years. Even decommissioning takes a long time.[17] Moreover,

- from blueprint to operation, lead times for many new energy projects may span up to a dozen years, and
- precisely because of entrenched public perceptions of energy needs, costs and environmental acceptability, the changeover system may be laden with inertia.

---

16 See in section 4 the problems associated with the timing of electric power delivered by Denmark's wind farms.

17 As discussed in Chapter 4, the dismantling and disposal of the Soviet-era nuclear plant in Lithuania is scheduled to take 25 years(!) and cost more than €1 billion ($1.5 billion).

While it is imperative to study new environmentally-friendly energy solutions in regard to their efficiency and its improvements over time, their longer-term assets and liabilities must also be documented, as this book has attempted. Very important also is to evaluate the efficiency of projects currently under way – which typically originated years ago and will dominate the scene for years to come.

Although there are many reasons for changing the status quo, it is wrong to think that change is possible at any price because the people pushing for it are, at the same time, its worst enemies when deprived of what they consider to be their 'energy endowments'. (In September 2005, for example, a spike in energy prices brought a swarm of British and French people to their respective streets in protest. First, European finance ministers vowed to stand together against fuel-price protesters who were demanding cuts in petrol taxes. But shortly thereafter several countries offered subsidies and tax rebates to favoured pressure groups.)

There is a silver lining to all this, which makes depressive reading for some people. The problems lying ahead provide excellent business opportunities for firms able to significantly improve energy efficiency and smooth the necessary transition. The market is vast both in developed and developing countries as both – most particularly the latter – have major energy efficiency gaps.

8. Capitalizing on Technology and Research Results

To convert a good part of its agriculture to biofuels, the European Union can use recent developments in technology, such as satellite monitoring. One of the major advantages derived from analyses based on real-time monitoring is that they compress time-lags. They are also effective in supporting large-scale solutions – an important contribution because, according to experts, the area of monitored farmland will increase as:

- the climate changes, and
- farmers can no longer rely on existing procedures if they want to be in charge of the future.

In planning for a rebirth of the European Union's agriculture it is also necessary to appreciate that crop prices are intertwined because they influence what farmers decide to plant. Higher wheat prices, for instance, are driving up the price of corn, which is even more sought after as an animal feed when wheat is too pricy to be a substitute; and, obviously, corn prices are also being pushed up by growing demand for biofuels.

Research on greater efficiency in the use of raw materials for energy and foodstuff has much to gain from cross-fertilization of themes and experiences. Both upsides and downsides must be considered. An example which comes to mind is the practice of feeding the by-product from corn ethanol distilleries to cattle, – a growing practice in the United States, particularly in the corn belt.

Some people say that this may be contributing to the spread of E.coli contamination in the food chain.[18] For instance, in early 2008 researchers at Kansas State University released

18 Strains of E.coli are commonly present in the guts of healthy cattle, but there are plenty of reasons to be attentive to the after-effects of new practices in dealing with foodstuffs.

a study showing that cattle have significantly more E.coli present in their intestines when fed a ration high in dried distillers' grains (DDG). One hypothesis is that DDG alters the acidity level in the gut, making it more conducive to bacterial growth.

Stockmen are also worried about the impact of wet distillers' grains. These are used for livestock feed in some locations and are favoured by ethanol plant operators who don't want to be burdened with the time and cost of drying it.

It is appropriate to note that feed made from biofuel waste is not the only subject connected with E.coli. As a bacteria, this is a lab favourite widely (and easily) manipulated for genetic engineering.

Scientists have established metabolic pathways to produce antibiotics, insulin and anti-cancer drugs. They are also researching into ways of creating something that doesn't exist in nature: long-chain alcohols. These are interesting because:

- they overcome some of the traditional limitations of ethanol, which has only two carbon atoms, and
- they have high energy density and low water solubility, so they can be optimal substitutes for current fuels.

Biologists and botanists also suggest that researchers should examine the second-generation biofuels and their raw material in terms of unintended consequences, such as developing an invasive crop strain which has the potential to escape plantations, overrun adjacent farms and create havoc.

The pros say that the risk of biofuel crops becoming a weed problem is overstated. This, however, does not mean that there is no risk. Laboratories have to look at the biology of the plant and the environments into which they are introduced and ask: 'Where are the issues to worry about? How serious can they be?' Every new industry has 'worry points' which have to be studied in advance. It would be irresponsible to do otherwise.

Nothing is, however, linear. While due diligence has to be exercised at laboratory and management level, it would be a mistake to ignore strategic and tactical moves by the competition. The European Union and the United States are by no means alone in a world in using more and more energy as a way of pressing the economic accelerator. Most of the developing countries, particularly China, India and Brazil, are doing precisely the same. And, unlike the West, they don't ask for anyone's approval about CO_2 emissions even though they lag behind in incorporating energy efficiency into their processes and systems of power production and distribution.

Curiously enough, with only minor exceptions, less developed nations have deliberately chosen to stay at the level of the now obsolete CO_2- intensive solutions to industrialization – as if any switch may turn back their timetables. The governments themselves:

- on the one hand, invite wastage by subsidizing energy prices, and
- on the other, turn a blind eye to the need for pollution control on the basis that this will slow their development.

Conveniently, but illogically, decision-makers in most developing countries forget that their people will greatly suffer from suffocating CO_2 (Chapter 11), as well as torrential storms and hurricanes whose intensity is allegedly greatly increased by environmental woes. No developing country has acid gas and CO_2 emissions anywhere near the top of its agenda, and politicians don't feel like making concessions in the direction of cleaner power generation.

Even among developing countries whose political leaders are considered to be environmentally-friendly, the typical reaction is: 'Environmental issues are very nice, but who's going to pay for them?' Nobody seems to have explained to them that, in a process of industrialization, increased energy efficiency means both:

- faster development, and
- significantly higher benefits to the economy.

Just as important is to appreciate that the able use of positive and negative incentives can provide a stimulus which classical political action lacks. It seems strange to think of putting a price on avoiding the use of natural resources, but now a number of countries are trying to do just that by assigning values to 'non-market benefits' aspects of the environment. (The downside is that these have no established market price and are therefore too theoretical and poorly documented.) Economists with an inclination towards environmental protection are estimating the value of things, such as:

- clean air,
- endangered species,
- preservation of quality of life, and
- preservation of unspoilt wilderness.

But values and cost taken out of a hat are worse than nothing. A better approach is to take a holistic view. *If* major environmental transmutations are a danger (which is true), *then* firm decisions with short to medium timeframes (not cloud-cuckoo 2030 and 2050) should be set and policed.

Adherence to commitments must be regularly audited and subject to global reviews – with merits and demerits attached to them. In the final analysis, no matter how one wants to look at these issues, the fact remains that people are simultaneously obstacles and resources.

CHAPTER 10 Producing and Trading Carbon Dioxide

1. A Man-made Ecological Collapse

Apart the extermination of dinosaurs and other sweeps of life on earth, which took place in remote antiquity,[1] the first known case of an ecological catastrophe which wiped out a civilization occurred in ancient Mesopotamia, in the valley of the Tigris and Euphrates. Archaeologists and historians are led to believe that this was a major event but, as usual, it has not been used to learn a lesson so that future catastrophes of similar nature can be avoided.

The Babylonian and Assyrian civilizations, as well as those of Ur and Elam which preceded them, were dependent on irrigation and therefore on an assured water supply. The great cities of ancient Mesopotamia were built following the invention of the plough, and agriculture was based on plenty of fresh water from these two great rivers. But as current experience with hydroelectric dams and irrigation canals helps document:

- dams and canals silt up,
- waterlogging and salt accumulation take their toll, and
- eventually the topsoil disappears, resulting in the land becoming infertile.

Life cannot be sustainable if the earth's resources have no chance to replenish themselves. The fertility of animals living on them must be below the regeneration rates of natural resources. *If* it exceeds it, *then* survival depends on radically downsizing the animals' reproduction rate (see also Chapter 13 on birth control). *Fertility* is the keyword, and the fertility of natural resources has a legion of enemies:

- from overexploitation,
- to land, air and water poisoning.

When this happens, the resulting disasters are definitely man-made, and dependence on technology or artifice to take care of abuse is ill-placed. Even modern societies with high technology at their disposal find it difficult to protect themselves from rapid deterioration of the environment in which they live.

Over the years, for example, California's Imperial Valley has experienced a fate similar to the one which wiped out the flourishing agriculture of Mesopotamia. Another similar case is the degradation of agriculture in the formerly fertile Nile valley, in part

---

1 Where lack of clear evidence mixes with various tentative statements and myths (like Noah's Ark).

due to Egypt's Aswan dam (Chapter 5) and in larger part to an unsupportable population growth.

The enemies of the environment do not have a unique profile. Some of them are not even looked as such, galloping urbanization being an example. The population of Cairo, I am told, is increasing by about 1 million people per year.

There is no doubt that many less developed countries are spoiling the habitat necessary for the survival of future generations, and the authorities – the local authorities in particular – are overwhelmed. While no other institution has as much leverage as the World Bank when it comes to giving the developing countries funds for their projects, it is not known to condition its loans to the recipients paying a great deal of attention to the preservation of forests and water resources, including lakes and rivers.[2]

Montreal, Rio, Kyoto, *Fiascopenhague* and other similar events notwithstanding, World Bank bureaucrats, as well as those of Western governments providing 'aid', do not seem to appreciate the importance of preserving global spaces free of increasing pollution. In sequel to business decisions which in practice make more out of ecological collapse (although they might have been benevolent in their original conception), the world is confronted by an ecological and societal disequilibrium, the latter of which includes behavioural and psychological aspects.

According to the media, there are many people around the globe concerned about the environment. Anecdotal evidence suggests that during the first decade of this century their number more than doubled. Few, however, ask the all-important question: 'Is man part of nature?' Or 'By wanting to dominate nature has man positioned himself against it?'

For some time it was thought that if finance were provided to help countries build up their own capacity to cope with economic development, environmental problems would somehow take care of themselves. It did not happen that way. Instead, the problems of widespread pollution and the earth's overexploitation got a big boost.

This does not mean that Western governments tried hard to put their own house in order. They did not. But even if they had done so, the problem is global; pollution knows no frontiers.

The huge production of carbon dioxide (CO_2, section 2), which can no longer be recycled by the earth's physical system, is an example. Oceans and plants absorb about 60 per cent of the CO_2 emitted as an consequence of human activities. The oceans do so because CO_2 is soluble in water. Plants absorb it because of photosynthesis. But this recycling by Mother Nature does not increase by leaps and bounds. At some point – which is not that well-defined, though it seems to have been reached –the forests and oceans become saturated, and the so-called *land sink* – essentially CO_2 absorption by oceans, forests and soils – goes into reverse gear.

This means that once the saturation point is reached, the land sink adds to net emissions rather than reducing them. In addition, scientists suggest that the size of land sinks depends on a long list of factors, many involving issues that have not yet been thoroughly investigated, as well as unknowns. These factors include:

---

2 Time and again the World Bank has financed projects without thinking about, much less examining, negative consequences. For this reason, during the 1980s critics nicknamed the supranational organization the 'World Destruction Bank'.

- concentration of CO_2 in the atmosphere,
- CO_2 appetites of plants and soil organisms,
- chemistry and circulation of the oceans, and
- the way in which they all vary in response to climate change.

These problems are not as new as they seem to be. Even though climatic change and its after-effects became a popular theme during the last decades of the twentieth century, the effects of large CO_2 emissions had worried scientists for more than a century. In 1890 Svante Arrhenius calculated the warming effect of carbon dioxide, but it was not till 1958 that research began on measuring CO_2 in the atmosphere at Mauna Loa observatory in Hawaii.

Decades of inaction in curbing atmospheric pollution saw to it that today, anywhere around the globe, society is faced with the same sort of environmental destruction confronted by the ancient Babylonians, Assyrians and other people living in the fertile crescent of Mesopotamia thousands of years ago.

- Both CO_2 and pollution at large are widespread, and
- plenty of unexpected consequences are waiting to show up as the earth's environment deteriorates.

Mexico City's air, for example, is fouled by a pound and a half of pollutants per day for every man, woman, and child in its ever-expanding metropolitan area. For children, breathing this polluted air is the equivalent of smoking cigarettes from birth, and newborn babies can expect a loss of five to ten IQ[3] points due to lead levels in the air.

Ozone levels in Mexico City have soared (see section 4), reaching a median level of about 80 per cent over World Health Organization's maximum standard. The average lead level in the blood of the city's residents is nearly four times that of Tokyo residents, and more than twice the levels found in Baltimore citizens. Similar levels characterize other king-size cities.

As if to add to these dreadful numbers which give a very negative quality of life, each day in Mexico City 700 tons of solid human waste are dumped, further polluting the air. This situation is made worse by the fact that 30 per cent of the population have no sewage service. It is *as if* human societies are making strenuous efforts to exterminate themselves. CO_2 is one of the problems – not the whole story.

2. The Scourge of Booming CO_2 and Other GHGs

Carbon dioxide (CO_2) and most other greenhouse gases (GHGs) have always been present in the atmosphere. In quantities acceptable to nature they have been keeping the earth hospitable for life by trapping heat that would otherwise escape into space. Essentially, this is what is called the *greenhouse effect*. The problem is that because of man-made environmental destruction it has gone out of control.

Since the Industrial Revolution, the concentration of CO_2 and other greenhouse gases in the atmosphere, from the burning of fossil fuels and other human activities, has very

3 Intelligence quotient.

significantly increased. Its effect is to trap more heat, amplifying the natural greenhouse process. Gases other than CO_2, released by man-made products and processes into the earth's environment and affecting it in a significant way, include:

- sulphur dioxide (SO_2),
- methane,
- nitrogen, and
- oxides of nitrogen (NO_X).

Sulphur dioxide, methane, nitrogen, oxides of nitrogen and other gases are GHGs, many of which convert to acids when they combine with water in the atmosphere. Combustion of fossil fuels, especially coal, accounts for more than 80 per cent of SO_2 and a large part of NO_X. (The nitrogen coming from the fuels and from the air combines with oxygen to form NO_X when combustion temperatures are high.)

For some unspecified reason, however, all attention is focused on reducing CO_2 which, although not wrong, constitutes an incomplete appreciation of the situation in which humankind finds itself today because of its own actions. Inversely, whatever succeeds in reducing the output of NO_X should also help reduce urban smog, a product of sunlight acting on NO_X, carbon monoxide (CO) and certain organic compounds.

In December 2009 the US Environmental Protection Agency declared that six greenhouse gases, including carbon dioxide, threaten the health and welfare of the American people. *If* this is true for the United States, *then* it is also true for all countries around the globe – and it is not going to get better in the future for reasons explained in Chapters 13, 14 and 15.

The public reaction, however, is no longer as alarmed as it used to be. According to an article in *The Economist*, by most pollsters' accounts voters 'are becoming less energized about global warming. Earlier this year [2010] Gallup found that 48 per cent of "moderate" Americans believe the seriousness of global warming "generally exaggerated", up from 35 per cent just two years before.'[4]

Ordinary people are waking up to the fact that CO_2 reductions don't come free of charge. In Britain, the Department of Energy and Climate Change put some numbers to this effect. In July 2010 it let it be known that 'Green policies will raise domestic gas prices in 2020 by 18 per cent and electricity prices by 33 per cent'. The impact will be up 43 per cent for non-domestic users, which will evidently find its way into consumer prices.[5]

As for the other GHGs and their footprint, so far no significant effort has been made to measure their emissions, let alone focus on wider issues including groundwater protection. Instead, all we hear are plenty of empty slogans like 'energy efficiency', 'greater use of renewable power' and 'geological carbon capture and storage (CCS)'.

Pollution is created by all these gases which, when released in big quantities, are poisonous. Unquestionably, they must be reduced, but this is not being done despite governments' propaganda about the 'accords' of Kyoto and other environmental tourism events (Chapter 14). Since 2000, CO_2 emissions from fossil-fuel energy – coal, oil and natural gas – has significantly increased, and this process is accelerating, especially in China, India, Indonesia and other developing countries that massively depend on coal

4 *The Economist*, 3 April 2010.

5 'Efficiency Drive', *The Economist*, 29 July 2010.

for most of their energy supplies. There are, of course, good reasons why fossil fuels are so popular.

- *Homo sapiens* has learned how to exploit them to produce the energy services it needs.
- A relatively simple technology of controlled combustion provides energy for a wide range of applications.
- Fossil fuels are accessible, in one form or another, in practically all regions of the world, which makes nations and people using them feel comfortable about using them.

There is also considerable flexibility in their usage. Fossil-based energy sources make excellent fuels for transportation as they are portable, which is not the case with most forms of renewable energy (Chapter 5). They also store a great deal of chemical energy, while the oxygen required for combustion is ubiquitous in the air.

Another reason why fossil fuels and their CO_2 emissions are a prized source of energy is that one form of them can readily be converted into another – for instance, from solid to liquid or gas.[6] All this used to be very good news for humanity, but suddenly, it has transformed into a fear that it is very bad news for acid deposition.

Humanity took a very long time to appreciate that, in large and uncontrollable quantities, CO_2 and the poison carried by other GHGs are the real enemy. This is not because of environmental warming because, at least according to some scientific opinions, changes in the earth's temperature may be due to the sun, over which we have no power.[7] Let's, then, get the target right.

- In large quantities CO_2 and the GHGs are potent poisons,[8] and
- these should be the primary target of a concerted global action to make human activities environmentally-friendly.

Unlike the poisonous effect of CO_2 and the other GHGs (as discussed in Part One of this book), the issue of global warming remains perplexing because, despite years of study, there are enormous uncertainties. All experts agree that certain gases put into the atmosphere by human activities trap heat and cause average temperatures to rise. But no one knows:

- how much or how fast temperatures will rise because of them, and
- whether the increase will make much difference in view of other even more potent reasons for climate change.

In an interview published in *EIR*, Dr R. Timothy Patterson explains in a comprehensive way the more potent reason:

---

6 Fossil fuels are also excellent feedstock for chemicals and plastics.

7 More precisely, the effect of changes in sunspot activity is that of cooling, not warming ,the earth's environment. We will return to this issue in subsequent chapters.

8 As a matter of fact, in big quantities everything – even the most benign substance (or attitude) – can be a poison.

> *There are different 'flavors' of sunspot cycles: There are the 11-year sunspot cycles, and the 88-year sunspot cycles, and the 200-year sunspot cycles, called the Glaissberg Cycle ... there is a correlation between sunspots and climate.*[9]

Dr Patterson's research involved the analysis of a 6,000-year record of laminated sediments whose X-ray provided very interesting patterns using time-series analysis. Another reference is a paper published by Jan Veizer and Nir Shaviv, which elaborates on cosmic ray amplifiers and how, with the clouds, they magnify solar effects.[10]

More evidence, based on different criteria, comes from Mojib Latif, professor and climate expert at Kiel University in Germany. His findings, presented at the 2009 World Climate Conference in Geneva, show that the mean global temperature has actually declined since 2001. Other predictions, some of them based on mathematical models, talk of further temperature drop over the coming decades. By contrast, media accounts 'spun' the story into the opposite scenario.

Climate change, including the melting of icebergs and rising sea levels, is a very complex issue not only because research projects like Patterson's and Latif's deny that CO_2 is the primary reason, but also because a natural increase in sea levels of 1–2 mm a year has been occurring for many centuries.

Taking these and similar references into account, as any scientific study should do, it looks *as if* all this environmental tourism from Rio and Kyoto to the Copenhagen circus (Chapter 14) is awfully misdirected. Like the invasion of Iraq by George W. Bush and Tony Blair, it looks like targeting the wrong enemy. CO_2 is indeed a subject of great concern (though less so for developing countries which have already taken certain measures), but it is not the only major environmental danger.

The real enemy of *Homo sapiens*, and of planet Earth itself, is the rapidly rising population which is unaffordable and unsustainable. Too many people are the reason for the booming quantities of CO_2 and the other GHGs, whether or not the sunspots are behind climate change.

- *If* we want to succeed in what we do,
- *then* we must identify the salient problem and put all our resources into solving it.

The salient problem is too many people depleting the earth's resources. Whether one wants to acknowledge it or not, the population explosion is the central issue, and the huge volume of CO_2 emissions is its result. Any management worth its salt tries to solve the salient problem first. Only one subject at a time is salient, although there may be many problems around – and one of them might become salient after the present one is solved.

9 'There's No Relation between CO_2 and Climate Change', Interview with Dr R. Timothy Patterson, *EIR*, 11 January 2008.

10 N. Shaviv and J. Veizer, 'Celestial Driver of Phanerozoic Climate?', *GSA Today*, July 2003.

3. The Vatican is Not Alone in Giving Indulgences

The volume of CO_2 emitted by industry, cars, other means of transport and households in Western nations is reasonably well known. So is the amount in the atmosphere, but not necessarily in the oceans. By contrast, the volume of CO_2 emitted by developing countries is largely unknown, and wild guesses can be way off the mark.

A big part of the destructive gases is methane from solid human waste by an exponentially increasing population. A major component of the salient problem lies in exploding urbanization, as in Cairo and in Mexico City.

Surprisingly enough, the eggheads who met and wrote up the Kyoto Protocol (Chapter 14) exempted developing countries from having to curb or mitigate their carbon emissions. This favoured a swarm of nations in a big way. As a result, China, India, Indonesia, South Africa and Brazil, among others:

- do not consider their inordinate contribution to CO_2 emissions as subject to curbs, and
- view polluting the earth as their non-negotiable 'right' – hence they do as they please.

The Vatican is not the only authority giving indulgences. From Tony Blair's time, Britain and the other EU members have adopted the same policy – except that they pay money, rather than receive it, for their carbon-offsetting indulgences

The United States neither gives nor receives indulgences for CO_2 emissions. Although Barack Obama likes to join the EU party, he nevertheless cut the subsidies allocated to alternative energy and energy as a whole in line with his federal budget presented on 1 February 2010. Being a novice, he was also trapped by the Chinese, Indians, Brazilians and South Africans in a unexpected night session on 18 December 2009 in Copenhagen (Chapter 14).

The developing countries have a ball in this Western comedy of errors. As Figure 3.1 documented, in 2011 the *big* polluters of the environment are China, India and the other developing countries, as their demand for energy is reaching for the stars. Worse still, this situation will have deteriorated even further by 2030, since they have assumed no obligations to control their CO_2 emissions.

- Rio, Kyoto, Copenhagen and the other environmental tourism gatherings have given an unprecedented number of indulgences to less developed countries.
- This Middle Ages invention by Pope Leo X, who issued them to get money to build St Peter's Basilica, has taken on a new life.

As an example, here is a brief list of the results of exempting developing countries from CO_2 caps. In November 2009 Maruti-Suzuki, India's leading car-maker, sold over 76,000 cars – a cool 60 per cent more than in November 2008. Statistics from China tell a similar story. China and India, the world's most populous countries, are also the world's biggest and fourth-biggest carbon emitters. As they are also the fastest-growing big economies, their CO_2 emissions similarly continue to grow most rapidly.[11]

11 China still outpaces India. Its emissions per head are around 5.5 tons. India's are 1.7 tons.

- At the time of Kyoto, China's and India's combined emissions accounted for 13 per cent of the world's total.
- Current projections indicate that in 2030 the proportion attributed to these two countries will reach 34 per cent.

During the Copenhagen circus of December 2009 India insisted that it should be less bound than other developing countries to take bolder mitigation steps.[12] Yet its government knows very well that this is against its own country's best interests, because it is already suffering badly from climate change.

According to environmental vigilantes, also at the top of the list of global polluters is another developing country: Indonesia. It emits 500 million tons of CO_2 a year, not including the consequences of fires ravaging, by omission or commission, its tropical forests.[13] But richer countries, too, are guilty and not only the much-criticized United States.

Among EU member states, Finland, Germany and Poland are respectively in 5th, 9th and 10th position in terms of CO_2 emissions from degrading peatlands. Indonesia, Russia, China and the United States, in that order, hold the 1st to 4th positions, Malaysia and Mongolia the 6th and 7th and Belarus, a relatively small country, the 8th position.[14]

Nor are the various EU governments embracing CO_2 quotas wholeheartedly. In May 2007 Slovakia and the Czech Republic said they would go to the European Court of Justice to appeal a CO_2 emissions quota imposed by the European Union. The Czech government added that abiding by this CO_2 quota was equivalent to swamping all the country's economic development for the next few years.

For its part, the Polish government stated that there was no way of converting out from carbon-fired power plants within deadlines imposed by the European Union. As for the United States, back in 1991, with the price of oil falling, the Bush administration decided against:

- setting federal efficiency standards for electric lights,
- exempting energy efficiency rebates from income tax,
- granting tax credits for electricity generated by renewable fuels, and
- creating a self-financing federal fund to advance loans for efficiency projects.

When the US government was developing its second energy strategy at the beginning of the last decade of the twentieth century, the pattern was one of heavy lobbying by industry and environmental groups, as well as arguments between the Energy Department and the White House. With all that, the basic theme became one of securing future supplies, rather than backing energy efficiency and conservation measures.

This position did not change at all during the eight years of the Clinton administration. Indeed, it was Bill Clinton who decided that America should make no commitment to the Kyoto Protocol and, at the end, he did not sign it. As for the next eight years, under the

---

12 Which is an oxymoron because no developing country participating in the Copenhagen circus accepted any responsibility to reduce, let alone control, its CO_2 emissions.

13 Which is very serious because tree-felling causes 20 per cent of man-made CO_2 emissions.

14 Statistics from Mongabay.com, 4 November 2009, at: http://news.mongabay.com/2009/1104-peatlands.html.

watch of George W. Bush, the least that can be said is that the control of CO_2 emissions was carefully kept in the time closet.

Regarding Barack Obama's early years in the White House, an interesting revelation has been that the power plant heating Congress produces more CO_2 than any other facility in Washington, DC. It is almost 100 years old and runs partly on coal.[15] It would have been a showcase for the Obama administration to make at least *that* plant ecologically friendly.

- The problem is that Obama, like so many other heads of state, talks about green issues because they are so fashionable.
- Each head of state boasts about his efforts to avert climate change, even if those efforts are on hold.

In the United States, the 2009 economic stimulus package allocated $65 billion in subsidies and tax breaks to the energy industry, most of which had at least a green colour. But really major reforms are still to come. Critics say that these billions were, more than anything else, a show for the gallery as the 'plan' must pass through Congress.

Critics also add that CO_2 emissions cuts selected by politicians are far more expensive than those chosen by the private sector. Some studies put the cost of every ton of emissions 'avoided' thanks to the renewables at between $70 and $140. Somebody must pay for that, and the taxpayer will be most probably asked to open his purse.

In the private sector itself, the financial crisis has curtailed new investment in green technology. Some people predict that once the recession is over, oil prices will rise again, making alternative fuels seem less expensive, and banks will start financial lending again (*if* and *when* abandon their financial gambles and resume their role in social service).[16] But all that is just hypotheses.

Market forces alone are unlikely to boost efficiency high enough to head off growing oil dependence, curtailing the poisonous CO_2 and other GHGs. Personal attitudes, too, are against change. Car buyers seem not to care about fuel economy once petrol prices drop below a certain threshold. In China GM's Buick is very popular. Society's interests are still far influencing consumer decisions, and therefore the destruction of the ozone layer continues.

4. Life, as We Know It, Would Not Have Developed Without an Ozone Layer

The holes man has created in the earth's protective *ozone layer* are one of the most flagrant examples of how ongoing policies of environmental destruction cause major deterioration in the environment. Many scientists repeatedly warned that chlorofluorocarbons (CFCs)[17]

15 *The Economist*, 14 March 2009.

16 D.N. Chorafas, *Capitalism Without Capital*, Palgrave Macmillan, Basingstoke, 2009.

17 Introduced by Du Pont in 1935 as Freon®, CFCs were seen for many years as a new wonder, quickly replacing toxic gases like ammonia in refrigerators and air-conditioners, including those in motor vehicles. They were also widely adopted as blowing agents for the production of insulation foams and as cleaning solvents for electronic equipment, including in semiconductor production.

might damage the ozone layer, but certain events, like the appearance of a hole in the ozone layer over Antarctica, have been unexpected.

- The ozone layer shields earth from ultraviolet radiation from the sun.
- CFCs (and volcanic eruptions) build up into polar stratospheric clouds, which act as sites for chemical reactions that turn chlorines into ozone destroyers.

At the beginning of the 1990s, for instance, the electronics industry accounted for around 20 per cent of CFC usage in the United States. Companies were using CFCs to clean circuit boards and other sensitive electronic components, but they moved quickly to phase out industrial cleaners. (In doing so, some electronics companies discovered that warm soapy water serves just as well as a cleaning agent.)

Ironically, like asbestos in the 1930s and immediate post-World War II years, CFCs started as a great benefit for human well-being and ended as a scourge. Ironically, the ozone we care so much about is the same substance which, at low atmospheric levels, constitutes a major component of smog and an irritant to the lungs.

As far as smog is concerned, according to the US Environmental Protection Agency (EPA), motor vehicles and industry contribute about 75 per cent of the millions of tons of nitrogen oxides released each year in the United States. Scientists say that it is not enough to control the big sources, like motor traffic. To make a difference, efforts have to focus on thousands of 'small sources' whose impact is rising because they are rapidly multiplying. These include:

- solvents used to clean the home,
- petrol-powered lawnmowers,
- starter fluids for backyard barbecues,
- hairsprays,
- windscreen cleaners, and
- even bakeries that emit alcohol vapour from yeast growing in bread dough.

For the majority of these smaller, but growing, sources, typically associated with quality of life, it is difficult to impose controls without all-out collaboration by the consumers themselves. As we saw in the first sections of this chapter, a similar statement is valid about CO_2; methane, for instance, is touted as an alternative fuel.

Along with CO_2 and methane, CFCs are greenhouse gases whose usage skyrocketed in the post-World War II years. By the late 1980s, CO_2 represented about 62 per cent, chlorofluorocarbons 24 per cent and methane the balance of greenhouse gas emissions. The prevailing scientific opinion now is that even if CFCs were completely banned (which is a weak hypothesis), degradation of the ozone layer will increase since the existing hole was caused by gases released before 2011.

- CFCs rising through the lower atmosphere take an estimated 7–10 years to reach the stratosphere, and
- once they are there they continue their demolition work of the ozone layer for as long as 100 years, as long as they remain active.

For the past few decades the hole in the ozone layer over Antarctica has been shifting with the winds round the continent, driving it faster and faster around the Southern Ocean.[18] Parts of the hole sometimes break off to create radiation hazards in Australia and New Zealand. The same could happen over the northern hemisphere if a hole forms above the Arctic.

Some years ago, a computer simulation indeed predicted that by 2030 the northern hemisphere would endure significantly more heat than the southern hemisphere. (According to an article in *Nature* magazine, this model was a new and improved version of a previous one, taking into account the ability of oceans to influence climate.)

Although there are still many unknowns accompanying such studies and forecasts, over the last three decades scientists have repeatedly warned that densely populated regions of the northern hemisphere and cities as far south as London, Amsterdam and Moscow will face solar radiation hazards with ozone depletion over the North Pole. An increase in radiation would not only have ecological effects, but also serious health effects. *Homo sapiens* will be more vulnerable to:

- skin cancer,[19]
- cataracts,
- immune system deficiencies and more.

Back in May 1990 the Bush administration was heavily criticized at a United Nations conference in Geneva because of its opposition to direct aid for developing countries in reducing their use of chlorofluorocarbons, to avoid depletion of the ozone layer. (The administration defended its stand on the ground that the World Bank already had the resources and structure to provide such aid.) Shifting responsibilities is no plan for the future. The then World Bank president, Barber B. Conable Jr, answered that his bank would have to create a new fund for such global environmental problems as chlorofluorocarbons, and a spokesman added the bank had insufficient funds for any kind of chlorofluorocarbon programme because:

- it was primarily devoted to internal domestic projects and
- most of its loans were committed.[20]

As usual, no money, no business.

By contrast, the then 12-nation European Community did act. Germany, the Netherlands and Denmark announced a deadline for banning CFCs while environmental ministers agreed to press for cut-off dates for the harmful chemicals throughout the European Union. The deadline exercised pressure on Europe's chemical industries, which had to step up production of substitutes needed for freezers, industrial refrigerators and other equipment.

18 See also the pending exploitation of Antarctica for minerals, in Chapter 11.

19 Studies by epidemiologists show that skin cancer rates significantly increase with greater exposure to ultraviolet light. Moving from low, cloudy Seattle to high, sunny Albuquerque, where ultraviolet radiation is twice as strong, raises the chances of getting skin cancers fourfold.

20 *International Herald Tribune*, 11 May 1990.

The slower pace of the US response was all the more surprising as, in April 1991, the Environmental Protection Agency (EPA) announced the results of a sobering study. Measurements from a NASA satellite revealed that the fragile ozone layer had shrunk by as much as 5 per cent over America in the preceding ten years. This was at least 50 per cent more than previously estimated.[21]

Beyond these studies and measurements, an international agreement known as the Montreal Protocol (Chapter 14), called for a phase-out by 2000 of CFCs and other chemicals that attack the ozone layer. Year 2000 has come and gone, but CFCs are still around in refrigerators, air-conditioners, aerosol sprays and plastics increasingly used in many countries, particularly in the developing world.

In a globalized world which cares much more about 'development' than its associated pollution, getting large numbers of nations to agree on anything, especially delicate environmental policy issues, is no easy job. Scientists have not been so successful in convincing policy-makers that the earth's ozone layer is in grave danger – even though the first warning was sounded by two chemists at the University of California, at Irvine, in 1973.

There is a collective absence of a universal conscience about the urgency of protecting the ozone layer from depletion. Nor is it properly appreciated that polluted air endangers all living species and that other animals should not be called on to pay for man's failure to respect his environment. Pollutants drift with the prevailing winds, descending in rain, mist and snow on the land, forests, lakes, rivers and oceans. The change in acidity kills trees, fish and other life which has counted for nothing in this man-made disaster.

5. Transportation: The Biggest CO_2 Polluter

There is plenty of factual and documented evidence that transportation at large, and most particularly motor vehicles, is the biggest single CO_2 polluter. This problem is going to get worse rather than better because transportation by plane, car or motorboat is also a status symbol and its increasing usage a sign of a higher standard of living. The curve of increasing CO_2 emissions resulting from transportation is difficult to bend, and any restrictions are seen as unpopular measures.

Statistics on the origin of emissions vary from country to country and, to say the least, they are seldom fully reliable. Nevertheless, some order of magnitude numbers in percentage of greenhouse gas emissions provide a pattern and the underlying causes make interesting reading. Chapter 6 made reference to California's gas emissions by major industry sectors as a proxy of the developed world. To avoid looking back at Figure 6.1, the statistics are as follows:

- transportation, 40.7 per cent,
- electric power, 22.2 per cent,
- industry, 20.5 per cent,
- agriculture and forestry, 8.3 per cent,
- other sources, 8.3 per cent.[22]

21 *Business Week*, 22 April 1991.

22 *The Economist*, 23 June 2007.

In terms of pollution, the effect of transportation and emissions figures are destined to worsen over time. As a news item on Al Jazeera had it on 12 April 2010, about 40 per cent of the world's goods are nowadays transported by air. It is pure hypocrisy, therefore, to talk of controlling CO_2 emissions.

Cars are still the biggest polluters in transportation, but air transport is catching up. On 16 April 2010 European airports cancelled 17,000 flights because of Iceland's volcanic ash cloud. This was said to represent 75 per cent of all daily flights, which means that in a single day some 23,000 flights carrying more than 200,000 passengers pollute the old continent's skies.

To appreciate what this means in terms of environmental effect, when flying from London to New York the average passenger is responsible for 500 kilos of carbon dioxide – in one single flight. That's more than the average African emits in a year. Yet, even Greens and other environmentalists are rushing to benefit from the supposed 'luxury' of affordable air travel or, more precisely, the lure of far-away places and convenience that air travel creates.

If the International Air Travel Association's (IATA's) projections are correct, the number of passenger departures will grow from 2.3 billion in 2009 to 16 billion by 2050 – by almost 7,000 per cent. It's 'cheap flights' that inflate the numbers by so much. In 2009 the biggest airline in the world in terms of the number of passengers carried was not British Airways, Lufthansa, Air France or American Airlines, but Dallas-based budget carrier Southwest Airlines. The fifth biggest was Ryanair, Southwest's imitator.

- These airlines, EasyJet included, fly short distances.
- Therefore, per passenger mile they pollute much more than BA and the other big airlines – which does not mean that the latter are environmentally-friendly.

In early 2009 it was astonishing to learn that 350,000 tourists were stranded at Thailand's airports and could not leave because the airports were occupied by political protesters. It is not precisely known how many of those stranded long-distance tourists were Greens or other anti-pollution protesters, but generally they seem to have no more conscience than the rest of us when it comes to catching a plane and travelling to any exotic location that takes their fancy.

Yet, pollution from air traffic is a real and present danger with some estimates indicating that each new super-jumbo jet will be pumping out CO_2 at the same rate as 5 million cars. Look at this in a different way. In terms of energy, flying a fully-laden A380[23] would give off the equivalent emissions to a 14-km queue of traffic on the road. And that is *just one* aircraft. Airbus reckons that in the next two decades 1,500 such planes will be in the air, while by then the total number of airliners is expected to have doubled to (hold your breath) 22,000.

Moreover, whereas cars are used for roughly an hour or so a day, each of the long-haul airliners flies for over ten hours per day. That adds a hell of a lot of CO_2 to that emitted by the 60 million vehicles produced every year, most significantly increasing the pollution from the more than 1 billion vehicles already on the world's roads.

Transportation at large is a god-size polluter, but not all modes of transportation have the same weight in CO_2. Back in 2006, CE Delft, a Dutch consultancy, made a comparison

23 The EADS superjumbo.

between different forms of travel. Assuming that aircraft are 70–75 per cent full and cars contain 2.5 people on average, since longer distances often involve the whole family:

- Long-haul flights of more than 1,500 km were 50 per cent worse than motor vehicles for each passenger kilometre.
- By contrast, short-haul flights were 300 per cent worse than cars running on petrol.[24]

Yet, not one of the UN conferences theoretically aimed at 'saving the environment' addressed the issue of CO_2 pollution from air transport – just like the United Nations initiatives have failed to examine the very negative impact of the population explosion (Chapter 13). If anything, as far as damage from air traffic is concerned, these UN tourist affairs have made the situation worse since, to attend them, thousands of people catch a plane. (At the 7–18 December 2009 Copenhagen circus there were *15,000* delegates.)

As for the slogan of the 1990s, 'communicate, don't commute', the score is not brilliant either. People have money both to fly to far-away places and to yakety-yak with one another uninterruptedly all day through social networking[25] and on mobile phones. Nowadays there are billions of mobile-phone subscribers, and their number continues to increase.

Communicate *and* commute has become second nature. Burgeoning numbers of airplane flights, as well as vehicles running on the world's roads, have a disastrous effect on global and regional environments. Air pollution threatens health, and much of it can be traced to emissions:

- carbon monoxide and nitrogen oxides, reacting with water to form nitric acid, and
- hydrocarbons, reacting with nitrogen oxides in the presence of sunlight to form ozone.

In developed countries – and pretty soon in less developed countries given their great love affair with the car – motor vehicles emit nearly half of the nitrogen oxides, two-thirds of the carbon monoxides and nearly half of the hydrocarbons. Moreover, where environmental controls are lax – a common occurrence – inefficient vehicles are a huge contributor to air pollution.

Technology being what it is and the efficiency of the internal combustion engine improving rather slowly, what really counts in emissions is the amount of car usage. Whichever way you count it, from private cars to trucks, traffic is clogging roads, highways and motorways. And while emissions are skyrocketing, the bureaucrats who are supposed to rein them in are just putting unrealistic goals on paper.

In January 2007, for instance, Germany's three big car-makers criticized a planned EU target to cut car exhaust emissions as 'unrealistic' and 'technically unrealizable', warning that it could make swathes of the European industry unviable. The chief executives of Volkswagen, DaimlerChrysler and BMW described the CO_2 emissions target proposed by the European Union as:

24 *The Economist*, 10 June 2006.

25 D.N. Chorafas, *Cloud Computing Strategies*, Auerbach/CRC, New York, 2010.

- a massive industrial policy intervention that would not only burden the entire European car industry,
- but also hit the German industry in particular, resulting in the elimination of numerous jobs at car producers as well as in the supplier industry of EU production centres.

Without due consideration to what can realistically be achieved by technological progress, the politicians and bureaucrats warming armchairs at the Brussels executive offices wanted car-makers to cut their vehicles' CO_2 emissions to an average of 120 grams per kilometre (g/km) by 2012. Car companies responded that they supported the European Union's drive to cut CO_2, but were struggling to meet a voluntary target of 140 g/km.[26]

(Looked at objectively, a 'typical' tank of petrol produces up to 180 kilos (400 pounds) of carbon dioxide, a major greenhouse gas implicated in global warming. While motor vehicles come up as winners in comparison to jet aircraft on 'lesser polluter grounds',[27] they remain big polluters because people use them so often for pleasure and leisure.)

It is, of course, people who drive vehicles, just as it is people who travel on jets and use products that contribute in one way or another to the environment's destruction. And, without doubt, it is also people who design internal combustion engines and manufacture planes and cars. The more the population explosion continues, the more the consumption habits of *Homo sapiens* pollute the earth and the bleaker is the people's future.

6. The Electric Car's Dark Prospects

One of the ironies of the European Union's *diktat* to European car manufacturers (mentioned in section 5) is that at the time it was issued, in May 2007, there had been three consecutive weekends of heavy motor traffic. A horde of people all over Europe (including EU employees and managers) took a short break by taking off a day between two public holidays. Consequently, millions of cars left for sea resorts or the mountains, adding their CO_2 to the prevailing pollution.

- This pattern of travelling long distances for an extended weekend is going to spread worldwide as less developed countries become more affluent, and
- as everybody is pursuing their own pleasure, it is utterly wrong to believe that policies to conserve energy and reduce vehicular emissions concern only the industrialized nations.

Today, developing countries present a much bigger menace to the environment. Electricity production and air traffic aside, an important factor driving demand for oil

---

26 Shortly afterwards the 120 g/km proposition was blocked as Angela Merkel, the German chancellor, warned Brussels against setting emissions targets that would hit the motor industry, particularly companies such as DaimlerChrysler and BMW that produce larger models with higher emissions.
Some political pundits said that the 120g/km target had nothing to with technical realities, but plenty to do with political infighting between EU member states about who would control the European car market.

27 See the aforementioned study by CE Delft.

and its accompanying CO_2 and other GHGs, is the rapid growth in ownership of personal motor vehicles. China, India and Brazil are examples.

The International Monetary Fund (IMF) forecasts that there will be some 2.9 billion cars in the world by 2050,[28] a 3,500 per cent increase, and the large majority will be owned by what are now developing nations. China and India are projected to have *half a billion* personal vehicles by 2030.

A good way of interpreting the IMF projection is that even if the various UN environmental conferences and self-inflicted carbon taxes (section 7) succeed in cutting demand for carbon fuels in the developed world, the much higher risk from the 'not yet developed' world will remain. And because of costs associated with the latest technological developments to cut emissions, developing nations will be given a *carte blanche* as CO_2 polluters.

The US Congressional Budget Office has estimated that a cap-and-trade (section 7), designed to cut emissions by 15 per cent, would reduce the average income of the lowest quintile of the population by 3.3 per cent.[29] The effect on the less financially healthy part of the world is not uniform.

- Not everyone drives a car, but those who do drive one spend a large part of their income on petrol.
- Although the poor may not all drive, they do heat their houses in winter and they do emit methane.

There are people who, unlike the tax-and-tax politicians, put their hopes for rapidly reducing CO_2 emissions on alternative means of transportation. But they are deceiving themselves. Industrial history teaches that the argument 'electric cars will take care of CO_2 emissions' is not tenable. Here are the reasons why.

Electric cars were not invented only yesterday, and a product that does not take off over a period of 100 years, is never going to lift off commercially. Electric vehicles have been around for a century and have never had great public appeal. Moreover, electric motors are never going to be able to power trucks, and the haulage industry is one of the major polluters.[30] Also, the success of individual means of transportation largely rests on its independence of action. There are no service networks for electric cars.

Much the same is true of hybrids, whose merits in terms of pollution control are hugely overblown. Nor should we ignore the fact that the electricity to charge their batteries must come from somewhere and, if this is a coal-fired power plant, the therapy may be worse than the illness.

Misled by politicians, the Greens and some car-makers, who see 'dollars now' in electric cars and hybrids,[31] the public thinks that such vehicles will make a significant contribution to improving CO_2 emissions. In light of what is stated in the preceding paragraphs, that hypothesis could not be more wrong.

---

28 'The Art of the Possible', *The Economist*, 13 November 2008.

29 *CBO Testimony: Approaches to Reducing Carbon Dioxide Emissions*, Congressional Budget Office, Washington, DC, 1 November 2007, p. 7.

30 Even if environmentalists choose to ignore this fact.

31 When, in the late 1980s, ISDN was promoted as a big step forward for telephony and data exchange (it was not), the joke was that the reason why it attracted the big equipment manufacturers was a novel reading of its initials: 'I see dollars now.'

Another basic argument against them is technology. As Robert Bishop, editor of *European Automotive Design*, pointed out in a feature article, it simply makes no sense to produce vehicles with two powertrains. Such a configuration is not sustainable either from an economic or environmental viewpoint. As Bishop sees it, 'In Europe they might, for the time being, have some small fuel economy advantage over conventional high-efficiency gasoline or diesel powertrains.' However, *if* the lifetime impact of their manufacture on the environment is included, as should always be the case, *then* this argument is unconvincing. Furthermore, Bishop adds: 'In North America the desire for dual powertrain hybrids which include huge internal combustion engines is just plain nonsensical.'[32]

Duncan Leslie, consulting editor of the same technical magazine, puts the same concept in a different way.[33] There is some great engineering going into hybrids and their performance. But, taking a life-cycle perspective, factoring in the extra costs and end-of-life issues, they don't look so impressive. There also the most crucial question: will the public buy them?

- When asked, they say they would be keen to consider it.
- But when they look at the price tag, they walk away.

Higher costs and resulting general public apathy can never make a mass market for electric cars and hybrids. Even as a niche market, they will have little impact, even assuming that the problem of producing electricity without CO_2 emissions is solved.

On the other hand, much greater benefits will come from advances in traditional internal combustion engines, coupled with lightweight structures and other improvements in the vehicle. Lightweight structures, however, mean less safety. Therefore, a global reduction in motor-vehicle traffic would be a great deal better than any electric car or hybrid strategy – which brings us back to the need to control the population explosion.

Probably constrained by the fact that, under current population patterns, a reduction in car traffic also means severe downsizing in employment opportunities, governments have failed to inform the public that our society's growing mobility does not present only advantages. Not only is growing mobility a major contributor to environmental pollution, but it is also severely taxing the transport system's infrastructure, including roads, highways, bridges and parking spaces:

- Transport routes, bridges and parking spaces cannot simply be 'just enlarged' when they reach the limits of their capacities, and
- the infrastructure of developed countries, designed in the post-World War II period, is decaying, while in many developing countries it is non-existent.

Year after year, more and more vehicles deliver increased mobility of people and goods and at the same time put severe strain on the existing infrastructure which is not always properly maintained.

The whole concept of mobility depends on numerous overriding conditions like incomes, energy prices, division of labour, consumer preferences, infrastructure and

32 http://www.automotivedesign.eu.com/magazine/Default.aspx , September 2008.

33 http://www.automotivedesign.eu.com/magazine/Default.aspx , June 2008.

more. There may be plans for improving infrastructural solutions, but they cost money and are not always put in practice – while major economic crises, like that of 2007–2011, turn what was thought to be a balanced economy on its head.[34]

7. Is There Any Sense in Cap-and-Trade CO_2?

For all their talk about saving the environment, politicians are decidedly silent on the vexed question of how responsibility for cuts in emissions should be shared between developed and less developed nations. The public is equally puzzled and sometimes takes to the streets, as often happens in Germany. Significantly, despite all the excitement, the roadshow of demonstrations and violence leads nowhere. Demonstrations and violence also accompany international conferences, and various 'initiatives' heralded by the media. But a fundamental study able to take a system view of the global problem of natural resources, pollution trends and population busts, as well as environmental decay, is not on hand. Ideally, such a study:

- should not target headlines but facts and figures in testing basic hypotheses about environmental projections, and
- should present each alternative solution's 'plus' and 'minuses' in a comprehensible way so that they can be debated in these conferences.

Those who still believe that CO_2 is by far the largest element at the origin of all the earth's troubles, even in the absence of full scientific evidence to that end, should make clear to themselves and to all others the results of their efforts so far, including their lavishly financed participation in the various 'save the world' conferences and congresses:

- What have they accomplished so far with their 'damn the CO_2' policies?
- How did they accomplish it?
- And at what cost to the economy?

To convince on the results of their action, they should also give an elaborate account of tangible benefits to the environment, both in absolute terms and in comparison to the goals they set when they started their action. In addition, they should come up with new goals *if* the older ones were attained. And *if*, after nearly two decades of effort, these have still not been accomplished, *then* they hardly worth pursuing.

Cap-and-trade (more on this later) is an example of a half-baked initiative to control CO_2 emissions, but the results are, at best, meagre. In late April 2010 Australia, one of its main proponents, shelved plans for an ambitious carbon emissions trading scheme for at least three years due to slow progress on a global climate pact. Kevin Rudd, the then prime minister, said the government would wait until the Kyoto pact expires in 2012 before pushing on with plans for a comprehensive regime.[35]

34 D.N. Chorafas, *Globalization's Limits. Conflicting National Interests in Trade and Finance*, Gower, Farnham, 2009.

35 There was also parliamentary opposition by a hostile Senate, where the socialist government was seven votes short of a majority.

That is not at all surprising. As stated in an article in *The Economist*, 'In the 1990s cap-and-trade ... was the darling of the green policy circuit ... And its great advantage was that it hardly looked like a tax at all, though it would bring in a lot of money.'[36] But now it looks very much like an extra tax on the economy.

This heralds that fancy idea's last hurrah, revealing it as a castle built on sand without proper foundations. Let it be an example of political initiatives taken without public debate, or on a debate based on scientifically incomplete (and often biased) data. Any environmental study that impacts on the economy must be scientifically based on reliable and verifiable data, including a stress test.

An example of the general outline, the kind of study that is needed and the structure it should take, is shown in Figure 10.1. Once the facts characterizing the evolution of disastrous man-made effects are available, and the hypotheses on cause and effect[37] are made, it would be relatively easy to agree on a global plan to cut carbon and other emissions and sign up to necessary initiatives. Nothing like that was done in advance of the December 2009 climate change 'summit' in Copenhagen.

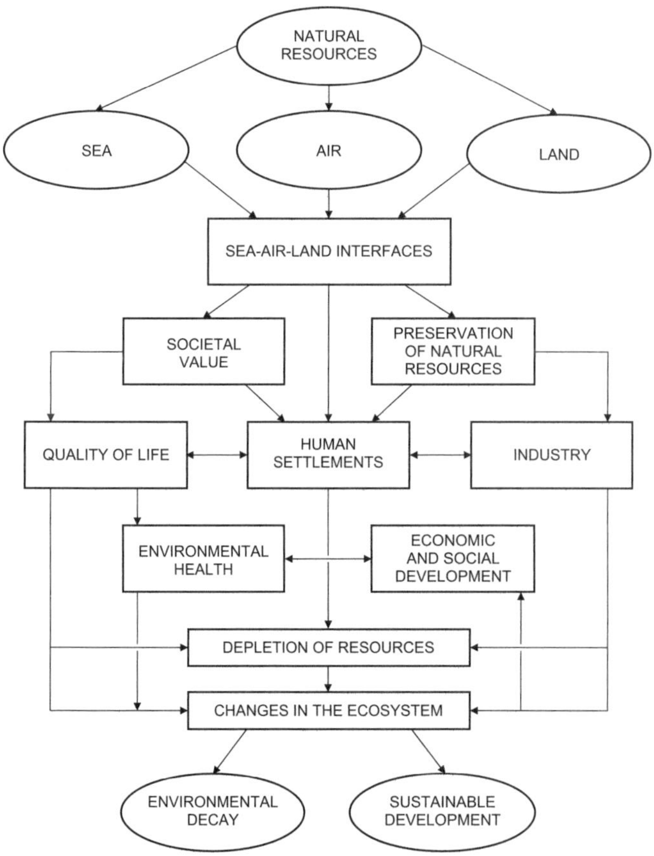

Figure 10.1 Human settlements and natural resources as an integrated system

36 *The Economist*, 20 March 2010.

37 As well as on the most likely future trend(s) under optimistic, realistic and pessimistic scenarios.

Moreover, it is absolutely necessary to have authoritative, empowered global enforcement mechanisms. Short of that, what is being agreed (if anything) will remain a dead letter. Such a mechanism has been utterly absent so far, even though there is a vague idea that there would be international monitoring of emissions to determine how well each country is doing in controlling the CO_2 and other GHGs it produces.

Short of a fundamental global system study, a firm plan and a fully empowered sheriff – who can apply heavy penalties – everyone can say what he wants and get away with it. This is totally different from the ill-studied cap-and-trade scheme, whereby:

- governments in developing countries place an overall blue-sky limit on emissions, and
- sell or give away an equivalent number of tradable rights to pollute the planet to companies of developed countries who have to pay for them.

Cap-and-trade is most definitely an ill-studied, cheap approach and a product of third-rate minds – even the thought of it is ridiculous. The whole scheme rests on the simplistic, dubious and not-so-ethical assumption that some people and countries (particularly the less developed) have the right to pollute planet Earth, and that they can sell that right to other countries and companies:

- who have been condemned as polluters,
- but have not been given an *a priori* licence for exercising their 'dirty trades'.

To make matters more ironic, but also to give some goodies to every embedded interest, all this unfunny business is nicely wrapped in a trading scheme that makes the big banks salivate. These are the same global banks who in the first decade of this century brought the American and British, then the global economy to the abyss with 'subprimes' and 'alt-As'.[38]

The principle of cap-and-trade sounds simple and straightforward: the quantity of emissions is limited (*cap*) and the right to emit is sold by those who are supposedly lesser polluters to greater polluters through a system of permits (*trade*). In reality, what lies in its murky background is that everybody has the right to pollute the earth. The 'right to pollute' is either God-given or bought with cash.

- The existence of such a 'right' is evidently highly questionable, and
- the scheme's distributed nature of emissions trade is full of loopholes.

The so-called *carbon credits*, essentially the merchandise being sold, are the brainchild of the Kyoto conference on environmental protection (Chapter 14). By buying carbon credits, Kyoto's decree has stated, industrial countries and their companies can meet their targets to reduce emissions by about 5 per cent by 2012.

Essentially this means that one is meeting CO_2 reduction targets by paying somebody, not by really meeting them. This is worse than Pope Leo X's gimmick of indulgences

38 D.N. Chorafas, *Globalization's Limits. Conflicting National Interests in Trade and Finance*, Gower, Farnham, 2009; and D.N. Chorafas, *Capitalism Without Capital*, Palgrave Macmillan, Basingstoke, 2009.

(section 3). With good reason critics say that most carbon offsetting does not even hold up in comparison to that medieval system of letting sinners off the divine hook because:

- in the medieval Vatican practice, sin was absolved through financial contributions,
- whereas now the earth's environment is definitely being destroyed by both those who sell and those who buy the indulgences.

This does not mean that cap-and-trade is not good business for those doing the short selling. Originally considered to be a lame duck because of too much hype, the market in carbon credits grew reasonably rapidly, reaching an estimated $30 billion by early 2007 – ten years after Kyoto – according to the World Bank, and it continues to grow.

This speaks volumes about how much money has been changing hands by selling and buying the right to pollute *à volonté*. During the December 2009 Copenhagen circus, Kyoto's cap-and-trade mechanism was seen as the new wheel of fortune, and many delegates were keen to expand it. They have reasons to be jubilant because this market is geographically one-sided.

A torrent of money has come from transactions under the European Union's ill-studied emissions trading scheme, and those who definitely benefit from the sale of carbon credits are the developing countries. But have these developing countries *true* carbon credits to sell? A study undertaken in 2007 showed that China was the biggest beneficiary in 2006, selling 61 per cent of carbon credits, while India took 12 per cent and Brazil 4 per cent. Britain was the biggest buyer of credits, purchasing 50 per cent of the world's supply thanks to Tony Blair, of Iraq fame, who promoted the scheme.

- This has been very bad for British industry, making it uncompetitive, and
- it has morphed into a most questionable transfer of 'carbon credits' and capital.

Even a blind man with black glasses can see that China is a huge polluter, and it is becoming increasingly so by the day. China and other developing countries simply do not qualify to be net sellers of carbon credits. They should rather be net buyers if there is anybody who has 'credits' to sell. What is happening is simply a mockery of environmental pollution control principles.

Critics correctly add that the cap-and-trade market as a whole lacks a generally acceptable standard and is therefore subject to a very significant *reputation risk*. This is true not only of the unregulated part of carbon credits, but also of what is said to be regulated emissions trading and carbon offsets. The implementation of ill-studied and one-sided solutions outweighs the risks of maintaining the status quo.

8. Carbon Tax versus Cap-and-Trade

Rather than cap-and-trade, many economists favour a different market-based alternative: *the carbon tax*, specifically aimed at inducing companies into renewing their technology. 'Tax', however, is a negative word in people's minds and therefore it has no political appeal.[39] Nevertheless, some countries prefer the solution of taxing carbon emissions or, as the French senate named them, 'carbon contributions'. The senate maintains that:

- carbon taxes are a better way of reducing greenhouse gases than cap-and-trade, and
- taxes deal with the problem more efficiently than do indulgences or permits, particularly in view of the uncertainty surrounding the control of carbon emissions.

(In economic theory terms, carbon reduction makes sense until the marginal cost of cutting carbon emissions is equal to the marginal benefit derived from the same process. By contrast, misjudging the number of permits can send permit prices through the ceiling or through the floor, with immediate and expensive economic consequences.)

As critics of the Kyoto trading invention point out, a fixed allocation of permits makes no adjustment for the business cycle – namely, for the simple fact that companies produce and pollute less during a recession. And because the price of permits can be volatile, cap-and-trade schemes have plenty of potential to cause economic harm.

For instance, even though it has never signed the Kyoto Protocol which originated the cap-and-trade scheme, the United States has used tradable SO_2 (sulphur dioxide – not carbon dioxide) emissions permits since the mid-1990s.[40] On average, the price of these tradable SO_2 emissions varied by more than 40 per cent per year. Volatility is always good business for dealers, but bad for end users. Moreover, given the importance to today's economy of carbon-emitting energy sources, such volatility:

- can significantly affect economic fundamentals, from inflation to business and consumer spending and
- can also deter firms from investing in green technology, which is evidently counterproductive.

Proponents of the US SO_2 scheme say that if and when they are designed and implemented *well*, emissions trading systems can reduce compliance costs compared to traditional command-and-control approaches. However, they also add that compliance costs may still be significant, and analysts must continue to compare these costs to the environmental benefits to determine the appropriate level of emissions reduction.[41]

To put it mildly, this is by no means what is happening with the UN Kyoto CO_2 cap-and-trade adventure. Among other reasons why this particular cap-and-trade system is not working as its enthusiasts predicted is because it is expensive, difficult to administer and is applied only to those polluters that can be identified – or have no inclination to pay what the French call *pots de vin,* the Greeks *fakelakia* and the Americans '*greenmail*'.

However, it is nevertheless a most profitable business for the sellers as evidenced by the fact that – according to what is known – the so-called *carbon credit industry* is struggling to keep up with demand. At the same time, there is a shortage of skilled technicians to:

39 Even if, when it gets down to basics, cap-and-trade is also a tax decreed by second-rate politicians.

40 The George Bush Sr created a programme to control sulphur dioxide (SO_2) emissions under the US Acid Rain Program – Title IV of the 1990 Clean Air Act Amendments. Research by the Center for Energy and Environmental Policy Research (CEEPR) in 1997 cautioned that the results of this programme, though positive, must be applied with care to other environmental problems.

41 *The MIT Report*, February 1998.

- monitor carbon reduction projects and
- verify the claimed emissions cuts are taking place.

The lack of skilled technicians, though not of carbon credit traders, also underlines the everyday fact that finding the correct price on CO_2 emissions requires huge amounts of detailed knowledge of the processes behind them. Part of the equation are the marginal costs faced by would-be or current polluters. Trial and error (including shortages and bias) is current currency. And it is not possible to appeal against high prices because, as a letter to *The Economist* put it:

- 'They are set by the market,' and
- 'The price that arises out of the market, then, is by definition the correct one' – which is a fake statement.

Other proponents of cap-and-trade argue that a carbon tax makes no sense because a 'one size fits all' taxation requires an impossible calculation of average cost of reducing emissions over a given period of time. By contrast, they say, an emissions-trading approach works on the free-floating marginal cost of abating emissions. This is a near-sighted argument, at best. Not only should a carbon tax not be 'one size fits all', but it also should not be permanent but phased out as the CO_2-emitting company proceeds with reducing its pollution quotient through technology and better organization – when compared to a pre-established level.

The carbon tax's weakness is that governments are prone to exempt key industries from it – for instance, those facing tough international competition for their products. (In January 2010, in France, the Conseil d'Etat rejected a carbon tax law because it had many exceptions.) Governments can also, if they please, turn a blind eye to the need for cap-and-trade by certain companies.

As for the argument made by cap-and-traders that carbon taxes would be levied locally which makes it impossible to 'properly administer them on a global scale', this can only be put forward by morons. Such an argument is simply for the birds, for three reasons:

- The concept of CO_2 trading is still foggy and easy to manipulate.
- The money changing hands disappears in the coffers of the indulgences' seller; it is not used to promote a greener industry.
- If pocketing these taxes is (hopefully) not the case, *then* the money should be used in the same country to reduce CO_2 emissions, rather than become another form of global giveaway.

All this is written in the understanding that the cap-and-trade market is unregulated, unaudited and opaque and therefore open to all sorts of Ponzi games. It is, however, possible that one day the market for carbon credits, including its sellers and buyers, will become respectable. Today it is not so. Theoretically, carbon offsetting is good for the world. In practice, it is a process overcrowded by problems of:

- false claims,
- double-counting,
- doubtful verification, and

- a total confusion of means and aims.

In addition, the goal should be to eliminate that market by taking care of heavy CO_2 emissions – not to perpetuate it as the cap-and-traders' dream.

To add a pinch of irony, according to the Greens and other theorists, paying for offsets is above all a challenge of eradicating poverty (whatever this means), rather than of saving the environment. And as Chris Goodall aptly suggested, all these arguments distract us from considering how economies dependent on growth without limits could be adapted to reduce pressure on the climate whose limits are proving perilously finite.[42]

42 *Financial Times*, 3 May 2007.

CHAPTER 11 Forests and Oceans: The Lungs of the Earth

1. Earth's Citizens: Their Responsibilities and Rights

A couple of decades ago, to prepare its January 1989 issue dedicated to planet Earth, *Time Magazine* invited 33 scientists, administrators and political leaders from ten countries. The group, which included experts in population, the preservation of species, climate change and waste disposal, met for a three-day conference[1] in Boulder, Colorado.

The participants did not limit themselves to explaining the complexities of interlocking issues characterizing planet Earth, and their problems. They also advanced a wide range of practical ideas and suggestions that *Time* fashioned into an agenda for environmental action. But the panel did not elaborate at all on two of the most important subjects of them all:

- family planning including birth control (Chapter 18), and
- the vital definition first of *responsibilities*, which always precede the *rights* of citizens.

'Your freedom of action ends where the nose of your neighbor starts,' said an American judge to a defendant who had rushed to justify what he did by invoking his individual liberties under the Constitution. With 6.9 billion of *Homo sapiens* on earth, a number still rapidly growing:

- the 'nose of the neighbor' gets closer and closer, and
- each individual's freedom of action shrinks and shrinks, as the people on earth multiply.

Individual or group 'freedoms' and individual 'rights' are non-comprehensive and unsustainable without the basic prerequisites of individual *responsibilities*. The latter come way ahead of the former. The older of these responsibilities are behavioural, partly (but only partly) defined by the law of the land. The newer responsibilities are global, and they concern:

- the preservation of the earth's resources, and
- their sustainable management, so that new generations do not inherit a derelict planet.

1 In November 1988.

The able management of natural resources today is much more important than it has ever been because, with globalization, what were once local environmental problems have taken on enormous proportions. The earth is literally being stripped of its wealth – while overpopulation, illiteracy and poverty are producing worldwide instability.

Everybody contributes to this run-down status. Examples are: emission of CO_2, SO_2 and other GHGs; overuse of chemicals; severe soil erosion; climatic change; and the accelerated extinction of animal and plant species. These facts are confronting the present and future generations with enormous problems that have not even been properly analysed, let alone find themselves on the way to a solution.

Some sociologists (and even a few anthropologists) have dreamed up the false notion that *Homo sapiens* alone, of all living things, has the capacity to view his surroundings on a more comprehensive scale than the span of a single life. That sounded fair enough, until we became aware that a long list of man's actions are destructive.

Examples of total lack of responsibilities include *tropical deforestation*, this chapter's theme. Cutting and burning permanently destroys as many as 50–60 million acres per year, particularly in the Amazon, Ecuador's Pacific coast, southern Mexico, Guatemala, Brazil's Atlantic coast, West Africa, Madagascar, India's eastern coast, Nepal, Burma, Vietnam, Malaysia, Indonesia, the Philippines and Australia's Cairn peninsula.

Although the rate at which the world's forests are disappearing has slowed over the past decade, according to the UN Food and Agriculture Organization (FAO), it remains troubling. Planning efforts in China, India and the United States helped to offset decreases in South America and Africa, but, despite the improvement, the yearly net loss of forest over the past decade amounted to an area the size of Costa Rica.

There is no great hope for change in these statistics, as African nations asked to be paid not to cut and burn their national wealth, and (unwisely) six Western nations pledged \$3.5 billion as a down payment on an effort to 'slow, halt and eventually reverse' deforestation in different countries.[2] Seeking rent for not destroying the earth's environment is indeed evidence of the abyss of current morals.

During the Copenhagen circus of December 2009, many people at the summit accepted the case made by friends of the forest that cutting down trees:

- contributes up to 20 per cent of global greenhouse emissions, and
- if forests are not better protected (according to the ongoing arguments), dangerous levels of climate change are inevitable.

A corollary to rampant deforestation is *desertification* (see section 6). Overstressed, very arid or fragile lands continue to be subjected to severe degradation. Here we are talking of massive irresponsibility which has spread from central Mexico to southern Chile, the north-east edge of Brazil (Belem), a wide spread west-to-east in central Africa, southern Africa, Madagascar, central Turkey, Mesopotamia, parts of Iran, Afghanistan, Pakistan, north-western India, western China, and north-western and south-western Australia.

2 'Touch Wood', *The Economist*, 17 December 2009. The benefactors are Australia, Britain, France, Japan, Norway and the United States. That money is a first instalment of the \$25 billion needed between now and 2015 to cut deforestation by a quarter – a vast and ill-studied plan.

Other great sufferers from man's irresponsibility are rivers, lakes, seas and oceans. They are exposed to severe pollution from big cities, industry at large and oil wells. In addition, urban wastes plague many semi-enclosed seas and find their way into oceans.

When in the early thirteenth century Ogedei Khan – son and successor to Genghis Khan and master of the empire – chose a site for his new capital, Karakorum, he applied principles way more far-sighted than those being used today. The place had to have:

- good wind to keep down mosquitoes, and
- ample water *sufficiently far away* that it would not be polluted by people living in the new city.[3]

What modern city is observing these principles? It's hard to find an example. By contrast, when it comes to spoilage by urban and industrial waste the examples become legion: the Gulf of Mexico, the eastern seaboard of the United States, Venezuela's coast, the North Sea, the Baltic Sea, all of the Mediterranean, the Red Sea, the Persian Gulf, the sea between Thailand, Sumatra and Borneo, the sea enclosed by Japan, Korea and China, and more.

Part, but only part, of the irreparable damage to the vibrant life that used to have its home in rivers, lakes, seas and oceans is the evidence of strip mining the sea. Huge drift nets trap and kill fish and marine mammals indiscriminately. Yet another common effect of man's destructive behaviour is coral bleaching. Many coral reefs from the Caribbean to the Pacific have been hit by a bleaching phenomenon that may be lethal.

Closely associated with the overall misery characterizing water resources (Chapter 12) are dying lakes. Thousands of acidified lakes can no longer support plant or fish life. (The Aral Sea in western Asia has shrunk by more than 80 per cent and has split into smaller lakes.) In addition, there is a global problem of groundwater pollution. Toxic wastes are seeping into aquifers in many locations, wiping out local life.

Another subject the *Time* panel of eminent scientists *did not* address – and a very important one, for that matter – was how to handle the pollutant aspects of *technology*. This is toxic waste at its best. The new materials of the information era – metals and synthetic ceramics used to construct machines and buildings – create toxic waste in much larger proportions than earlier, now obsolete materials.

Technology has not yet developed efficient means to cope with its own waste. Ironically, in early January 1989, when the aforementioned *Time* issue dedicated to planet Earth appeared on the stands, emergency crews rushed to a semiconductor factory in Silicon Valley in response to alarms indicating a leak of potentially lethal arsine gas. Hundreds of workers were evacuated while fire crews wearing oxygen masks entered the contaminated plant. The incident turned out to be relatively minor, but it served as a reminder of the potential dangers faced by people who work for, or live near, factories specializing in the semiconductors miracle.

Although previously widely regarded as clean and pollution-free, the high-technology industry, manufacturing semiconductor chips and other materials, has now come under increasing scrutiny, as highly toxic chemicals are an essential part of it. Arsine and phosphine gases are used to dope silicon in order to produce the electrical properties needed to create semiconductors; hydrofluoric acid is employed to clean wafers; organic

---

3 Jack Weatherford, *Genghis Khan*, Three Rivers Press, New York, 2004.

solvents such as trichloroethylene, which has been linked with cancer in animal tests, are used to remove impurities from the wafers. The semiconductor industry is also a big user of chlorofluorocarbons (CFCs), which are damaging to the earth's ozone layer as we saw in Chapter 10.[4]

Most of these matters were skipped over by the November 1988 *Time* conference, as well as by the Rio de Janeiro, Kyoto, Copenhagen and other environmental tourism gatherings. In particular, the most important – such as those relating to population control – are taboos and therefore untouchable. Yet, without control over all of the critical issues of the earth's ongoing destruction, any effort to reign on the future of this planet will be half-baked.

2. The Holocaust of Forests

President Ronald Reagan once said: 'If you've seen one ... tree, you've seen them all.' Think of a world where only one tree is left, with 9 billion *Homo sapiens* waiting to see it. Even the thought of it is not funny, yet it is no less true that the enemies of forests are numerous and some of them masquerade as 'dear friends' of humanity.

Continents have changed in configuration, glaciers have advanced and retreated, seas have been elevated and mountains submerged, because of the forces of nature. But, with forests, the primary force taking them off the map is man's irresponsibility which upsets nature's delicate balance.

Destroying tropical forests at the rate of millions of acres annually exacerbates the greenhouse effect. This annual sweep includes tropical and other rainforests, while plenty of forests are also dying from acid rain and burned by fires – as in Spain, Italy, Greece, Australia and elsewhere.

Forests are cut down to make place for housing, as *Homo sapiens* continues to multiply; others are erased for logging, a case being the beautiful rainforests off British Columbia, which have become a memory; still others are turned into grazing grounds. There is a guesstimated 1.4 billion cattle spread across the earth's continents, occupying nearly a quarter of the total land space and consuming about a third of the world's grain harvest.

As the careful reader will recall, this passion for beef used to characterize the developed countries. Today, however, it has become status symbol to the rising middle classes of developing nations, which means a couple of billion people. Cattle raised to satisfy such a huge market are stripping away vegetation, destroying natural habitat, eroding soil and contributing to desert-like conditions in Asia, Africa and the Americas.

Forests and oceans are the lungs of the earth. Their steady downsizing, combined with pollution of the water (Chapter 12), is among the principal reasons for the annual diminution of the volume of oxygen in the atmosphere by an estimated 10 billion tons. Curiously enough, however, saving the forests has only been very timidly discussed at *Fiascopenhague* (December 2009), two decades after environmental tourism started.

Yet, the evidence has been present for many decades. According to experts, nearly 20 per cent of the Amazon rainforest has been cleared for cattle-crazing, despite unsuitable soil conditions. According to a 2009 World Bank study, the loss of another 2 per cent

4 In the domain of unexpected consequences also comes polybrominated diphenyl ethers (PBDEs), a 30-year-old flame-retardant chemical widely used in furniture, carpeting, bedding, textiles, plastics and electronics.

Amazon forests could start to trigger dieback in the relatively dry southern and south-eastern sections.[5] If a global temperature increase is added to this, the joint effect may well be to release the equivalent of 50 gigatonnes of carbon into the atmosphere.

The obsession with cattle is also leading to water shortages. Nearly half the water consumed in the United States goes to grow feed for cattle and other livestock.[6]

Deforestation to raise cattle, expand the big cities' slums or for other man-made 'initiatives' is what interests us in this chapter. To continue with the previous example, the vast Amazon region generates half its own moisture, and deforestation would have dramatic effects on this giant rain-producing machine. In addition, the Amazon basin contains around 1 million known insect species and more than half of the earth's 10 million animal and plant species, which will be exterminated as the forest disappears.

- For several decades the Brazilian government did almost nothing to stop the extinction of Amazon rainforest.
- Grand plans to halt its destruction have come and gone over the years, with little or no success because of political instability and lack of will.

It was only in 2009 that Brazil's federal government seemed to come to recognize the importance of stopping Amazon rainforest's slide into oblivion, and even that is not certain. Hopes rest on a new plan, bringing together state governors and federal officials, and aiming to regularize the titles to 80 per cent of the private land holdings in Brazilian Amazonia by 2010 to 2012. It is too early yet to talk of expected results.

Critics say that even if some rules come out of it, enforcing them will be difficult because Brazilian governments are not renowned as implementers of rules which are far-sighted but unpopular. Despite its pronouncements, even the government of Luiz Inácio Lula da Silva has shown an indulgent attitude to violations of property rights, let alone:

- the right of trees in the Amazon to continue being upright, and
- the right of an incredible variety of species living on them to continue their existence.

There are also Indian tribes in risk of extermination as their traditional habitat in the Amazon rainforest is burned to the ground. The Ticuna (who have a reservation at Novo Paraiso near Brazil's borders with Colombia and Peru) had their first encounter with environmental plunderers when they had the misfortune to encounter Portuguese raiders centuries ago. Then came the robber barons, as rubber drew the Amazon into the list of hinterlands that could be profitably exploited.

- Other invaders followed the early raiders, and
- the latest breed are the beef industry's farmers and their political friends.

Politicians could not care less about the after-effects of the invaders on the rainforest. Supposedly to improve the lives of Brazilians living in the Amazon, the government has

---

5 *The Economist*, 25 September 2010.

6 Cattle also account for a sizeable portion of all methane emissions, thus contributing directly to global warming, under the hypothesis that this is a man-made effect.

come up with a policy known as Plano Amazônìa. It envisages an expansion of road-building in the forest, along with big hydroelectric projects. These are loathed by people who want to preserve what remains of the trees.

- Theoretically, Plano Amazônìa also contains measures to slow deforestation.
- In practice, everyone knows that these are just the icing on the cake, and they will be hard to enforce even *if* there was a political will to do so.

Nor is the Amazon an isolated case. According to several studies, human activity worldwide is causing more than one species per hour to become extinct. It is not just 'burn trees and you get carbon dioxide', or 'the forests that are cleared are no longer breathing in carbon dioxide and converting it back to wood'. We are talking about obliteration of other species because of man's irresponsibility.

In Africa, deforestation may partly account for shorter rainy seasons in the western part of the continent. Even where vegetation is relatively sparse, the removal of bushes, hedges, shrubs and other ground cover affects local weather. Some scientists suggest that, once exposed, land reflects more sunlight and sets in motion atmospheric processes that reduce rainfall by drawing dryer air into a given area.

Africans should have been aware of these facts before initiating the policy of decimating their forests and turning parts of their countryside into a lunar landscape. After all, they and their families depend for a living on the nature surrounding their homes. Therefore, it was appalling to hear in *Fiascopenhague*'s December 2009 meetings that Africans have been asking for money *not* to cut down *their* forests. (Several years ago, Brazilians were making similar noises to allow the survival of rainforest and species in the Amazon.) This request by the Africans resulted in one of the worst acts of what became known as the Copenhagen circus (Chapter 14).

- To put it mildly, it is repulsive to pay people not to burn down their own habitat, and
- burning, cutting and depleting forests is an act of barbarism worse than the 'professional' pirating of ships by the Somalis.

People who destroy their habitat through deforestation and desertification (section 1) should be taught that the absence of vegetation hardens the soil and cuts back the amount of cloud that returns the nourishing moisture that has evaporated from the land. As an example, such processes have arguably contributed to the 10 per cent drop in rainfall over Florida during the past 30 years. Letting this happen in the country in which you are living is sheer lunacy.

3. Game Slaughtered by Words Cannot be Skinned

The Mongols have a saying: 'Game slaughtered by words cannot be skinned.' Game slaughtered by words, and only by words, is exactly what the different heads of state and the United Nations top brass are doing about protecting the earth's environment, its forests, rivers, lakes and oceans: words, words, and words.

Fiascopenhague had set the end of January 2010 as the deadline for firm commitments by the over 100 UN member states who took part in the circus. When this deadline passed, it was said that there were plenty of agreements with the vague principles, but firm quantitative commitments in reducing CO_2 and the other GHGs (Chapter 10) were as rare as angels.

In short, the participants to the Copenhagen circus did exactly what they were expected to do: nothing! They did not square up to the need for salvaging the earth's assets, at least those that are still salvageable. Carbon dioxide emissions have nothing to fear from environmental tourism and its participants.[7] By contrast, forests have plenty to lose, because:

- their survival depends on global support and *Homo sapiens*' ability to control their worst habits, and
- this control is not forthcoming, because none of the parading dignitaries has learned that a leader should demonstrate his thoughts and opinions through his actions, not through his words.

The Amazon rainforest and Africa's tropical forests (section 2) are by no means the only areas in the world where rampant deforestation is having its heyday. Somehow, governments looking for rapid cash fail to appreciate that deforestation is rarely in the economic interest of the country itself, its people and its governors. Practical evidence suggests that this lack of responsibility is due to a combination of:

- bad policies, and
- rapid population growth which promotes poverty, *not* a higher quality of life.

In some parts of the world, like the highlands of Bolivia, Peru and Nepal, and in the countryside surrounding many fast-growing cities in Africa, trees are lost because people use wood for fuel and governments don't care if whole forests go up in smoke. (A basket-case is Haiti, the now bare western part of formerly forest-covered Hispaniola.)

Commercial logging is also a big cause of deforestation. Demand for industrial timber has significantly increased and is expected to accelerate. Developing countries with hot climates consider that they have a competitive advantage in this market, simply because trees grow much faster than in temperate climates – but they fail to think about the long-term damage they inflict upon themselves by rapid deforestation. Nor do they care to cultivate trees for commerce. Yet, plantation forestry can be as profitable as chopping trees from virgin forest, and it has the advantage that growers can choose which species to cultivate. This leads to the opinion that the main reason why virgin forests are being cut down is a set of perverse economic incentives, including personal greed by those in charge of governments. It is as if:

- today everyone milks the forest for what he can get, and
- tomorrow everyone will pay for it in terms of CO_2 poisoning and the extermination of small game and other species making up the ecosystem.

---

7 Who enjoy themselves at taxpayers' expense.

In Asia-Pacific, for example, swathes of moist tropical forest continue to be cleared. Studies by the FAO suggest that Indonesia alone loses 2 per cent of its forests every year, representing an area of 19,000 square kilometres – bigger than the state of Connecticut.[8]

Loggers wrench out the most valuable hardwoods that grow there – the dipterocarp family which reaches up to 50 metres high and provides a protective canopy for the plants and animals below. They leave behind a shadeless so-called 'secondary' forest, which dries out and becomes vulnerable to fire, releasing huge amounts of carbon dioxide.[9] Also:

- Many blazes are set deliberately in a two-stage drama.
- Loggers move in first to chop the tall trees, followed by farmers who burn the rest to clear the land for planting.

Indeed, deforestation in Indonesia caught the world's attention in mid-1998 as the dry season spread smoke across much of south-east Asia, producing a catastrophic smog. But by then, however, the decimation of Indonesia had already been going on for more than 20 years. As in Brazil, the government had an official programme encouraging millions of people to move from the crowded islands of Java and Bali to less densely populated, but heavily forested, islands such as Kalimantan, Sumatra and Irian Jaya. Settling them there became an ecological catastrophe.

Also as in Brazil, property rights in Indonesia's forests are often ill-defined, leading to violent conflicts between locals, migrants and forestry firms. But whereas Brazil seems to have abandoned government policies that explicitly encourage deforestation, Indonesia lags behind.

In addition, there is conflict of interest. Over several decades the Indonesian government has levied high export taxes on unprocessed logs to help the domestic wood industry. This has kept domestic timber prices below world levels, providing forestry firms with an implicit subsidy estimated at over $2 billion a year, but also encouraging them to use logs inefficiently.

While Indonesia's intensive logging of its rainforests has long been identified as a big contributor to the world's emissions of greenhouse gases, more recently attention has also turned to the soil beneath the trees, and especially to peat (see also Chapter 10). According to some estimates:

- the top two metres of soil contain three times as much carbon as the entire vegetation on the planet, and
- soil degradation, such as the burning of peatland, is the main cause of Indonesia's high level of CO_2 emissions.

Other Asian countries, too, like Myanmar and Cambodia, are losing tree cover at high rates. Particularly damaging is illegal logging, often carried out by 'untouchable' cronies of politicians and other protectors. This looks like being unstoppable, because it continues to be a profitable business.

8 'Falling Here, Rising There', *The Economist World in 2008*, January 2008.

9 Statistics on the extent of illegal logging in Indonesia's forests are not exactly reliable. By one estimate, 73 per cent of the $1.6 billion-worth of forest products imported by the European Union from Indonesia in 2008 came from illegally felled timber. *The Economist*, 12 September 2009.

None of what has been said in the preceding paragraphs means that people in Western countries are not plunderers of forests and of the wildlife living in them. According to the US Center for Biological Diversity, there were 283 endangered species in the United States in 2005.[10] Despite various state and federal forms of protection, two-thirds of these species continue to decline.

Back in 1973 California's legislature passed a law to assure the sustainable management of its forests. However:

- timber companies have replanted new species instead of replenishing existing forests, and
- have cut too often to permit the forests to regenerate themselves.

That attitude, which also prevails in many other areas of the globe, has sparked increasingly bitter court battles and legislative debates between the forces of conservation and the defenders of the logging industry. Environmental activists have rightly demanded protection for the northern goshawk in Arizona and New Mexico, and the grizzly bear in the northern Rockies. In Alaska they have forced the government to re-examine timber sales from the wild Tongass National Forest.

Here is precisely where there is a major difference between developed and developing countries. In the West there are laws which have to be respected. When this is not the case, it leads to public action and it is brought to the voters – not only to the courts. In 1992 George Bush said that in a second term in White House he would not be bothered about the spotted owl – and he was voted out of office.

4. Man's Abuse of the Oceans

The holocaust of forests by *Homo sapiens* is matched by their abuse of rivers, lakes, oceans and other water resources. The ever-larger volume of carbon emissions does not stay in the air forever. Roughly half of it contributes to the greenhouse effect. Most of the other half is absorbed by the oceans where it forms carbonic acid.

This increase in acidity of seawater does not only affect individual species on which man depended for his Blue Revolution (Chapter 7). It destroys the whole food chain, starting with the destruction of the plankton and proceeding to species feeding on it, which in turn pass the poisoning on to their predators. Acidification is making life difficult, if not impossible, for marine organisms with calcium-carbonate skeletons or shells – from krill to crabs.

The krill plays a critical part in the food chain, as it is a staple food for other marine animals, and passing the poison to higher levels destabilizes the entire ecosystem. Acidification also destroys the coral reefs which are already suffering from overheating and pollution. The consequences can be dramatic as coral reefs are home to a quarter of all marine species. Massive amounts of man-made CO_2 also affect seas and oceans in other ways, particularly through global warming.

- The oceans expand as they warm up.

10 Press Release, Center of Biological Diversity, Washington, DC, 8 November 2005.

- They are further swollen by melting glaciers, ice caps and ice sheets.
- Swathes of red algae blooms, plagues of jellyfish and dead zones appear, in which only simple organisms can thrive.

This unprecedented destruction of the oceans is associated with the growing number and intensity of stresses that *Homo sapiens* inflicts on the entire marine ecosystem. Rivers, lakes, seas and all other waterways, such as canals, are steadily used as waste baskets – while fertilizers from the land drain into rivers and estuaries and find their way to the oceans where they poison the fine balance of life.

As if chemical poisoning was not enough, the use of rivers, seas and oceans as waste baskets has led to a mass of discarded plastic that swirls round in clots. Two such clots have been observed in the Pacific and, reportedly, each is as large as the United States. The way an Australian cartoon had it, floating containers, all sorts of plastic, tyres, drums and every other sort of refuse have become synonymous with civilization.[11] Man is destroying his own habitat and that of other species by throwing all this rubbish on to the land and into the seas – from sewage to toxic waste and from rubber tyres to plastic packaging, bunker oil, heavy metals and other pollutants. These bad habits have become global, and there is no evidence that they will cease.

In 1989 the *Exxon Valdez* dumped 11 million US gallons of oil into Alaska's pristine Prince William Sound. This is only one out of many oil spills caused by negligence, unseaworthy tankers or both (more on the damage caused by oil spills later).

If tanker accidents happen are only occasional, the discharge of all sort of waste by big cities is steady and even more damaging. In New York, where the Hudson River empties into Hudson Bay, pollution is propagated and sustained by daily human and industrial discharge. Other areas of the East Coast tell a similar story. Lobsters and crabs off Rhode Island have been exhibiting, and continue to exhibit, marks resembling burns (as they do in Europe's coastal areas).

- Throughout New England, the southern states, the Gulf of Mexico and Pacific coast more than a third of ocean acreage is closed to commercial shellfish harvesting.
- Pesticide contamination and oxygen depletion are rampant all along the coast of California, the Gulf of Mexico and several states on the eastern seaboard.
- High levels of toxic chemicals can be found off the coasts of Maine, New Hampshire, New Jersey, Oregon and Washington State as well as in Boston harbour, Long Island Sound, Galveston Bay, Santa Monica Bay, San Diego Bay and San Francisco Bay.

In northern Europe the Elbe, one of Germany's main rivers, is full of chemicals. As a popular joke puts it, 'Elbe's waste is an elixir of Life. Drink it and you never become old.' In southern Europe ecological damage makes the Black Sea bleak. For more than two decades biologists have been worrying about the growing tide of human, farm and industrial discharge pouring in from the six Black Sea countries: Turkey, Bulgaria, Romania, Ukraine Russia and Georgia. Surveys show that every town, industry and mine along its shores uses this land-locked formerly pristine sea as a dump and sewer.

11 *The Bulletin*, 20 November 1990.

In antiquity, the Black Sea produced a wealth of marine food. Like the Caspian Sea, the Black Sea (which used to be a sweet water[12] lake) was renowned for its caviar, and many people thought this would never end. It did, as the sea, which is also a highway between Europe and Asia, is suffering extensive damage. Poison comes from more than 60 rivers and streams delivering waste from a catchment area of 160 million people.

- The larger rivers like the Danube, Don, Dnieper and Dniester arrive here after sweeping through some of the most polluted areas of the European continent.
- Every year these rivers carry an increased amount of toxic materials including pesticides, oil, lead, nitrates, chromium, cadmium, phosphorous and more.

This has led to major changes for the worse in the ecosystem, including the collapse of fisheries. But poison is not the only enemy of marine life. Overfishing is another. This is by no means only true of the Black Sea but also of the:

- Mediterranean,
- north-west Pacific,
- south-east Pacific,
- eastern Indian Ocean, and so on.

In many countries coastal pollution is exacerbated by hydrocarbons, many of which are carcinogenic. Those found in asphalt and tar are often released by changing weather conditions which break down the pollutants contained in them. High temperatures contribute to this process, causing toxins to seemingly boil up from the roadways.

Another pollutant, whose after-effect is rarely discussed, is friction from car tyres which loosens the tar and asphalt deposits on the surface, combining them with additional surface pollutants from sources like car exhausts and road dust. Tiny particles are washed into the soil, and nearby waterways by this polluted liquid known as 'urban run-off'.

- Carcinogenic particles settle into soil and sediment where they are absorbed by plants and eaten by fish, hence entering the food chain.
- At higher temperatures carcinogenic particles adopt a gaseous form, finding their way into soil, streams and rivers through air and rain – and from there to seas and oceans.

Shipping itself is a major cause of pollution. The International Maritime Organization says that sea transport accounts for less than 3 per cent of total emissions. But an unpublished report by the Intergovernmental Panel on Climate Change puts that figure at nearly 4.5 per cent, about twice as much as the share of air transport.

While shipping is rather in decline, by 2020 emissions are expected to rise because ships burn 'bunker oil', a dirty fuel considered to be responsible for about 60,000 deaths from chest and lung diseases, including cancer, year on year. It is interesting that most of these occur near coastlines.

Accidents involving tankers in the open seas and in sounds are not the only ones contributing to pollution through oil spills. Russia has more than 1.5 million kilometres

---

12 Low salinity.

of gas and oil pipelines, many of them poorly maintained. Every year, up to a fifth of the country's total oil production is lost, much of it through leakage. The spills are occasionally cleaned up, but all too often they pollute the waterways that drain into seas and oceans.

5. Man's Destruction of Antarctic Wilderness

The systematic pollution of rivers, lakes, seas and oceans imposes a great cost on the next generations of *Homo sapiens* because much of it is irreversible. It will take not just a discontinuation of pollution, but also tens of thousands of years for ocean chemistry to return to a condition similar to its state prior to the Industrial Revolution.

In addition, many scientists fear that some changes are reaching thresholds after which:

- further degradation may gain a critical mass, and
- from that point on accelerate uncontrollably.

Pristine areas are as vulnerable as those that have already gone beyond repair. A case in point is the destruction of the ecosystem of Antarctica, the only continent which still remains reasonably unscathed by man's destructive acts, although it suffers from a growing ozone hole (Chapter 10).

At the Antarctic Station of Uruguay, waste water and sewage are stored in underground tanks, presumably forever. There is no evidence available about disposal plans, but there is an open-air dump which contains metals, glass, plastics, drums, timber and other waste.

The sinking of the Argentinean supply ship *Bahia Paraiso* off the Antarctic Peninsula and its subsequent 170,000-gallon oil spill in 1989 caused the death of many penguins and seabirds. Four weeks after that accident, the same area was menaced by a new oil spill from the Peruvian research vessel *Humboldt* that went aground off King George Island. A Chilean navy spokesman said at the time that there was a 140-yard oil slick floating towards the coast.[13]

At the Bellinghausen Station, belonging to the former Soviet Union, sewage was typically dumped into the sea. Waste was piled near the base, while wood and paper were burned in open fires. In 1997, the rubbish was finally cleared; it totalled 1,500 tons.

At Palmer Station of the United States, sewage used to be discharged directly into the sea along with metals, plastics and even laboratory chemicals. Britain's 'G' Base has been officially designated as unoccupied; in practice, this apparently means abandoned with no clean-up. Items left behind include equipment, buildings and fuel.

All this man-made destruction took place while Antarctica was simultaneously no man's land and everyone's land. Drawn up in 1959, the Antarctic Treaty, which went into effect in 1961, did not even attempt to resolve the often conflicting claims of seven nations (at the time) that staked out territory on the continent, or the dozen countries with scientific stations.

Some countries like Argentina and Chile considered their projection into the Antarctic continent as *their* territory. The Antarctic Treaty's did not challenge such pseudo-statutory

13 'Peruvian Ship Runs Aground in the Antarctic and Spills Oil', *The New York Times*, 1 March 1989.

rights. Its aim was much more limited: to create a framework in which the so-called consultative parties, engaged in what was (incorrectly) labelled 'substantial scientific research activity', would meet periodically to sort out issues about Antarctica's future.

- This has been a confirmation of the status quo, but hardly a blueprint for action, and
- while these meetings attracted little interest as models of international cooperation, they became horse-trading sessions.

The bureaucrats and diplomats who attended them were largely engaging in pseudo-scientific tourism under the pretext of discussing claims on sovereign rights for future exploitation of Antarctica's riches – which continue unabated to this day.

For ordinary people, a trip to Antarctica became the 'in' thing. By sea and air 'exotic' Antarctic tourism thrived. In the mid-1970s Qantas and Air New Zealand offered half-day overflights of the seventh continent. Flying extremely low over the icy landscape, the airplanes offered spectacular rides, but they also poured fumes into the pure air and posed an ever-present risk of a crash.

These flights ended in November 1979 after a DC-10 smashed into Mount Erebus, killing all 257 people aboard. Seaborne tourism, however, continued and, before too long, it became a permanent feature of a landscape full of penguins – despite on-and-off discussions of banning it.[14]

When mining starts, which some experts consider unavoidable, Antarctica will become like any other continent. Theoretically, the aforementioned Antarctic Treaty prohibits mining and oil-drilling unless virtually all the Treaty signatories agree. This leaves the door open to wide interpretations guided by conflicts of interest.

Indeed, mining is currently being contemplated by several of the original 'consultative' parties, and (apart many other negatives) it would inevitably create quantities of dust and waste that would be carried by the winds over a large area. The ice would grow darker, absorb more heat and melt. The level of the oceans could rise, perhaps enough to inundate coastal cities around the world. Global warming might do this eventually, but at slower pace.

Even prospecting for oil deposits would be disastrous. Geologists normally use explosive charges and seismic refraction to find potential oil deposits. On ice under stress, an explosion could engineer the dislocation of a substantial part of the ice sheet. Plunging into the sea, that ice could provoke enormous tidal waves.

- From an environmental standpoint there would be untold casualties.
- But this does not seem to be enough to discourage the would-be plunderers, nor are the UN events on 'environmental protection' concerned with Antarctica.

Back in December 1990 the self-appointed custodians of the world's last great wilderness met to develop a new charter. Anecdotal evidence, however, suggested that

14 To govern general behaviour there are a 'Code of Conduct on Waste Disposal', 'Measures to Regulate Antarctic Tourism' and 'Recommendations on Environmental Impact Assessment'. All these are nothing more than a few sheets of paper.

they were salivating when considering Antarctica's hidden resources. It was reported at the time that a formula for a 50-year moratorium was blocked by Japan.[15]

A curious position was taken by the Bush administration, with a US State Department official saying that the US government had decided not to submit the minerals convention to the Senate for ratification because of objections to it. The United States was supposed to propose a long-term moratorium on mineral activities – but it opposed a total ban because future scientists might discover 'some rare mineral that could be extracted without damage to the environment'.[16] (No kidding!)

Precisely because the preservation of natural resources was not really part of the signatories' agenda, opponents of the convention argue that it is making even easier to kill Antarctica – and this is in contradiction of the interests of humanity as a whole. The interest in preserving Antarctic wilderness nearly intact derives from the fact that this most isolated corner of the globe is one of the truly vital components in the earth's climatic machinery.

- It contains some 90 per cent of the world's natural ice, and
- this endowment represents most of the planet's freshwater reserves.

The ice at its deepest is 4.7 kilometres thick. At the waterline, Antarctica sloughs off enormous, flat icebergs, the biggest of them roughly the size of Belgium. Yet across most of its land mass there is rarely more precipitation than there is in the Sahara. The ice pack is the accumulation of light snowfalls from the last 100,000 years.

In conclusion, whether we talk of the forests, the oceans or Antarctica, the need to conserve the earth's resources is plain. The remedies are not hard to grasp – at the top of the list is keeping man's lust and greed under lock and key. Politicians, however, lack backbone. Few of them are prepared to stand up to powerful lobbies. Yet they know very well that mass extinctions cannot be ignored – including that of *Homo sapiens*.

6. The Earth's Disappearing Land Assets

According to some estimates, the destruction of the earth's soil has already eliminated 10 billion hectares of fertile and possibly fertile land. This disregard for, and squandering of, a precarious asset like soil runs contrary to humankind's own interest and its chances of survival. It also speaks volumes about current ethics.

Since remote antiquity, the science of geography has had its own built-in morality and ethics. This was based on an assumption that the earth harbours a force for life that should not only be appreciated, but also become a pivotal point of humankind's behaviour. Today, the issue of protecting the earth's soil causes great acrimony and tension between:

- those who foresee the catastrophic consequences of current policies, and
- the plunderers who could not care less about poisoning the soil, exterminating the forests and polluting seas and oceans.

15 *International Herald Tribune*, 6 September 1990.

16 Ibid.

Many people are imprudent with natural resources because they do not have an understanding of basic concepts of physical and cultural geography. For example, it came to my attention that, in a high-school test, more than one student out of five responded that removal of vegetation would *increase* wildlife populations.

- This crazy answer has much more to do with IQ than with a disrespect for nature.
- The real reasons for the growing rate of degradation are much deeper, and (most unfortunately) much more lasting.

Take soil erosion as a first example. Over the ages, soil erosion, as a natural process, has formed some of the world's richest river valleys. This has taken millennia. But with agriculture, even with relatively careful agricultural practices, plenty of topsoil is lost every year. The difference is that with sound practices what is lost is at about the same rate, or slightly more, than that which is created. By contrast, the widely practised misuse of croplands accelerates erosion which:

- diminishes the soil's ability to retain water,
- removes important nutrients,
- stunts plant growth, and
- destroys the topsoil well beyond the earth's ability to repair the damage.

All that happens in one go. Subsequently, eroded soil blown away by wind or carried off by water cripples the land's long-term productivity and harms waterways. This is happening all over the globe, but most particularly in developing countries where erosion is acute, leading to desertification which constitutes another form of land degradation.

As briefly discussed in section 1, *desertification* is the loss of productive land for farming and grazing in dry regions of the world.[17] It is not always appreciated that the cause for desertification is not climate change but excessive stresses like overgrazing. This hardens the land and destroys its ability to absorb water. The end result is a soil that is:

- bone-dry,
- devoid of vegetation, and
- virtually useless for agricultural activities.

It goes without saying that this is a perilous prospect for humanity. Today, desertification threatens about one-third of the earth's land. According to the United Nations Environment Programme, the livelihoods of more than 850 million people are being affected by it, but it is severe damage of a self-inflicted kind.

Land degradation, another scourge, is a result of both misuse and overuse of soil. As we have already seen in section 2, it can happen when large areas of trees are cleared to make way for farmland. But it also takes place when areas of land are sown year after year with only one type of crop. Therefore, this, too, is self-inflicted damage. As a rule:

- the more the *Homo sapiens* on spaceship Earth,

17 Some years ago, a UN study estimated that in Africa 40 per cent of the world's fertile land was at risk of desertification.

- the greater the damage they inflict, not only because of lust and greed but also because of a 'don't care less' attitude and lack of conscience.

In many developing countries farmers think that once the soil is worn out in one area, they can move on to better conditions elsewhere. That is just one outcome of the population explosion (see Chapter 13) – developing countries fast running out of suitable farmland and their governments looking the other way as the people destroy the land in a desperate effort to survive.

Desertification and deforestation can be seen as twins. In some areas of the world – for example, in parts of Sudan – what was previously forested land is now desert, unable to be used for anything. Here and there, trying to restore some sense of balance, local groups have planted windbreaks and grasses. But this effort to retain what little soil there exists is not systemic – and it is therefore doomed to failure.

There are also other ways in which land degradation manifests itself – for instance, nutrient depletion. Salinity, which happens when trees, which normally soak up much of the rainwater, are cleared, allows increased amounts of rainwater to soak into the ground. The result is a rise in the water table (more on this later).

To compensate for lack of nutrients, and quite often without a real need to do so in a massive way, people use fertilizers which eventually poison streams, rivers and lakes, asphyxiating local fish. They also employ a vast panoply of pesticides in an effort to protect the crops and increase land productivity.

As will be recalled from Chapter 7, no part of the globe has escaped the reach of fertilizers and pesticides – nor their severe after-effects. In practically every country in the world there are testaments to how improper disposal of hazardous chemicals can harm farmers and their communities, not only the environment.

Pesticides containing DDT (which after World War II was a wonder chemical, but is now banned in many countries) and mercury compounds have created dangerous poisonous concentrations found in plants, birds, fish, and even mothers' milk. With plenty of reason, people who care for the protection of the soil, water resources and health standards are demanding rules that would force farmers to dramatically cut the use of chemicals. They also want growers to:

- draw up plans for the future use of agricultural chemicals, and
- make public records of the different pesticides being employed, banning the most destructive.

Politicians whose re-election depends on farmers' votes, however, are not ready to oblige. True enough, particularly in developed countries, some farmers already have amended their practices to protect the land and reduce chemical use. But environmentalists say that such efforts are not enough. They are asking why agriculture should not begin to do what every other industry has started – that is:

- cut pollution, and
- protect the environment.

That 'every other industry' does so is, by any measure, an overstatement. But the fact that the destruction of the topsoil, desertification and the wide use of pesticides and fertilizers should be subjected to careful scrutiny and rethinking is one of the most urgent issues confronting mankind.[18]

So far, an imaginative approach to this very serious problem has not been found. While different ideas float around and discussions go on without clear winners, other challenges concerning soil preservation are showing up as well. One of those not generally expected is irrigation.

For millennia irrigation has been considered to be a very positive practice. It has transformed much arid land into fertile soil and increased crop yields in marginal tracts. Less appreciated is the fact that it can also pose a threat. There is plenty of evidence that poorly managed irrigation systems actually reduce yields over the long term by:

- waterlogging the soil, and
- building up deposits of salt and other minerals in the topsoil.

Salinity, of which we have briefly spoken, seriously lowers productivity on about 7 per cent of the world's irrigated land. In South America, for instance, an estimated one-third to one-half of all irrigated land has been harmed by salinization. In the United States, 20–25 per cent of all irrigated land, much of it in the south-west, suffers from salt-caused reductions in crop yields. In parts of the European Union damage due to salinity is not that different.

In conclusion, the domain of Goddess Demeter – more precisely the earth's capacity to produce food and sustain life – is being seriously jeopardized by man's activities. Practices that foster soil erosion, deforestation, desertification, salinity, the pollution of waterways and lakes or coastal areas, as well as disposal of vast amounts of waste (section 7) are depriving *Homo sapiens* of their living space – and this cannot continue forever.

7. Man-made Mountains of Waste as Tall as Mount Everest[19]

Immediately after World War II the story was heard in Europe that the amount of food an American city threw away could feed a European city. Times have changed. Today, the food thrown away by an American or European city could feed a city in a developing country. While the United States still holds the upper ground in the mountain of waste, *waste* has become a scourge of the earth worldwide.

According to the US Environmental Protection Agency (EPA) more than 6 billion tons of domestic, commercial and industrial waste are produced each year in the United States – a huge amount which threatens not only the environment, but also public health. Each year the United States generates over 200 million tons of municipal solid waste alone. This roughly corresponds to two kilos per day per person of which roughly:

- 40 per cent is paper,
- 18 per cent is yard waste, and

18 The point has been already made that an alternative to wide use of pesticides is genetically engineered grains.

19 Most of the examples in this section are 20 years old or more. Their inclusion in the text is deliberate, aiming to demonstrate that though many challenges to the environment are well known, noithing has been done to solve these environmental problems.

- the rest consists of food, plastics, glass, metals and other materials.

On average, each European produces 550 kilos of garbage per year, which corresponds to 1.5 kilos per day, including Sundays and holidays.[20] That's not much better than two kilos per day, and it is a terrifying statistic.

All this trash has to go somewhere. In the United States roughly 80 per cent of it is put into landfill, 10 per cent is incinerated and 10 per cent is recycled. Landfilling poisons the soil and waterways. Moreover, like their European counterparts American municipalities are running out of places to continue with such practice.

In the European Union most of the member states and their municipalities simply don't have the room to dispose their mountain of garbage in 'the old way'. But even big land masses, like the United States and China, do not allocate endless space to dispose of their waste. Therefore, all nations should consider drastic measures to manage this phenomenal growth in domestic waste.

As should be expected, opposition to the gas given off by landfill sites has grown among local residents who do not want to smell the acrid fumes of society's discarded and rotting waste. Apart from considerations of raping the countryside, the opening of new dumping sites has been hampered by tougher government regulations in the West:

- correctly aimed at protecting the environment, but
- providing no solutions other than incineration, which entails problems of gas emissions among others.

Some local communities are beginning to make waste a political issue. They are launching mandatory recycling programmes to force residents to separate paper, glass and metals. Shipping of garbage to other, poorer nations is neither ethical nor easy, because of the growing unwillingness of less developed countries to become the dumping grounds for richer nations.

In reality, no country is free from the environmental nightmare created by waste problems. In Soviet times, in Siberia's Lake Baikal basin, together with the Ministry of Timber, Pulp and Paper, the wood processing industry erected the Baikalsh pulp factory. This managed to pollute the majestic lake which holds 80 per cent of the country's freshwater and 20 per cent of the world's supply. It was also a disaster for marine life.[21]

Siberia is laced with wastes from steel, chemical and coal industries. Dead sturgeon float atop the polluted Volga River, threatening Russia's caviar supply – a big foreign-exchange earner. As of 2009, resorts along the Black Sea banned swimming after the Ukraine government's warning that the waters are contaminated with dysentery and typhoid germs.

With the rush towards urbanization, in many developing countries individuals and families live amidst the garbage pile, and they often feed on it. Rubbish dumps are now supporting a lot of destitute people who forage for food and saleable goods that have been thrown away. Occasionally on the TV bulletins we see hungry people in Manila fighting among themselves to access the waste freshly delivered to a dump.

---

20 *Bloomberg News*, 9 April 2008.

21 Ibid.

But living amid piles of waste is not just a characteristic of less developed countries. The throwaway society of Naples provides an EU example which made headlines in 2009. In Campania, the province of the Italian South, which includes Naples, each person produces somewhat below the average European amount of garbage – roughly 1.3 kilos per day – but:

- for years a big part of the daily production over of 5,000 tons of garbage produced by 4 million people piled up in the streets of Naples, and
- when the crisis came, 60,000 tons of garbage had accumulated in parking places, along the road and in depots, massively reducing the citizens' quality of life.

Successive Italian governments, as well as city administrations, have been unwilling, unable or both to solve the garbage problem. So, it has lingered on and rumour has it that the local Mafia have contributed to the deadlock because it has traditionally made profits from garbage disposal and is afraid of losing its curious privileges.

The local authorities feared that reopening the garbage dump of Chiaino, on the outskirts of Naples, would bring social revolution because the local citizens were opposed to it and would fight rather than live with it. The showdown was supposed to happen on the weekend of 25 May 2008 as garbage trucks were scheduled to pass the last kilometre from the city to the dump.

Molotov cocktails with inflammable chemicals holding glue and petrol were thrown at the policemen along with stones and insults. But the Berlusconi government stuck to its guns, refused to negotiate with the insurgents (or the Mafia) and called in the army. Suddenly the resistance fizzled; despite the Mafia's *agents provocateurs*, the citizens in revolt suddenly discovered that it was better to go back home than become statistics.

There exists a long list of concerns regarding environmental pollution from waste. For instance, only 5–15 per cent of municipal landfills in Western nations have the equipment necessary to prevent contamination of water supplies.

In the United States the EPA says that fewer than 30 per cent of its landfill sites have some type of groundwater monitoring system (EU statistics on this issue are missing, but anecdotal evidence suggests that the percentage is much lower). Indeed, the EPA has identified 30,000 sites holding the potential to release hazardous substances that are toxic, flammable, corrosive or reactive. A significant proportion of these substances is often carelessly discarded and enters water supplies – a scenario just as valid in the European Union.

In addition, some 5–6 million underground storage tanks in the United States contain petroleum products or hazardous chemicals, and the EPA estimates that 2 million of them may be leaking. To make matters worse, the US government, too, has no model of environmental protection.

As an example, from 1946 to 1970, the US government dumped some 47,500 barrels of nuclear and industrial waste in or next to the Gulf of Farallones National Marine Sanctuary, just off the coast of California. By 1990 these 55-gallon barrels had corroded and burst, releasing plutonium and cesium into a sanctuary that supports the largest population of seabirds south of Alaska and an abundance of fish, invertebrates and marine mammals.[22]

22 Prudential Bache Securities, *Strategy Weekly*, 13 June 1990.

An article in the October 1989 issue of the *Bulletin of Atomic Scientists* started this way:

> *The jackrabbits, burrowing owl, and rattlesnakes that inhabit the arid expanses of the Hanford Nuclear Reservation are radioactive ... So are many of the shoreline and underground springs discharging groundwater into the mighty Columbia River that bends around the northeast perimeter of this desolate, guarded place in eastern Washington.*

The article noted that:

> *... deadly wastes from the plutonium production and extraction processes were dumped into soil and trenches or poured into million-gallon steel tanks. Some of the tanks cracked and leaked after the liquids were mixed with chemicals in an effort to neutralize the most radioactive materials.*

Critics said that neither the government nor business firms cared about the management of waste because it cost money.

Japan provides a more positive example of waste control. An estimated 50 per cent of wastepaper, 55 per cent of glass bottles and 66 per cent of beverage and food cans are recycled. Much of the remaining garbage is turned into fertilizers, fuel gases and recovered metals.

Along the same frame of reference, in the United States, too, millions are paid out every year to recyclers who collect billions of aluminum cans. The economic value of recycling has been nicknamed mining 'urban ore', as contrasted to extracting more of the earth's raw materials.[23] But with all recycling there is a hitch.

On paper, producing a ton of paper from recycled fibre saves some 3,700 pounds of trees and 24,000 gallons of water. Melting down a ton of aluminum cans saves 8,800 pounds of bauxite. Such statistics make recyclers look like friends of natural resources. Recycling projects, however, use energy and, most importantly, employ a large amount of chemicals which pollute the environment. Conservation is a much better bet.

Nor are all countries tuned to the concept of recycling. In Argentina, for example, every year 100,000 tons of electronic gear is thrown on to the rubbish dumps without any recycling. (Technology, too, makes a major contribution to the mountains of waste.) And in no country that I know of is the auditing of waste management an essential government and corporate strategy.

If governments want to be in charge of the proliferating waste problem, *then* eco-auditing should become institutionalized on a global scale, with all major projects subjected to environmental impact assessment. The development of strategies on waste management auditing should include a roadmap on how local and national government, as well as industry, address the need to protect the environment and assure that the mountains of garbage are significantly downsized – which means changing the policies of the throwaway society in which they live.

---

23 Recycling can also be a business opportunity if one comes up with good ideas. An example is TerraCycles which offers some 100 branded 'upcycled' products, from plastic garbage cans to pencil cases and backpacks. The company started in 2001 by selling plant food made from worm waste products.

CHAPTER 12 The Scarcity of Clean Water and its Impact

1. Water and Energy: Strategic and Scarce Products

Water, like energy, is a most critical natural resource, indispensable in maintaining life. It is therefore quite worrisome that water shortages, particularly of clean water, are now confronting most parts of the world and that they are getting worse over time.

- Rapidly rising populations and pollution are now doing the damage which, in other times, conquerors did, and
- in the long march of history, few events have had so much impact on the course of humanity as the disappearance of water resources.

'The migration of the Turks to the Middle East and from there to Europe has been due to the desiccation of Central Asia. Therefore, the fundamental study in searching for historical causes is hydrography,' said Bertrand Russell.[1]

To subdue the countries which resisted his conquest, Genghis Khan depopulated extensive areas of land through the laborious destruction of their irrigation system. 'Without irrigation, the villagers and farmers left, and the fields reverted to grazing land,' suggests Jack Weatherford. 'This allowed large areas to be set aside for the herds that accompanied the army and were kept as reserves for future campaigns.'[2]

The trigger for other major events, too, has been water. Since the beginning of history one of the main problems with water in some parts of the world was its lack or scarcity. Statistics on water resources are therefore vital, but because the definition of clean, fresh water is not cast in iron, percentage estimations vary. According to prevailing estimates, as an order of magnitude and not necessarily as exact statistics:

- all but 1 per cent of water on earth is ice locked up in glaciers or comes in the form of salty seas, and
- *if* water in glaciers is added to freshwater, although much of it is not easy to reach, *then* this share rises to 2.6 per cent. The rest is salty.

Even these 1 per cent and 2.6 per cent estimates are unevenly distributed. Africa, Asia and Europe have less freshwater per head of population than North and South America. Many countries in North Africa and the Middle East suffer chronic water scarcities. And

1 Niall Fergusson (ed.), *Virtual History*, Basic Books, New York, 1999.

2 Jack Weatherford, *Genghis Khan and the Making of the Modern World*, Three Rivers Press, New York, 2004.

because water usage increases so much, so rapidly, in many parts of the world underground aquifers have been drained.

Growing populations are the number one cause, and the situation is getting worse, not better. According to various studies, in the not-too-distant future big and medium-sized cities will be starving for water. More than half of China's 600 cities already suffer water shortages, and in parts of the North China plains water tables are falling by as much as a metre a year.

Even in areas of the world where this fall in water tables is not a critical problem, pumps and aqueducts that carry water to the cities are creaking with age, threatening the water supply (see sections 5 and 6). Water is a particularly big issue for sprawling megapolis. Their insatiable water consumption will not destroy agriculture, but whether it will permit the continued cultivation of the crops that the land currently produces is questionable. It is not unthinkable that, as the situation deteriorates, big cities, like big banks, may have to be broken up.

An inheritance the twenty-first century received from the mid- to late twentieth century was the dubious (in terms of feasibility and affordability) assurance that all *Homo sapiens* 'must have access to sufficient supplies of clean water'. This has proven difficult to uphold as consumption rises to the stars. According to the UN Commission on Sustainable Development, demand for water:

- will double by the middle of this century, and
- increase at a rate of 50 per cent with each subsequent generation.

Providing clean water supplies at that rate is clearly unsustainable. Africa, the Middle East and Asia are likely to be most vulnerable, and it is estimated that the portion of the world's population that will not have access to clean (or even sufficient) water will increase from one out three to one out of six people.

The earth's water resources are being abused to the point that a heavy price will eventually be paid by humankind itself. In Punjab, India's and Pakistan's bread basket, from early 1970 to late 2010 the water table dropped by 10 metres, and its rate of decline is accelerating. Big drops have also been registered in the aquifer in the north-western Sahara desert, which serves Libya, Algeria and Tunisia; in the Nubian sandstone aquifer serving Egypt, Libya, Chad and Sudan; and the Iullemeden aquifer which sustains life in Mali, Niger and Nigeria. Soil salinity has also increased all over these areas. Nor are other parts of the world free from the depletion of their water resources. In South America the Guarani aquifer that benefits Brazil, Paraguay, Uruguay and Argentina is also being depleted.

This disregard for the sustainability of water resources through overexploitation is very serious because groundwater provides about:

- a fifth of the world's water needs, and
- half its drinking water.[3]

'In effect, we are on countdown,' says Erik Peterson, director of the Center for Strategic and International Studies, at a symposium of the New York Academy of Sciences organized

3 *The Economist*, 9 October 2010.

in collaboration with the Johns Hopkins Bloomberg School of Public Health.[4] Plenty of reasons underlie this projection, and in the background of all of them is the explosive growth of *Homo sapiens*. The proportion of people living in countries chronically short of water,

- stood at 8 per cent (or about 500 million) the turn of the century,
- but is set to rise to 45 per cent or 4 billion people by 2050.[5]

Moreover, to feed these rapidly and irrationally rising number of mouths (Chapter 13), farmers are increasingly demanding more water. Over and above that comes people's desire for better-tasting food. It takes nearly four times more water to produce a kilo of beef than a kilo of chicken, and five times more water to produce a glass of orange juice than a cup of tea – which used to be the staple drink.

As people move to urban areas, including shanty towns adjunct to the big cities, their direct and indirect water consumption rises. City people also have to be fed, and not everybody truly appreciates the amount of water required to produce what has come to be considered essential foodstuffs. For example:

- 1 litre of milk requires 2–4 litres of water.
- 1 litre of beer requires 3–7 litres of water.
- 1 kilo of sugar requires 10–13 litres of water.

Factories planted in or around the city also need plenty of water for their produce. For example:

- 1 kilo of steel requires 30–50 litres of water.
- 1 kilo of paper requires 50–100 litres of water.
- 1 kilo of plastic material requires 200–500 litres of water.

As the examples for milk, beer and sugar show, water is a most essential ingredient in many of the products that line supermarket shelves. (Some experts reckon that, on a global basis, the five big food and beverage multinationals – Anheuser-Bush, Coca-Cola, Danone, Nestlé and Unilever – consume almost 575 billion litres of water a year, enough to satisfy the daily water needs of every person on the planet under present conditions.)

This is a different way of saying that the shift in people's diet, particularly of city-dwellers, has presented many unexpected consequences. The water intensity in food increases very rapidly as people begin to climb out of poverty because they begin to eat more meat and consume drinks that use plenty of water. In the aftermath:

- The good news is that living standards in the poorest countries are starting to rise.
- The bad news is that water use is soaring, and there is no way of reversing that trend short of global thirst and famine.[6]

4 *Sciences Magazine*, Winter 2010.

5 *The Economist*, 22 May 2010.

6 Conservation alone will not be enough. Besides, the concept of water conservation is totally alien in most countries.

The demands for water posed by the traditional manufacturing industry – for instance, steel, paper and plastic – are by no means the highest. High technology is even thirstier. It takes about 13,000 litres of freshwater to produce a single 200 mm semiconductor wafer. (Chip-making is thought to account for 25 per cent of water consumption in Silicon Valley.)

Energy production is also water-intensive. Year on year, roughly 40 per cent of the freshwater withdrawn from lakes and aquifers in America is used to cool power plants. Who would argue against industrialization? But ignoring its negative effects, such as its contribution to water shortages, is a prescription for future problems.

Many of the challenges confronting human societies correlate. This is the case, for instance, with the availability of plenty of water and hydroelectric energy production. When rivers and reservoirs are low, so too is the output of hydroelectric plants, which are touted as a major domain of alternative energy (Chapter 5) and today generate about one-sixth of the world's electricity.

All these examples lead us to the conclusion that water resources, which for millennia tended to be taken for granted, are under strain. Water pollution comes over and above all that, significantly reducing the aforementioned 1 per cent ratio of freshwater. Some years ago, a report for the European Commission argued that in the European Union:

- growth could be restrained by a lack of freshwater caused by industrial and agricultural contamination, and
- this was not a worst-case scenario because, on average, water resources in the European Union are better managed that in other parts of the world.

Particularly in developing countries, industrialization, urbanization and uncontrollable population growth are leading to economic and socio-political pressures accompanying the growing demand for water. Their governments are largely inactive on the water front, as if they have failed to notice that freshwater resources are scarce and not easy to sustain.

Nor are they easy to reach. Water held as ice or snow at the poles is not transferable to urban centres,[7] and freshwater that is not frozen is poorly distributed across the surface of the world.[8] Therefore, raising water conservation standards is a 'must', but it is costly and correlates inversely with the population boom. According to some estimates,

- the world's use of freshwater has increased nearly sixfold in the last 70 years, and
- water recycling by nature is no longer able to satisfy rising demand, raising the likelihood that *Homo sapiens* may be just a temporary species.

Current and projected water scarcity poses a great dilemma to policy-makers. Projections are bleak and solutions are not obvious. The sustenance of water supplies at a health-conscious level, let alone at a level of water purity, requires major investments

---

7 Some years ago, promoters convinced a Saudi Arabian prince to bring icebergs from the South Pole to his country. But, as was to be expected, nothing came out of it.

8 About a fifth of the world's freshwater resources is held in Lake Baikal in Siberia, roughly the size of Belgium. It suffered severe pollution from the timber industry during the Soviet era.

which few countries can afford.[9] For decades the price paid for water has been rock-bottom. Now it is rising, but it is not rising fast enough to allow loans to be taken out and depreciate such investments, because the price of water is a political issue. Therefore, clean water scarcity will, by all probability, become a key phrase in the coming decades.

In conclusion, what makes the future so uncertain in terms of water supply is that water resources are being depleted, many are polluted, and demand for water is rising because an estimated 90 per cent of population growth in the next 40 years will be in the cities where water leaks and wastage are legion. With cities mushrooming, and with them industry, their appetite has no limits. But there is no way that water can be diverted from farms to towns, because then city-dwellers would not have enough to eat.

2. Confronting the Hydra of Clean Water Scarcity

'Water scarcity' is an odd phrase for something that covers almost three-quarters of the planet's surface. Theoretically, water is plentiful, but, as highlighted in section 1, most of it is saltwater. The proportion that could be called 'freshwater' is a tiny 2.6 per cent and only 38 per cent of it (1 per cent of the total) can be tapped. Even that, however, is an impressive figure. Every year an average of over 6,000 cubic metres of water per person flows into rivers and underground channels. The problem is that:

- the rain does not always arrive where and when it is needed, and
- almost a quarter of the world's countries are dangerously short of freshwater.

Canada, for example, has 26 times more freshwater per capita as Mexico. Also within some countries, there may be lush forests in one region and deserts in another. Be what may, the importance of water in sustaining life is so great that the randomness of its distribution has determined the map of civilizations.

Cities located on rivers and oceans developed faster than others. Human populations eventually became concentrated along rivers, lakes, seas and oceans. Waterways turned into highways for trade and greatly facilitated not only merchandizing in all sorts of goods, but also cultural interchange.

- The more advanced civilizations watered their cities through an elaborate network of aqueducts. But they eventually decayed after suffering grievous damage from toxic heavy metals diluted in their water supplies.

Confronting water scarcity is a many-headed problem, like a hydra, involving not just supplies but also wastage, pollution and unexpected consequences. Many ecological catastrophes have been provoked by dwindling and altered water supplies, including the collapse and disappearance of waterfowl caused by dams and other hydroprojects (see Chapter 5 on alternative energies).

An estimated 75 per cent of the world's available freshwater is used for irrigation; in the less developed countries, this proportion is a good deal higher.

9 Some years ago the US Environmental Protection Agency (EPA) estimated that at least $150 billion must be invested to bring the nation's water systems up to federal water quality standards.

- Irrigation improves agricultural productivity, but, at the same time, much of that water is wasted.
- The poorer countries use about twice as much water per acre as the richer ones do, but achieve crop yields that are only one-third as high.
- In addition, poorly managed irrigation can lead to salinization and degradation of agricultural soils (Chapter 11).

Significant improvements in irrigation methods are one way of confronting the hydra of water scarcity.

Solving the water resource problem requires both general principles and case-by-case analyses. Many countries pursue water policies that are inefficient and hard to sustain. These have been chosen mainly (if not purely) for social or economic reasons, such as:

- preventing migration from the countryside to the towns,
- supplying ample water resources for irrigation, or
- providing electrification with poorly studied consequences.

Critics do not fail to point out that there are fallacies associated with these reasons. Edward Goldsmith and Nicholas Hildyard have argued that: irrigation schemes often created no more new land than they destroyed; resettling people to build dams caused huge economic and cultural disruption; and dams frequently failed to control floods.[10]

These references press the point that society has not yet found a way of getting on top of its water problems.[11] For all practical purposes, we are in a corner. We must use water to maintain our current, comfortable standard of living, but we know that water scarcity and pollution are almost inevitable.

Catastrophes in civilization have been propagated, and at times instigated, by polluted water resources. According to some estimates, diseases associated with filthy water kill more than 5 million people a year. The increase in pollution in many countries is further aggravating water shortages. Backed by the World Health Organization and other UN agencies, a global monitoring scheme suggests that as many as 10 per cent of all rivers being monitored are polluted.

Much was said in Chapter 7 about pesticides and fertilizers. Their massive usage has a great deal to do with water pollution in both developed and developing countries. For their part, industrialized countries have the problem of high levels of poisonous heavy metals like mercury.[12]

Technology tries to remove pollutants at their source, but this is often too expensive and quite often ineffectual. Desalination which is often touted as a remedy, is, in many cases, a half-baked solution to the shortage of freshwater because of the energy demands it poses. As for the level of pollution to be tolerated in water supplies, this depends on the use to which the water must be put:

- drinking needs,

---

10 Edward Goldsmith and Nicholas Hildyard, *The Social and Environmental Effects of Large Dams*, Wadebridge Ecological Centre, Camelford, 1984.

11 The same can be said of its energy problems (Chapter 1).

12 The rivers Rhine and Meuse are considered to be two of the most polluted in Europe.

- factory usage,
- farm irrigation,
- commercial fisheries,
- general recreation,
- navigation, and more.

All these activities demand water of greater or lesser quality. End-use requirements must evidently be taken into account in determining acceptable pollution levels, but even that is not objective because of conflicts of interest. Particularly important in drinking water supplies is an understanding of purification levels. In Nagoya, Japan, expert systems[13] are is used to help separate water into:

- upper-level for drinking (tap water), and
- lower level, for other activities.

To save the relatively pure water resources for drinking, the Japanese increasingly use recycled drainage for lower-level water. Intelligent buildings in Tokyo and other Japanese cities are employing recycled water for toilet-flushing and similar purposes. These are commendable solutions and should be applied to all new buildings.

To be effective, technological advances must successfully integrate with governmental regulations. Likewise, regulations should be tuned to take advantage of tools and methods made available through scientific effort, as well as to account for negative factors affecting water quality. This is the purpose of water-pollution regulations which, however, are:

- by no means global, and
- not always observed, even where they exist.

In early 1992 the British government lost a long fight with the European Commission, when the European Court declared that Britain had failed to bring its drinking water up to EC standards.[14] This was the first time in Britain's (then 20-year) membership of the European Union that it had been found guilty of disobeying environmental laws. In fact, Britain was the second EU state to be condemned on water-related issues. Belgium had lost a court case over lead levels in drinking water in 1990, two years previously. Over the years, every member of the European Union, except Portugal, has faced Commission prosecutions over drinking and bathing water. Portugal escaped because, in a show of silly European Commission 'generosity', it was not required to meet the regulations until 1993.

Yet there is a real and present need for water quality rules and standards. A short time before the aforementioned events, a survey by the Japanese government had found that 4 per cent of almost 8,000 drinking wells nationwide contained unsafe levels of tetrachloroethylene, while 3 per cent were contaminated with trichloroethylene. Both are carcinogenic chemicals. (Tetrachloroethylene is commonly used by dry cleaners, and trichloroethylene is employed as a solvent in the manufacture of computer chips.)[15]

13 Computer-based, knowledge-engineering artefacts.

14 *The Times*, 22 January 1992.

15 *The Japan Times*, 29 March 1989.

This survey, which covered thousands of factories, also documented that 9 per cent of businesses using tetrachloroethylene and 3 per cent of Japanese factories employing trichloroethylene were discharging unsafe levels of the chemicals into the waterways. While less developed countries are often pinpointed for having contaminated their water supplies, as the above examples demonstrate, developed countries have not necessarily taken all advisable precautions. Water quality rules count for nothing if they are not regularly policed.

3. Water: Prime Ingredient of Life and Daily Necessity

The importance of clean water can better be appreciated if we look at its origin as a vital ingredient of life. Water was formed as product of volcanic activity, heat and swirling gases on early earth. The human body[16] synthesizes water in the process of metabolizing its food – at the rate of about two quarts per week. In terms of its effects on society, and on living organisms at large, clean water means good health.

Scientists believe that solar events affect water, and by extension human molecules, as water is large part of human body. This belief further emphasizes water's vital importance to the evolution of life on earth.

Chemically, water is a compound of hydrogen and oxygen whose unusual angle of atom-bonding keeps it liquid at ordinary temperatures. As a chemical, water is unique. In its pure, clean state it is odourless, colourless and tasteless. It is also incredibly stable and a remarkable solvent.

- Water molecules cling together more tenaciously than certain metals, and
- water can absorb and release more heat than most common substances.

Until 200 years ago, water was believed to be an indivisible element. Now we know it is not so, but tremendous energy is needed to break water apart. By contrast, it is very easy to pollute water, and, as already discussed, this is what people are doing routinely day-in and day-out. They do so because only very few appreciate the key role played by water in the survival of the species.

Those who appreciate water's role think of it as the blood of the land that nourishes crops, grasslands and forests. But the majority of people seldom think about water in these terms until it is threatened, or gone. A person can live for weeks without food; but without water, barely a week. Also little appreciated are some special characteristics of water:

- It can creep uphill.
- It can disintegrate the toughest metal.
- It supports life ranging from single-celled bacteria to elephants.
- What is more, unlike oil or coal in energy resources, water has no substitute.

Living organisms originated in water, and their biochemical systems are all adapted for functioning in it. Our society has followed on this dependence. We each use water for

16 Indeed, the body of all animals. Water is locked in practically all living organisms.

drinking, cooking, bathing, laundry, domestic and industrial waste disposal, irrigation, the cooling of electricity-generating plants and many other activities.

Despite these important roles, no one really knows how much water people can safely use from the available resources. It is certainly not all of it, because other species, too, have to live off the earth's water. In many places the maximum may well be less than one-fifth, which seems to be the average for Asia as a whole. This maximum quantity depends on many factors, two of which are:

- how much water is returned to the natural system, and
- how much is taken from underground aquifers.

Nobody either really knows how long the earth's clean water supplies will last. According to the latest hypotheses there is some patchy evidence that, given current patterns of use and abuse, the amount now being withdrawn and employed is dangerously close to the limit of safety and the natural system's ability to recycle no longer suffices for 7 billion growing into 9 billion *Homo sapiens*.

Certainly, the signs are not positive. Freshwater fish populations are in rapid decline. According to some estimates, in the four decades since 1970 fish stocks in the world's lakes and rivers have fallen by roughly 30 per cent. To a large extent, the population explosion (Chapter 13) is responsible for this:

- directly, through overfishing, and
- indirectly, by polluting the earth's water resources.

Water is also a fundamental component of alternative energies and not only for hydroelectric and geothermal plants (Chapter 5). In February 2008 city officials in Tampa, Florida, were surprised when a local firm building the state's first ethanol production factory put in a request for 1.5 million litres of water a day, making that factory one of the city's top ten water consumers.[17]

In Missouri, residents went to court to stop a $165 million biofuels factory being built, because it was projected that it would draw 4.5 million litres of water per day from the Ozark aquifer.[18] Alternative energy may be solving some problems but, as we can see, they are creating others which may be more serious.

Water conservation, like energy conservation, is the answer but this goes against the habits of people and the unaffordable status quo of water usage. Statistics published in 2007 by Veolia, one of the two major French water companies, dramatize how much water could be saved by paying greater attention to its usage. According to these statistics, in the typical family:

- drinking represents 1 per cent of consumed water,
- cooking, 6 per cent,
- washing the dishes, 10 per cent,
- clothes washing, 12 per cent,
- flushing the toilet and sanitary use, 12 per cent,

17 'Don't Mix', *The Economist*, 28 February 2008.

18 *The Economist*, 1 March 2008.

- baths and showers, 39 per cent,
- various domestic duties, 6 per cent,
- washing the car and watering the garden, 6 per cent.

Taking a bath uses 150–200 litres of water; a five-minute shower uses 60–80 litres. The toilet is a major water waster, as each flushing toilet throws 6–12 litres of drinking water down the drain.

That sort of water usage has become second nature. Therefore these statistics are hardly appreciated by the majority of the population who are wasting water resources. In 1999 Americans objected strongly to a new federal law which imposed a limit of 5.7 litres per toilet cistern instead of the then prevailing 10–12 litres. Consumers complained by phone, letters and e-mails against the new norms.[19]

Yet, new norms on water usage are urgently needed. So, it is good news that in developed countries several governments are busy instituting them in regard to the use of water resources. But in less developed countries governments have taken no action to reduce the wastage of water, and, overall, world water consumption shows no sign of abating.

Goldman Sachs has estimated that global water consumption is doubling every 20 years, which is obviously an unsustainable rate of growth. In addition, climate changes alter the patterns of freshwater availability in ways too complex to analyse and forecast, and these changes can lead to unexpected consequences.[20]

No government has paid enough attention to the fact that population explosion (specifically in less developed countries), increased industrial production, intensive agriculture and activities closely connected to harvesting, storing, treating and delivering water are putting water resources under stress. Consumers should think about this next time they turn on the tap.

4. Downstream Quality Rights and Water Wars

'Life as we know it is at stake,' argued Maurice Strong, a Canadian businessman and secretary-general of the 1992 UNCED meeting in Rio de Janeiro.[21] In the background of his statement was the fact that although the Cold War had ended, population pressures and environmental decay might lead to conflict among nations over *food*, *water* and *land* – as well as massive illegal immigration.

Sixteen years later, Andrew Liveris, chief executive of Dow Chemical, expressed his thoughts in an equally eye-opening way: 'Water is the oil of the 21st century.'[22] As we saw in the preceding sections, water is a critical lubricant of the global economy, just like oil, and supplies of water are coming under enormous strain because of:

- the rapidly growing global population,
- the different countries' industrialization,

19 *Nice Matin*, 4 August 1999.

20 'Running Dry', *The Economist*, 21 August 2008.

21 *Business Week*, 11 May 1992.

22 'Running Dry', *The Economist*, 21 August 2008.

- water-intensive agriculture, and
- a huge amount of spoilage all over the world.

The water crisis has crept up while the political leadership has been distracted by global warming, ozone depletion and other emerging environmental threats. Curiously enough, the various environmental gatherings and social pressure groups failed to take account of the fact that there simply is not enough water around for several billion men, women and children.

In fact, nearly everybody seems to be living under the false assumption that if left to their own devices, water problems will take care of themselves. The time has come, however, to carefully examine a serious hypothesis which can be briefly stated in two bullets:

- *If* today countries are poised to go to war over oil,
- *then* the catalyst for future armed conflict could very well be water.

Egypt and Ethiopia, Israel and Jordan, India and Bangladesh, and a host of other pairs of neighbouring countries have serious disputes over water supplies.[23] Today's unilateral act of taking hold of water supplies at the expense of one's neighbours closely resembles the protectionism and competitive devaluations of the late 1920s which brought about the Great Depression.

For instance, Turkey has had the power, but not the legal basis, to curtail the water flowing into Iraq through the Tigris and Euphrates rivers. Starting in January 1990, Turkey diverted the Euphrates to fill a reservoir behind its new Ataturk dam, cutting off the flow of water into neighbouring Syria and Iraq. Yet Syria relies on the Euphrates for 90 per cent of its surface-water supply and has itself built the massive Al Thawrah dam on Lake Assad.

The Euphrates was one of the four rivers of paradise in the Book of Genesis, and some 10,000 years ago it nourished the world's first civilization. Today, it may be nourishing frictions. After Turkey's unilateral action, one of its tributaries near the Syrian village of Jarablos (right on the Turkish border) became a series of semi-stagnant pools. The water became stained green-blue from sewage, while dead fish floated by.[24]

From Syria's viewpoint this case exemplifies what has become known as *downstream water quality* rights – a new issue which creates conflict not only between neighbouring countries, but also among neighbouring communities. An interesting American example dates back to the early 1990s.

This case concerned Champion International's paper mill in Canton, North Carolina, its 2,000 workers, $100 million annual payroll and the waters of the Pigeon River, which winds through Canton on its way to the Tennessee. The people of eastern Tennessee,

---

23 Similarly, in the United States there exist water disputes among neighbouring states like California, Nevada and Arizona. Problems are also present in California. In 2007 in the Golden State a federal judge ordered the huge pumping stations of the Sacramento Delta, the largest estuary on the US West Coast, to reduce the water delivered to two aqueducts that run south to San Joaquin Valley and southern California by a third. His reason was the delta smelt, a fish that lives only in the delta and is considered endangered under federal law. 'Of Farms, Folks and Fish', *The Economist*, 22 October 2009.

24 Turkey's control of the sources of the Euphrates and Tigris rivers, both of which rise in its central highlands, has long soured relations with its Arab neighbours. Over and above that came Turkey's irrigation project, known as Gap.

some 42 kilometres downstream, did not like what had happened to their river. They wanted Champion to:

- clean up its act,
- rid the river of what they called 'the stench of rotten eggs', and
- restore the clarity of its water by removing the source of the coffee-brown colour it had acquired from the factory's effluent.

Health and social interaction are at stake when neighbouring communities or states are squabbling over downstream water quality rights. The two are intermingled and decision-makers should have the authority to police clean water. Is the downstream state authorized to control water resources, or does it have to accept what the upstream authorities judge as fair?

When such differences in opinion arise in the same country they are resolved in the courts. But when the downstream water quality deteriorates and associated rights concern different jurisdictions this leads to cross-border disputes between sovereign countries – and war is the time-honoured solution.

If Turkey's dam and wide-ranging irrigation project near the Syrian and Iraqi borders brought Syria, Iraq and Turkey only to the brink of armed conflict, some historians have found water-control reasons in the roots of Israel's six-day war with its Arab neighbours. (In Israel much of the country's underground water supplies lie beneath the West Bank, and experts say that the challenge of water rights could be crucial to a political solution of the Palestinian question.) And the late King Hussein of Jordan once admitted that water rights could drive him back to war.

In this narrow stretch of the Middle East where three religions and many states rub noses, Israel shares relatively scarce water resources with Jordan, Syria and the Palestinians. Unsurprisingly, fast-growing populations have been putting a strain on those resources. And as for replenishment of water resources:

- average rainfall in the region has not changed greatly,
- but rain showers have become shorter and more intense, so that more water runs into the sea instead of recharging the aquifers.

Consequently, the Jordan River is a trickle of its former self, and the level of the Dead Sea, which it replenishes, is falling.[25] The Middle East is but one example of several around the globe – short of water and getting shorter all the time.[26] The region also faces drought every four years or so. (Desalination is an option – more on this in section 7)

Because a continuing supply of clean water is so important, there are plenty of reasons why, in years to come, the primary focus may not be on boundary lines, but on one of the most fundamental of human needs – water. Historians point out that these problems are not new, but they have been aggravated over the centuries. Since biblical times the competition for water, and fierce protection of access to it, has been a source of tension and conflict between peoples and nations. But the greatest battle still lies ahead. Experts warn that:

---

25 Farming swallows at least 80 per cent of the region's water.

26 Lebanon is the only tiny place that has enough.

- the region is running out of cheap, uncontaminated water supplies, and
- as competition for this vital resource mounts, both within and between nations, further conflicts are bound to arise.

In conclusion, a booming population, relatively higher standards of living, industry, agriculture, overexploitation, spoilage and poor management of water resources create an explosive situation not only in the Middle East, but also in many other areas of the world. Over and above that comes the violation of downstream water quality rights by direct and regional neighbours, without any mechanism in place to resolve water disputes in a relatively peaceful way.

5. For Business, the Silver Lining is Infrastructural Renewal

Most people find it difficult to appreciate what it means to guarantee pure water resources, yet humankind has settled in the cradles of river systems for millennia. A close relation-ship has been built around the flow of water which brings freshness and change with every passing moment. Leonardo da Vinci wrote: 'When you put your hand in a flowing stream, you touch last what has gone before and the first of what is still to come.'

Aristotle studied the behaviour of water, including its condensation and precipitation. Over the ages scientists and philosophers looked at water as the cradle of life. But today we know more and understand less. This is why we are in the process of converting mighty rivers into sterile lakes (look at Aswan), and living lakes into polluted ponds.

Pollution and the rapid surge in population are affecting not only exploding urban areas like Mexico City, Cairo and Bombay, but also formerly pristine countries like Nepal in the Himalayas. In Katmandu, Nepalese mothers now have to filter water for their children through cotton rags as most of the drinking water is highly polluted.

Although most of India has abundant monsoon rains, its water supplies are strained by unchecked population growth and rapid urban development. The overpumping of underground aquifers and the relentless felling of trees, which trap and store water, have upset the hydrological balance, turning the subcontinent into a fugue of floods and droughts.

- This strains the country's agriculture which must feed nearly 1 billion people, and even has dreams of huge exports.
- And what water is available is sometimes contaminated by sewage and industrial waste.

There was a drought in India in 2009, and, because none of the country's previous five big post-independence droughts caused famine, there seemed little cause for people to worry. Indeed, the government said that, after two bumper years in 2007 and 2008, it had enough wheat and rice in store to prevent serious food-grain price inflation as a result of the latest drought.

Yet, there is plenty of reason for the government to worry. India's extremes of hydrology, poverty and population explosion present huge difficulties for water conservation and management. The monsoon rains, which supply over 50 per cent of India's annual precipitation in just 15 days, are predicted to become:

- more contracted, and
- significantly more unpredictable.

As if this was not enough, the melting of Himalayan glaciers is set to deprive the great rivers of the Indian sub-continent – the Indus, Ganges and Brahmaputra – of their summer source of water. All this adds up to the prospect of longer dry seasons with low river levels and unpredictable volumes of water during the monsoons.[27]

China provides another example of the infrastructural investments necessary to confront and avert future water famine. At a time when China's population of about 1.4 billion[28] is enjoying a harvest of plenty relative to any other time in its recent history, the country has been running out of clean water. Some years ago, a study by World Bank scientists[29] warned that the increasing pressures on the country's water resources to feed, house and meet energy needs was:

- rapidly destroying 'whole ecosystems', and
- threatened to put the brakes on the economic boom.

Although China has been facing a water crisis for a number of years the lack of serious infrastructural improvements exacerbated the associated problems. Accidents as a result of widespread pollution bring to a wider stage several disturbing trends about China's water resources, many of them connected to:

- the booming chemicals plants that line the country's rivers,
- severe water shortages in certain agricultural areas, and
- the secrecy with which the government sometimes treats accidents, suggesting that they are rather severe.
-

As a consequence of pollution an estimated 60 per cent of China's rivers cannot be used as sources of drinking water. Anecdotal evidence suggests that many rivers are not even clean enough for industrial use.

In addition, little or nothing is being done to control CO_2, SO_2 and NO emissions which sees to it that air pollution morphs into water pollution. With the rapid pace of Chinese industrialization, the consumption of aluminium, copper nickel, and iron ore has grown to an extraordinary level – and, as we have seen, intensive industrial activity requires massive amounts of water.

The Chinese government has tried to compensate for this water shortage by making use of once-abundant groundwater; but this supply is gradually disappearing. To avoid a daunting water resource crisis China, like many other developing countries, has only one road open to it:

27 'When the Rains Fail', *The Economist*, 10 September 2009.

28 They were 1.1 billion just 20 years ago.

29 All references and statistics attributed to the World Bank in the following paragraphs were obtained by the author during a meeting in Washington, DC.

- Adopt effective measures without further delay.
- Renew and restructure the country's water infrastructure.
- Train human resources in water efficiency and avoidance of spoilage.

In India, in China and in practically all developing countries there is a present and urgent need for restructuring the whole domain of water management. Everything should be negotiable, although infrastructural works and intense training are the pillars.

- Rebuilding the water infrastructure and making it much more efficient requires plenty of engineering ingenuity, and
- the World Bank says that billions must be invested just to avoid severe water shortages, not just in India but in all developing countries.

Over and above skills, and the huge sums that will be required, tough environmental standards will be needed *if* and *when* such standards are formulated. Moreover, developing countries must urgently build up water management skills. This is easier said than done, not least because most developing countries' water industries are still owned and run by governments, which means political patronage.

Many governments often treat their water utilities as job-creation opportunities for political friends. According to the World Bank, the number of employees per 1,000 water connections in West European water utilities is two or three. A corresponding figure for many Latin American utilities is 10–20 – which speaks volumes about prevailing inefficiency as well as opportunities to cultivate connections for contracts.

Ironically, the fact that in developing countries many water firms are beset by political scandal favours investments (although it works against greater efficiency in water management). Bribery in the award of water-supply and purification contracts pushes governments to find the money, but income from the water network is low because water[30] is hugely subsidized and therefore underpriced. Some years ago, a survey of municipal water projects financed by the World Bank showed that the price charged for water covered only around 35 per cent of the average cost of supplying it. The shortfall is made up by outright subsidies and by allowing the water infrastructure to decay. In the aftermath:

- Heavy rain, as well as floods, overwhelm existing controls.
- Reservoirs do not store enough to get people, fields and plants through longer droughts.
- Leaking pipes cause the water distribution networks to lose a large proportion of water.

Damascus suffers nightly water stoppages, partly because the city loses 30 per cent of its water from old pipes. The situation in Egypt is worse. Nile basin farmers use almost twice as much water as they need because of poor irrigation methods, while an estimated two-thirds of the water allocated to Egyptian cities and towns is lost through leaking pipes, taps that do not shut and malfunctioning toilets.

30 Like petrol in oil-producing countries.

6. Investment in Sewage and Water Treatment

At the dawn of the twenty-first century, UN member states set a goal of ensuring clean water for all people worldwide. Totally missing from the UN plan was a hint as to where the financing for such a colossal investment will come from.

Allowing for the number of years this will take and for the expected growth in population – most of it in cities of less developed countries – this UN plan implied providing clean water and basic sanitation to about half the population of the earth by 2015. We are at 2011 and, not surprisingly, nothing has been done.

Around the same time as this great UN pronouncement, a study chaired by Michel Camdessus, the former managing director of the International Monetary Fund (IMF) and former governor of the Bank of France, forecast that spending on water and waste projects in developing countries would need a budget of about $80 billion a year.

Camdessus also pointed out that to achieve full 'water security' by 2025, a twin UN goal, would need at least $180 billion a year for the next 20 years or more.[31] Even this may prove insufficient, as practical examples in this section will demonstrate.

In the background of this huge financial requirement is the dilapidated water system found all over the globe and a horde of difficulties and challenges stemming from a combination of the growth of the world's population as well as its increasing concentration in cities. For instance:

- Shanghai's population growth exceeds 1 million people a year.
- Rabat, Morocco's capital, is growing by 100,000 a year – as are several other cities in developing countries.

Some lessons can, however, be learned from past successes and failures, particularly in investments in sewage treatment. In 1989 a study by sewage consultants Camp, Dresser and McKee recommended that the Sydney Water Board, spend AU$4.5 billion (about US$ 3.6 billion, at the time) to improve Sydney's sewage system.[32]

Re-evaluated to 2011 prices, this represents about US$ 6 billion – and we are only talking of *one* city. On a global scale – if a project like that was at all feasible – it would not be surprising if the budget were in excess of $6 trillion.

Moreover, the Camp report recommended doubling the capacity of Sydney's sewage system and vastly improving treatment facilities. It's advice which will perfectly suit, for example, all large and growing cities around the globe that are still without appropriate sewage treatment and dump sludge into the Atlantic, the Pacific and the Mediterranean, as well as other seas and plenty of rivers. (Many countries and cities consider sludge-dumping in seas and oceans as being relatively cheap. It is indeed a 'cheap solution' but it is not low-cost. It cost New York City $27 million to acquire a fleet of ocean-going sludge barges, while towing and other costs exceeded $20 million a year.)

Another example is provided by Boston where in 1988 a Massachusetts state authority broke ground for the most ambitious water-cleansing system in US history, a $6.1 billion

---

31 *Financing Water for All*, Report of the World Panel on Financing Water Infrastructure, March 2003.

32 *Business Review Weekly*, 22 September 1989. As noted at the beginning of this chapter, it has been a deliberate choice to include references and examples dating back 20 or more years. This helps on two counts. First, clean water and sewage treatment troubles are not a recent phenonenom; second, even though so many years have passed by, nothing has been done to correct the roots of the problem.

complex (the final clean-up bill exceeded $8 billion). Residents were largely paying for the project themselves, as Boston harbour had become a metaphor for the polluted environment during the 1988 presidential campaign, when George Bush embarrassed Michael Dukakis on the clean-up issue.

- Some 500 million gallons of partially treated waste water, and
- 500,000 gallons of toxic sludge, equal to 70 dry tons,

poured daily into the harbour, producing a so-called 'slumgullion' of raw sewage, ammonia, phosphorous, pesticides, toxic trace metals like copper and mercury, carcinogenic organic compounds, petrochemical solvents and other poisons. The Boston harbour clean-up became the biggest public works project in New England history, and it offered a sharp lesson for other countries around the globe grappling with the once mundane, but now vital, issue of the safe disposal of a city's nastiest wastes.

Nor is money, and the lack of it, the only handicap. Skills are in short supply, and the number of companies able to undertake such projects is rather limited. Also, as section 5 brought to the reader's attention, the large majority (estimated at more than 90 per cent) of water supply and sewage operations around the world are public bodies, most of them known for their:

- inertia,
- high levels of corruption, and
- limited know-how in regard to new water technologies.

Many water utilities worldwide are overstaffed. In 1992 a consortium led by Lyonnaise des Eaux, a French water firm, won a 30-year contract to run water and sewerage services in Buenos Aires. By 1995 it had cut the labour force of the Buenos Aires water entity from 7,500 to under 4,000, and renovated thousands of kilometres of pipes.[33]

Political patronage, low prices for water and too many people doing nothing works against investments in sewerage and clean water systems. Even in developed countries the application of technology to water management is uneven. To my knowledge, only 15 per cent of the US waste stream is being recycled, compared with 50 per cent in Japan.

Recycling water is necessary, but this does not mean that it has no negatives or unexpected consequences. It does. Treating sewage with chlorine and aluminium salts makes the water unsuitable for drinking and poisons the soil when the residue of solid waste is dumped in a landfill. There were several surprises after the US Water Pollution Control Act of 1972 set a policy of zero discharge to cleanse every last pollutant from the nation's navigable waters. Critics said this was an impossible goal given:

- the state of waterways, and
- problems associated with a wide reversal of policies.

33 *The Economist*, 21 March 1998.

The water of the Great Lakes, for example, was full of dioxins and PCBs, even though the surrounding communities had been using it for their freshwater supply. The greatest problem associated with recycling is not the mandatory chlorination, but the chemicals invariably found in it. This is better understood if we keep in perspective that on the American side of the Niagara river is the largest petrochemical agglomeration in the world.

Account should also be taken of an irony associated with success in the battle against water pollution. The better the water treatment gets, the worse the residual sludge problem becomes. Incineration is one option. But the US Congress has voted strongly against sludge incineration, so permits have been denied. In Switzerland where the burning of sludge has been investigated for heating purposes, but the results were not convincing and it has been rejected.

Some experts want to see sludge recycled into fertilizer. But people who are likely to live adjacent to a sludge processor angrily oppose such a practice. In addition,

- on paper, sludge fertilizers are great stuff,
- but farmers are wary of negative responses to their produce by consumers.

There is no shortage of ideas on what to do with the mountains of sludge, except that none really works as expected. In Britain officials at Southern Water, seeking ways to use sewage when marine dumping of sludge became illegal in 1998, examined a Japanese process that recycles sewage into a material called metro-marble. Sounds great? It was said at the time that the Sewage Bureau of Tokyo (the body which handles waste) even made pieces of jewellery out of it. One was a light dark grey stone set in a silver pendant that reportedly 'looked like onyx and pearl'. The other was a gold tie-pin with a brownish stone the size of a one-penny piece. The way modern alchemy wanted it, jewels can be made out of garbage. Nowadays nobody talks anymore about onyx, pearls and brownish precious stones made out of sludge.

While the sewage systems of Boston, New York, London, Paris, Berlin and other major Western cities are generally in better condition than those of developing countries, they are far from perfect. At least part of their waste disposal systems were designed at a time when their populations were much smaller and the throw-away society had not yet taken root. Also, the technology was much less advanced.

On the positive side has been the fact that the building of the Western clean water supply infrastructure was accompanied by the development of judicial and organizational mechanisms to deliver a standard which public authorities and water system operators had to observe. Such standards are rarely found in the developing countries' megacities.

In conclusion, there are two major challenges facing *Homo sapiens* in connection to the sufficiency of water resources. The one is water conservation – as water has joined energy in being a scarce resource. The other is the renewal of the water management infrastructure that is often more than a century old. Expanding and upgrading water and sewage systems in big cities, small towns and the countryside is not an option. It is a basic necessity.

7. Desalinated Water: An Alternative Clean Water Supply?

If land-based water resources have been exploited and are no longer in a position to meet the ever-growing demand for freshwater, why not use the water resources of seas and oceans?[34] Desalination looks like being the answer. It is technically feasible, but far too expensive to be a general solution. It also requires significant amounts of energy and presents other problems (as discussed later). The separation of water from the salts it contains is a difficult process mastered through distillation. The usual methods are:

- thermal, and
- reverse osmosis.

A thermal process involves multiple-stage flash, seawater heating, freezing and electrodialysis. In the course of electrodialysis, salt and water are separated under an electric field using a semi-permeable film. Freezing capitalizes on the ability of saline solutions to perform pure crystals of freshwater ice at a particular temperature. This, in a nutshell, is the basic thermal technology introduced in the 1950s.

A reverse osmosis process is energy-intensive requiring about 6 kWh of electricity per cubic metre of water, using high-pressure pumps to force saltwater through fine filters. It goes without saying that the higher the salt content of the water, the more intense and costly is the process.

If there are two proven methods for desalination, why not use them? At least theoretically, the seas and oceans offer a virtually limitless supply of water. John Kennedy had that in mind when in the early 1960s he said: 'If we could ever competitively, at a cheap rate, get fresh water from salt water, that would be in the long-range interest of humanity, and would really dwarf any other scientific accomplishment.' As a good politician, however:

- Kennedy started his statement with the word '*if*' and
- also inserted another condition – 'at a cheap rate'.
- This makes the Kennedy pronouncement a double '*if*', which is not going to be fulfilled too soon.

Apart the capital investment required, the major operating expenditure of a distillation process is the correct choice of energy sources which represent much of the operational cost. It is not by accident that of the estimated 13,000 (or more) desalination plants of various sizes currently active around the world about half are in the Middle East[35] where oil is so abundant.

According to various educated guesses, these various plants have the capacity to produce up to 56 million cubic metres of drinkable water a day. This is not necessarily of truly *clean* water, because a great deal depends on exactly where they take the seawater from.

---

34 The first known concerted efforts to produce drinking water from seawater date back to the sixteenth century when explorers on long sea expeditions began to install simple and inefficient desalination equipment on their ships for emergency use – boiling and distilling seawater above a furnace or stove.

35 In terms of capacity, the concentration is in Saudi Arabia, the United Arab Emirates, Qatar and Kuwait where energy is available nearby and there isn't any choice.

For instance, it was appalling to learn during a cruise in September 2009 with one of the huge 60,000-ton new-generation cruise ships, that all freshwater, including drinking water in the cabins, was straight out of the ship's desalination plant. This meant that:

- seawater was taken up and treated at any place the ship found itself, including filthy harbours, and
- although the desalinated water was treated with chemicals before being pumped into the passengers' cabins, what the passengers ultimately drunk was filthy water plus chemicals.

Proponents of this process would say that this does not happen with land-based desalination plants, because they are built at clear corners of the sea. But it does happen: first, because polluted seas don't have 'clean corners'; and, second, with pollution being so widespread, what might have been clean turns to dirty over time. Desalination plants are major investments; they are tied to location and cannot be moved at will to another 'clean corner'.

Energy costs and purity of seawater intake aside, the cost-effectiveness of desalination plants and their water depends on how much their contribution is vital to who builds them. In Abu Dhabi desalination plants supply 95 per cent of all water used. By contrast, in the United States less than 5 per cent of water comes from such plants.

In Israel, the Ashkelon plant produces 105 million cubic metres of water a year – roughly 15 per cent of domestic water demand, excluding agriculture and industry. Ashkelon has tried to control energy costs by building its own power station, but a dedicated electric power plant also costs a significant amount of money.

To better appreciate this statement let's go back in time. In the remote past, merchants from Bahrain in the Persian Gulf dived to the bottom of the sea to fill their goatskin bags with freshwater that flowed out of the depths through ruptures and fissures in bedrock. Freshwater springs of this kind often played a decisive role when cities and fortresses were defended against the enemy.

About 2,000 years ago, the Greek geographer and historian Strabo described a remarkable island-city situated in the arid eastern part of the Mediterranean. It was an inhabited rock washed from all sides by the waves. During wartime the people living on that island received water from a channel that ran from the coastal region and got its regular supply from a powerful submarine spring of freshwater.

- A pumping installation with an upturned wide funnel made of lead was submerged in the channel.
- Attached to the end of the funnel was a pipeline that came out on to the surface and was connected to leather bags.
- Water from the spring flowed into these bags, after having traversed the whole aforementioned system.

Such submarine sea sources of freshwater exist near the shores of several countries around the world. They can be found near the coasts of Florida, close to the Yucatan peninsula and in the Pacific near the Hawaiian and Samoa islands. Although these are examples of freshwater sources at sea, they are also clean-water cases.

Exploiting such undersea clean water sources offers no solution to modern society's hugely increased need of water because they are tiny. The only way to make use of sea and ocean water is by desalination, but it has environmental consequences.

A desalination plant takes in seawater and discharges brine and this increases the saltiness of coastal environment even though salinity tends to return to normal levels at a distance of about 500–700 metres of the plant's discharge outlet. In short, waste brine alters the environment when put back into the sea.

In addition, thermal saltwater desalination plants are prone to corrosion, and may shed traces of copper and other heavy metals into the waste stream. For their part, reverse osmosis plants employ chemicals during the pre-treatment and cleaning of the membranes, which also negatively affect sea environments.

Other technical problems have to do with membranes. New technology provides better possibilities, but no assurance. Currently being considered is a process of making membranes out of carbon nanotubes, which consist of sheets of carbon atoms rolled up into tubes But this is still at the laboratory stage, and experts think that it will be a decade before it is ready for commercialization – and, therefore, for judging its performance.

Some laboratories are looking into unconventional desalination techniques, including so-called solar desalination. This uses natural convection in vertical chimneys, which draw the outgoing heated water vapour past condenser plates.[36] In short, there is no 'ideal' solution so far.

The better strategy is to make desalination part of a system approach to water management, particularly focused on relieving drinking water shortages. Seas and oceans don't hold surprises like that of Beijing where a third of the wells have gone dry, and the water table drops steadily year after year. But massive desalination is *not* a solution, not least because poorer countries cannot afford it.

36 Small-scale plants have been built, but there are several problems connected to inefficiency and cost.

PART IV The World's Salient Problem Has Not Been Solved

CHAPTER 13 Human Fertility: The Number One Environmental Enemy

1. The Self-decimation of *Homo Sapiens*

In his keynote address to the 26th Progress International Seminar,[1] Professor Grzegorz Kolodko of Kozminski University, Warsaw, made the point that there is no world government to confront and coordinate policy response to problems arising from globalization, including its negative aspects such as money-laundering, terrorism, global crime and the black economy of drugs, slavery and arms. Then he added that 'The greatest threat is from the growth of world population to 9 billion by 2030, which will greatly affect the planetary climate.'[2]

The Four Horsemen of the Apocalypse – war, pestilence, famine, and death – still ride freely across the globe as they have since the beginning of time. To them, however, has been added a fifth horseman, who is man-made and an even greater nightmare: the population explosion.

This *fifth horseman* of the apocalypse specializes in both selective decimation and (in due time) mass extermination of the human race. The number of female babies being killed in China, India, Bangladesh, Afghanistan, Iran and Pakistan is reportedly skyrocketing. Census data are evidently missing, but reliable sources indicate that these countries now have a combined deficit of some 77 million women.

These are not just estimates made by Western countries. A late 2009 study by the Chinese Academy of Social Sciences highlighted the growing sex imbalance in China. It reported that 119 boys are born in the country for every 100 girls, and that, by 2020, 24 million men of marriageable age[3] might find themselves without wives.[4] More pessimistic estimates put that deficit at 44 million.

'Men only' is indeed a curious aspect of the ongoing explosion of *Homo sapiens*. Sociologists say that the motor power behind such an unstoppable trend is the market value of infant labour. Children are put to work at early age, but it is the boys, not the girls, who are more in demand.[5]

---

1 Held in Glion sur Montreux, Switzerland, on 15–16 April 2010; organized by the Geneva Association in the context of its programme on regulation, supervision and legal issues.

2 *Progress*, No. 51, June 2010, The Geneva Association, Geneva.

3 Liberally interpreted to mean men who are reaching, have reached or are briefly past marriageable age.

4 'China Faces Growing Gender Imbalance', *BBC News*, 11 January 2010, at: http://news.bbc.co.uk/1/hi/8451289.stm.

5 According to a research by ARD, the German national television station, 85 per cent of carpet-making in Iran is done by children. In India 20 million minors work in places as diverse as stone quarries, carpet manufacturing and prostitution.

The other side of the population explosion is that the number of people in poor countries is growing so fast that improvements in agriculture, health services and all-important water supplies have failed to keep up with the number of newborns. Employment opportunities are abysmal, and beyond that the poorer countries also have the world's worst environmental problems. Whether one acknowledges it or not, the population explosion is a major contributor to:

- food challenges,
- energy shortages, and
- environmental pollution.

Yet, during the Copenhagen circus of 7–18 December 2009 (see Chapter 14), the poorer countries who participated massively did not ask for assistance in controlling their population explosion. All they asked for was *more money* – which, as usual, will find its way into the pockets of corrupt, well-connected and well-fed politicians.

Along with the *fifth horseman* and his population explosion comes the surge of extra-large cities. 'The centralization of population in great cities exercises of itself an unfavourable influence,' wrote Friedrich Engels in 1844. 'All putrefying vegetable and animal substances give off gases decidedly injurious to health, and if these gases have no free way of escape, they inevitably poison the atmosphere...'[6]

- In Engels' time this reference to booming cities referred to then rapidly industrializing London.
- Now, as Figure 13.1 suggests, the population explosion has altered the geography, so that the wave of booming cities is located today in Asia, Africa and Latin America.

Engels was indeed prophetic. While richer countries – including developing ones like Brazil, China and India – pollute the environments through the smokestacks of their industries and heavy motor vehicles, the poorer countries are not far behind. The mountains of excrement, as well as disgusting drainage associated with their exploding populations, infects the people's own dwellings. Methane is more polluting than CO_2.

Many socialists, communists and Greens quote Engels and Marx, but conveniently forget Engels' wise words about the dreadful consequences of extra-large cities. Judging from the delegates' performance in Copenhagen 2009, one gets the message that they have no concern for environmental issues connected to overpopulation. At best, they consider them as being a luxury only the rich world can afford to talk about and that wealthier nations should *pay* those less developed to:

- reproduce as fast as they can,
- cover their basic needs for food and shelter, and
- provide them with not-so-basic commodities like mobile phone networks, cars and television.

6 Friedrich Engels, *The Condition of the Working Class in England*, Penguin Classics, London, 1987, p. 128.

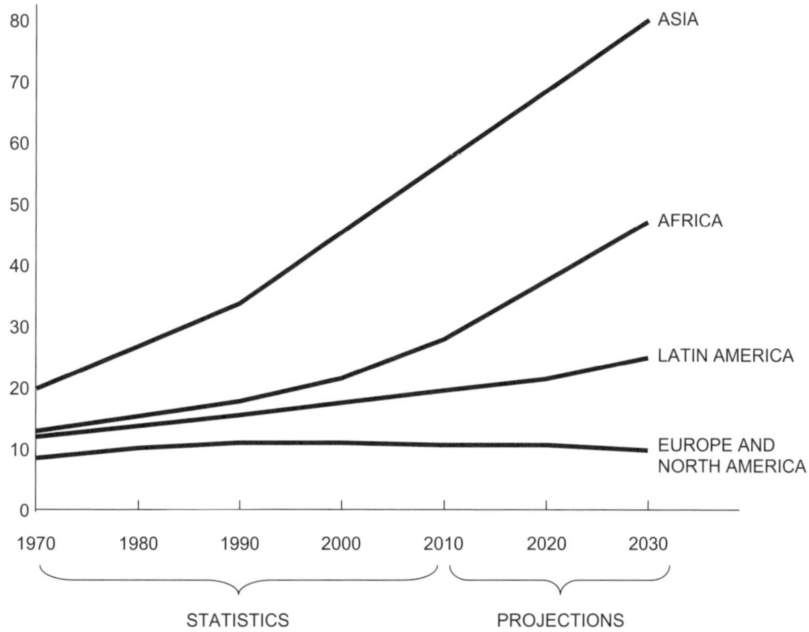

Figure 13.1 Number of urban centres with an exploding population of 5 million residents or more

It is quite curious that none of the developed countries' political leaders – from Barack Obama, who features a brand-new Nobel Prize for 'War and Peace',[7] to the Europeans – found the courage to tell the less developed nations what their salient problem is that first and foremost they have to put their own house in order, birth control being the crucial pillar of a sound, sustainable policy.

Priority should be given to quality, not to quantity. Putting quantity first is like putting the carriage before the horses. Today over a billion people have no access to clean water (Chapter 12), while water contaminated by sewage is a worldwide killer, particularly of children. It is both kinder and more rational not to give birth to children than to condemn them to that fate, because their country lacks not only the ability and willingness to promote family planning, but also the resources and skills to create sound living conditions.

Because of the population explosion, as well as pure greed, throughout Asia, Africa and Latin America forests are disappearing causing long-term concern about climate change as well as immediate economic damage. As we saw in Chapter 11, sporadic forest fires in Indonesia create a huge blanket of smog that envelopes much of south-east Asia and burn, in the process, other species which pay for *Homo sapiens'* irresponsibility.

Wide-ranging mismanagement is bound to happen when the world's population is increasing by an estimated 90–95 million every year, equivalent to one and a half times the population of today's Britain. According to UN projections, the pace of growth is so

7 But failed to keep his pre-election promise to substantially beef up American assistance to developing countries on urgently needed birth control and family planning.

fast that it will push global population above 9 billion by 2050, with the lion's share of that growth in developing countries which can the least afford it.[8] Nobody seems to be asking the most crucial questions:

- Can the earth's resources stretch to feeding an extra 3 billion people?
- Can rapidly increasing urbanization be handled without severe health damage to the city-dwellers?

The UN expects that by 2025 the number of people living in urban areas will exceed *5 billion* and that 4 billion of these will be in developing countries.[9] This rampant urbanization represents an unprecedented and unplanned migration of people into the cities, as in Africa and Asia, where it predominantly takes place, more than half the population still lives in the countryside (compared with only a fifth in Europe and North America).

In addition to energy shortages, *favelas*[10] and water pollution, air pollution is a major and growing concern. Interestingly, the sort of air pollution thought to cause the most damage to human health in developing countries comes from small particles of less than 10 microns in diameter. Known as PM10, they are caused by:

- coal burning,
- vehicle exhausts,
- dust stirred up by vehicles, and
- smoke from factories.

Such particles easily find their way into people's lungs, not just by respiring dirty air outside the house, but also through indoor air pollution which – according to medical practitioners – is as damaging to health as chain-smoking cigarettes. Quite often, air pollution indoors and outdoors is the same.

Providing dwellings for skyrocketing urbanization is yet another challenge. Anecdotal evidence suggests that in Asia's big cities a quarter of the urban population lives in slums. Usually, poor migrants arriving from the countryside are unable to afford housing in the city and they squat on land where they meet least resistance from landowners: steep hillsides, river beds or railway cuttings.[11]

It is no less evident that countries which have been utterly imprudent by failing to control their exploding populations are at a loss in providing basic food for the multiplication of mouths – let alone jobs. As for clean water and clean air, these belong to the realm of miracles. (The World Bank estimates that, in China, air pollution causes about 200,000 premature deaths among city-dwellers and some 1.8 million cases of chronic bronchitis annually.)

8 According to a press release issued by the UN Population Division, New York, 11 March 2008.

9 Michael Pacione, *Urban Geography: A Global Perspective* (3rd edn), Routledge, Abingdon and New York, 2009, p. 71.

10 A Brazilian term for spontaneous shanty-town settlements which spring up on the outskirts of cities and are populated by migrants from the countryside.

11 I was told in a research meeting that in Mexico City squatters make regular payments to local politicians in return for promises that they will not be evicted. But they do not necessarily benefit from municipal services such as rubbish collection, water supply, sewage and electricity.

In the West, environmental groups are now actively trying to change centuries-old practices with regard to cemeteries because the fluids from decaying corpses are polluting the water table. But in less developed countries there is no movement against polluted air and polluted water *as if* it doesn't matter overmuch in what dismal way the children who never asked to be born are going to die.

2. The Correlation between Birth Control and Quality of Life

Principle number one in terms of quality of life is that a world which tries to *banish death* by extending human longevity must also *banish outsize birth rates*. Birth control or, if you prefer, rigorous and enforceable family planning, has traditionally been a taboo subject. People are afraid to even talk about it because it is anathema:

- to the religious top brass,
- to big chunks of the general public,
- to all sorts of politicians, and
- to many of the so-called 'social scientists'.

'Death is not an event in life: we do not live to experience death,' said Ludwig Wittgenstein (1889–1951), 'If we take eternity to mean not infinite temporal relation but timeliness, then eternal life belongs to those who live in the present.'[12] This raises the all-important question: 'Have we sorted out our priorities on how to live in a healthy and sustainable way in the present?'

A factual answer to this query is vital because, as section 1 has documented, failure to exercise birth control is synonymous with an abysmal quality of life. Whichever way one looks at it, the earth cannot afford the population explosion. Shortages abound, from food and energy to education and jobs. Today there are *260 per cent* more people living on this planet than at the end of World War II. These billions of *Homo sapiens*:

- keep pumping more pollutants in the atmosphere,
- practise intensive, fertilizer-based agriculture which poisons the soil and water,
- continue to pile up more trash than can be handled by landfill, and
- are driving most other species – on land, in the air and in the seas – to extinction.

In short, they are damaging their own habitat beyond repair.

There is much speculation today on whether the earth's overexploitation and pollution (including air, land and water resources) has passed the point of no return. However, most people agree that by all likelihood we have already reached the limits of growth, while on average the human population as a whole is becoming more illiterate every day.[13]

- The visible result of unaffordable population levels is that 1 million children are prostituted every year in Asia alone.

12 Ludwig Wittgenstein, *Tractatus Logico-Philosophicus*, Cosimo Books, New York, 2007 (first published in 1922).

13 D.N. Chorafas, *The Social Cost of Business, Education and Employment in the European Union*, Gower, Farnham, 2011.

- At the same time, as competition from countries paying much lower wages intensifies, the Western world is being hit by savage and persistent unemployment.

Nearly everybody thinks that he or she has a right to reproduce. But this is not a right but a *privilege* which can be revoked, just like civil liberties are revoked in major wars and under siege conditions. Let's face it, the earth is under siege not by extraterrestrials but by *Homo sapiens* who have inhabited it for millennia.

It is utterly naive to think that this salient problem of humanity – *rampant reproduction* – will somehow take care of itself. Or that skyrocketing birth rates, quality of life and food, education and energy 'for all' can coexist. Yet, judging from the daily news reports, this irrationality seems to be the prevailing hypothesis.

In addition, some parts of the population are totally uninterested in taking their own future into their hands – and the gap between those still able to think and those who don't is widening. Half the babies in Texas are born to Latina mothers,[14] even though Latinos make up less than 40 per cent of that state's population, and they are in a more precarious financial condition than the others. This phenomenon is not present only in Texas. In Chicago's public-school district, 41 per cent of the students are Hispanics.[15]

On the other side of the world, more than 15 per cent of Indonesia's 240 million people are dirt-poor despite the fact that this has been a relatively rich oil-producing country. As with Latinos in the United States, the Indonesian poor are the section of the population multiplying the fastest. Not unexpectedly, unemployment is high, as the workforce is growing more rapidly than in any other Asian country apart from India and China.

It is therefore no surprise that inequality is consequently extending its reach and there is no prospect of reversing this trend. In addition, for the first time more than half the Indonesian population is living in urban areas, implying a further boost to the ills of urbanization and pollution. After more than four decades of being an exporter of oil, Indonesia is now a net importer of energy.

The dual effect of mismanagement of resources and of population explosion sees to it that other oil-producing countries, too, have been confronted by an energy squeeze, including power shortages. In early February 2010 Hugo Chavez, the CEO of oil-rich Venezuela which nevertheless depends on hydroelectric power for 70 percent of its electricity production, declared an 'electricity emergency' as a result of drought. Customers who did not cut their power usage faced steep increases in their bills.[16]

In Turkey and Iran more than half of the population is under 30. Worse still, in India 40 per cent of the population is under 15.[17] The Indian prime minister wants to turn this youthful population into an asset by nurturing an educated class. This, however, requires a massive training effort, significant improvements in educational infrastructure, faculties whose quality and numbers must be ahead of the population explosion curve, and great deal of money. Contrary to these stated objectives:

---

14 There is no uniform way in America to describe people of nationalities 'south of the border', in an all-inclusive way. The most frequently used terms are Latinos, Latin Americans and Hispanics. In this text they are used interchangeably because the quoted sources employ all three terms.

15 'Reshaping Politics', *The Economist*, 9 January 2010.

16 'Chavez Puts Venezuela under "Electricity Emergency"', *BBC News*, 9 February, at: http://news.bbc.co.uk/1/hi/8505906.stm.

17 *Sciences Magazine*, Winter 2010.

- the unprecedented surge in birth rates has been accompanied by falling educational standards, and
- because of the brain drain, mainly to the United States and Britain, India is confronted by a shortage of innovative scientists and engineers.

The drama is that the reactionary forces of society who are dead-set against birth control but care little about these children's quality of life, don't give up. The so-called *Tea Party* movement in the United States offers an example. In 2010 the star speaker was Sarah Palin, the defeated Republican vice-presidential candidate of 2008 whose 15-year-old daughter had an illegitimate child.[18] Despite that, because of her ultraconservative views Palin became the darling of religious fundamentalists.

What all these occurrences have in common is the sense of a total lack of political leadership, with the result that the long-term drivers of population growth remain in place. The earth as a whole and, with it, quality of life, pays the price. Excuses are easily found, one of them being divine will and another being the uncertainty of action on birth control. But it does not take a genius to appreciate that, despite the uncertainty associated with any action, the price for inaction is much higher.

3. Man against Himself[19]

Estimates made by biologists, physiologists, botanists, medical doctors, disease specialists and other scientists point to a high probability that *the human race as a whole, whatever its colour or creed, is on its way to total extinction*. And *Homo sapiens* will not go out of this world alone. Like Sampson in the Temple, the human race will take many other species with it to the mass slaughter – although some highly resistant forms of life like the arthropods may survive and become the new master species of the earth.

When is this God-size genocide, for which we are all responsible, going to happen? Estimates and answers vary, but some scientific responses indicate that with the current rate of accelerated reproduction and environmental degradation, there is a distinct probability it can happen in the next 50 to 100 years *if* the ongoing pace of overpopulation and self-destruction continues.

The question we should ask ourselves is: 'Is this the future me and my family are preparing for our children and grandchildren? If not, what shall we do to avert such catastrophe?' An answer I often hear when I ask this question is that 'technology will take care of that.' (How wrong this answer is will be explained in Chapter 15.) In the 1960s the answer to a similar question about prospective food shortages was the Green and Blue Revolutions but, as we saw in Chapter 7, they turned out to be huge mirages.

There are good reasons why some people – and scientists are prominent among them – take a rather pessimistic view of technology's contribution. First is the fact that science and technology have not provided, by themselves, the means of satisfying society's

18 According to the newborn's grandmother, this was a gift from God. It was probably brought by Archangel Gabriel in his second descent to earth.

19 In the early 1950s Dr Karl Menninger wrote a terrific book *Man Against Himself*, largely about self-destruction. The six decades following its publication extended the perspective of Menninger's reach.

expectations for a better life without inflicting undue hardship on the environment. In fact, they have traded the one for the other.

Slogans, of course, come and go. In the 1970s, as satellite networks started circling the earth, the slogan was 'Communicate, don't commute'. Working from home was thought to be both environmentally-friendly and efficient, but we now know that the proliferation of computers and communications has had major unexpected consequences, from the surge in energy demand to pollution.

This is true even of the life sciences. Optimists thought that advances in technology would play a decisive role in the reduction in birth defects, only to find out afterwards that in reality such defects are increasing because of pollution and environmental poisoning. Science has *not* entered a strange netherworld. Rather, people misjudged the nature of its deliverables.

In Paris, in the early 1970s, Professor Maurois found that 10 per cent of newborns had a defect, and that 10 per cent of *that* population was seriously defective. Over the years, these numbers have increased, particularly in the less developed countries where famine associated with overpopulation creates an appalling number of brain-damaged children. George Bernard Shaw once observed that:

- Man has invented nothing new in the science of life.
- But, by contrast, in the art of death he had surpassed nature itself.

Nor should we lose sight of the fact that many of man's discoveries and inventions eventually turned against him. We have seen lots of evidence why man-made chemicals and the massive way in which they are being put to use can bring more extermination than scarcity of food has ever brought.

True enough, chemistry and pharmacology have made available medicines which helped to control infant mortality and more than doubled the average human life over a period of several hundred years. But excesses have started to reverse this trend,[20] and this relapse is still in its infancy. Biotechnology has created major advances in the life sciences and may well deliver more. It is no less true, however, that the tools of biotechnology can be used to create monsters – and ethical standards, which might have stopped this happening, are not in place.

In other words, our age is full of contradictions. There is the possibility of rejuvenating brain cells and thereby deferring the onset of senility, but at the same time, in wide areas of the globe, the starvation of those who should never have been born damages their brains beyond repair. There are new materials to use in prosthesis, and wars worldwide are mutilating scores of people (about 6,000 American soldiers were killed in the Iraq War, engineered and mismanaged by George W. Bush, but over 100,000 were wounded and mutilated for no purpose).

Several routes have been taken towards a reduction in the incidence of heart diseases, but both heart disease and cancer are on the increase. The American Cancer Society estimates that 45 million Americans now living will eventually develop cancer. This means that malignant disease will strike the majority of US families. Statistics on children and terminal illness are even more disturbing.

20 In Russia, for example, the average human lifespan is now getting shorter.

- Half a century ago, cancer in children was considered a medical rarity.
- Now, more American schoolchildren die of cancer than from any other disease.

A significant number of malignant tumours are discovered clinically in children under the age of five, but it is an even more sobering fact that an increasing number of such malignant growths are present at or before birth. The prevailing hypothesis is that they are created by countless new cancer-causing chemicals with which the father and mother of the unborn are continually in contact. Are these carcinogenous unborns our society's goal? Is it what the different religious movements try to protect through their opposition to birth control?

Curiously enough, man has become ingenious in finding ways which ensure that he will not be the dominant species on this planet for much longer. We are imposing many forms of environmental degradation on ourselves and we face a long list of associated challenges – scientific, energy-related, health-connected, industrial and agricultural – over and above political and social problems. In the aftermath, it is hard to know where to start in reversing the downturn.

- Mind-boggling disasters are predicted in regard to the environment.
- Energy and food solutions, which superficially look brilliant, fall under their own unsupportable weight.
- Proposals with little chance of being adopted are pouring out at a bewildering rate, while the political leadership around the globe has neither clear focus nor priorities.

Despite the media's increasing coverage of these issues over the past few years, Americans, Europeans and people in all other countries are still ill-informed (and generally unconcerned) about the ongoing crisis. Partial solutions do not answer the *real* 'better life' requirements which relate to the way people live, the education they receive, the food they eat and the energy resources they consume as well as:

- the creative activities in which they engage, and
- their respect for the environment in which they are born, work and die.

To test what has been stated in the preceding paragraph, ask your priest where he stands in terms of family planning and environmental protection. Is he so preoccupied with the next life – from which nobody came back to tell us the whereabouts – that he forgets that we live *now* on earth? Is he so blind to the issue of birth control that he wants to see the newborn deformed and eventually the whole human race extinct from the face of this planet?

Above all, ask your own conscience. Why are you bringing children into this world? Do you wish them to suffer at a young age because the world around them is becoming increasingly polluted? Do you wish to have deformed grandchildren because you did not take the necessary steps to safeguard the precious resources you inherited but did not bother to maintain? To put it bluntly, have you joined the club of Man Against Himself?

4. Fertility Will not Fall without Licensing the Newborn

One of the basic principles I never tire of repeating is that a world which tries to banish death by extending human longevity must necessarily banish birth or at least it closely regulate it. *Licensing the newborn* is an effective way for doing so. We give licences to people who drive cars – why should we not do so for activities that are so much more important for humanity's future like:

- child-bearing, and
- raising children – an act much more complex than giving birth.

Just because *child-bearing* has been traditionally thought to be granted by nature doesn't mean that everyone is fit to be a parent, and even less that everyone has the character, serenity and financial means to do so. Is it fair on the newborn to be brought into the world by parents living *on rent*,[21] people who are so busy receiving social-security handouts they can't occupy themselves with:

- what it takes to raise their children, and
- make them into worthy citizens of society?[22]

In an inspiring lecture at Washington State University back in 1967, when I was a member of the faculty, Dr Kenneth Boulding advanced the suggestion that every newborn should be given a licence to reproduce, say, up to 1.05 offspring. This licence should not only be issued but also policed. Boulding saw the licensing of new births as a necessary regulatory activity, whose absence would lead humanity into lots of trouble.

The elapsed four and a half decades provide ample proof that Dr Boulding was right. The unwarranted overcrowding of this planet with billions of *Homo sapiens*, unsupportable by the earth's resources, is a vote of confidence in the licensing scheme. In fact, some self-respecting states have adopted it. For nearly two decades Singapore's enlightened leadership has drummed the motto 'Two is Right', putting incentives and disincentives into motion to ensure that it is observed by every couple living in the city-state. Far-sighted Mao placed the bar at one child per family (although his successors removed it). These enlightened policies are, however, the exception. The majority of the world, and most particularly the developing countries, ignored Boulding's call. As a result, they are the parties responsible for the current strangulation of available resources and the ongoing destruction of planet Earth.

- Measures now have to be taken that are much more dramatic than would have been necessary in the late 1960s, and
- the metrics to be applied have to be more stiff than those required 44 years ago.

To actually shrink the human population inhabiting spaceship Earth to manageable numbers – eventually at or below the 1 billion mark worldwide – the reproduction licences

21 By abusing endowments as well as on all sorts of benefits, charity-as-a-policy, and the like. 'On rent' does not mean in rented property.

22 And, even worse, some have children in order to hold on to their benefits and family allowances.

to be issued from Day 1 can no longer be 1.05, as Boulding suggested. They have to start well below the 1.0 level per newborn.

For instance, a 0.7 birth licence would amount to 1.4 children per family, and, as Ken Boulding aptly suggested, 'the free market will be playing its role to bring the coupons up to an integral number'. Those who can afford to bring up children would buy the difference from those who cannot. Heads of state who flocked to Kyoto, Rio de Janeiro, the Copenhagen circus and other environmental entertainments want to trade CO_2 licences.

- So, why not trade licences for the newborn?
- After all, it is people who produce CO_2 – and the methane – and not only by burning coal.

In Dr Boulding's view, such a free market mechanism will place the rearing of families where it is affordable and better managed. This means the more stable families with higher net incomes, better positioned to rear children and give them the necessary education – the luggage they most need for their future.

There is a precedent to what I am saying. In 1991 the Board of Supervisors in San Mateo, California, took the unprecedented step of requiring pet owners to acquire a reproduction licence or sterilize their pet. Why not to apply the same principle to *Homo sapiens* whose population explosion is so much more of a world disaster? Let's not be so arrogant as to think that man is an animal superior to all others.

It should also be noted that some parts of the globe have already applied a voluntary standstill. In 1900 Europe accounted for a quarter of the world's population. Projections indicate that by 2050 it may account for between 7 and 8 per cent, and almost a third of these people will be more than 65 years old. This has several important consequences, one of them being economic.

Tough economic situations are well-known in less developed countries, and they are made worse by skyrocketing birth rates. But taboos blind governments, making them unwilling to tackle one of the biggest barriers to sustainable growth: redimensioning the population with a longer-term perspective. Zimbabwe provides an example where decades of *laissez-faire* in population explosion has left the country with over 50 per cent of its people:

- poorly educated,
- out of work, and
- young and hungry.

The population explosion fans answer that the divine doctrine is inscribed in Genesis 1:28: 'Be fruitful and multiply and replenish the earth and subdue it; and have dominion over the fish of the sea and over the fowl of the air and over every living thing that moves upon the earth.' That's high-handed, of course, but at the time the Genesis book was written it might have made sense. Not today.

In this modern age, 'be fruitful' and 'multiply' contradict one another – while uncontrollable multiplication is the population equivalent of the hydrogen bomb. As John Bongaarts, vice president of New York-based Population Council wrote in a letter to *The Economist*, although much of the developing world now benefits from declining fertility the demographic prospects for sub-Saharan Africa (in which 'multiply' has become sort of a religion) are grim.

- Its average fertility is more than double that of Asia and Latin America, and
- its population is expected to grow by 1 billion over the next half-century, with high levels of unplanned child-bearing because many women lack access to, and information about, contraception.[23]

Population growth might have been necessary when *Homo sapiens* was uncertain about whether it could secure a place on this planet. By contrast, today in order to accommodate the minimum needs of an exponentially expanding world population, it is necessary to build houses, hospitals, schools, ports, bridges, agricultural domains, factories and all other kinds of facilities. This needs to be done in numbers that almost equal all the construction work undertaken by human society to date.

- It does not take a genius to understand that when birth rates are exploding this is an impossibility, and
- by consequence, from education to food, energy and hygiene, quality of life is diving and will continue descending into the abyss.

What many people don't seem to have understood, and what political leaders are too apathetic or afraid to talk about, is that without licensing the newborn and policing whether the licence is thoroughly observed, the only way human society and the environment can go is to hell. There are now simply too many people for the earth's fragile ecosystem, and voluntary constraints don't work.

The time to stop and ideally reverse the population explosion is *now or never*, and by several accounts it is already too late. As an article in *The Economist* had it, the abstract notion of celebrating falling fertility is like congratulating the captain of the *Titanic* on heading towards the iceberg more slowly.[24]

5. Facing the Challenges of Downsizing

More than four decades after the religious revolution, and its accompanying exploding birth rates, at long last Iran provides a good example on downsizing – because the people themselves now believe that fewer people are better. When the clerical regime took over in 1979 the mullahs wanted their flock to go forth and multiply, and they abolished the country's family planning system.

- Fertility rose, reaching 7.0 in 1984.
- But in the 2006 census the average fertility rate had fallen to 1.9, and just 1.5 in Tehran.[25]

23 *The Economist*, 21 November 2009.

24 *The Economist*, 31 October 31.

25 'Go Forth and Multiply a Lot Less', *The Economist*, 29 October 2009.

Unfortunately, there are not many stories like this to brighten the book's pages. In China, Mao was successful in bringing fertility sharply down because he saw that his country was sinking under a population explosion. Other political leaders, however, are not that far-sighted. Or they simply lack the guts to explain to the people what needs to be done and take the necessary measures – in short, *to lead* because, like any important enterprise, population downsizing has basic prerequisites. To summarize:

- The first basic prerequisite is family planning.
- Educating young and old is another 'must'.
- Providing the information and means necessary for avoiding accidental births is still another basic condition.

Brazil did that much when its government distributed 55 million condoms for just one event: the 2010 carnival. No matter that it might involve actions forbidden by taboos, or be objected to by those with vested interests. Leaving things that are going downhill to their own devices is disastrous, as documented by lessons learned from the events which followed the Spanish conquest of the New World. While the white and mixed population of the Americas had continued to grow, the Indian population of Mexico shrunk, scourged by terrible epidemics in 1545–1546 and again in 1576–1579. Implosion brought it from some 11 million at the time of Mexico's conquest in 1519 to little more than 2 million by the end of the sixteenth century. A similar fate overtook the native population of Peru.

Hell broke loose because health conditions had become abysmal, with important economic consequences. The labour force on which settlers depended was dramatically reduced, raising a number of challenges to which solutions had to be provided before it was too late.

- The great building projects were abruptly halted.
- The problem of food shortages was met by a drastic agrarian reorganization (see also Appendices A and B).
- As it became increasingly difficult to find labour for the mines, society had to review its priorities.

With muscular power still the basic energy source and slavery the order of the day, labour was brought from Africa to replace the Indians, but it, too, proved to be vulnerable to diseases. The century that followed the great Indian epidemic of 1576–1579 has been called 'New Spain's century of depression'. It was a time of economic contraction, during which the New World closed in on itself.

Forecasting, planning and controlling were not the sixteenth century's *forte*, and the consequences spread far and wide. During that time of change, the New World's markets came to require less of Europe – or at least of Spain. Moreover, then as now, European luxury products found themselves competing with the products of the Far East carried to the Americas in the Manila galleons, while several basic goods were produced by the colonies themselves.

- Mexico had developed a coarse cloth industry.
- Peru was producing grain, wine and oil.

These were exactly the goods which had bulked so large in the cargoes from Seville during the preceding decades. The staple Spanish exports to the Americas were ceasing to be indispensable to the settlers, and in 1597 Spanish merchants found it impossible to dispose of all their wares. For the first time, the New World's market was overstocked (an event repeated with the late twentieth century's globalization).[26]

The changing demands of the New World's market presented the Castilian economy with problems of readjustment which it was ill-equipped to tackle – something that happens these days with American and European products. During the preceding decades there had been a total failure in reversing the economic trends apparent during the later years of the reign of Charles V. Neither the Spanish industry nor its agriculture was able to meet the challenge of:

- changing demand, and
- increasing foreign competition.

Late sixteenth- and seventeenth-century text reads like a page ripped out of the 2010 Almanac, and what it teaches must be taken most carefully into account. Downsizing a booming population can be managed, but we should carefully study the process and its implications, as well as experiment with the best way of how to do it. Downsizing will affect all societies. Passing the test of an inflection point is always difficult, but it can succeed:

- *if* it is done by absolutely first-rate, ethical people in every sense of the word, and
- *if* it is managed by forward-looking leaders of the high quality the world needs in times of crisis – Churchill and Roosevelt being examples.

The fact is that there are no alternatives to downsizing. Not only is the world's population still increasing, but it is doing so at an accelerating pace. The preceding 12 chapters have documented that this can do a lot of environmental damage before it peaks at just over 9 billion in 2050, as various experts are projecting.

Any plans made today to stop (let alone reverse) the destruction of the earth and its environment must fully account for the fact that, from now on, more damage comes from the poor, fast-growing countries which try to follow the economic trajectories of the rich world, but have neither skills nor the means to do so. In addition, if the developing countries try to emulate the rise in standard of living currently established by Europe and America with an additional 3 billion people:

- from energy to food, they will create an irreversible degradation of the earth's resources with grim consequences for themselves and everybody else, and
- this will happen in violation of any good sense, since the parts of the world where the population is growing fastest are also those most vulnerable to climate change, water shortages, declining food yields and poor energy supplies.

26 D.N. Chorafas, *Globalization's Limits. Conflicting National Interests in Trade and Finance*, Gower, Farnham, 2009.

An extremely dangerous idea has been floated by silly leftists during the Copenhagen circus: that there should be a way to accommodate 'climatic refugees'. In other words, the very people who are now causing the mass destruction of the earth through ignorance, taboos, lack of leadership and/or means to implement birth control,[27] or misguided religious beliefs must be 'accommodated' by providing them with residence in the West – all expenses paid.

6. Establishing the Right Priorities for a Turnaround

Section 5 has explained the need for downsizing the population of *Homo sapiens*, with the goal of having fewer but better educated and healthier people, economizing on the use of the earth's remaining resources, and making life on this planet last longer. These are both goals and basic prerequisites for survival, and are therefore priorities.

They are also only the beginning of what is needed for a turnaround. Under currently prevailing conditions, even an actual reduction in the number of *Homo sapiens* alone will not reverse today's highly destructive policies in terms of overexploiting natural resources. We have to develop and instil a totally new culture which looks at spoilage as criminal offence (see also Appendix C).

'The cause of all these evils was the desire for power which greed and ambition inspire,' wrote Thucydides, the great historian of the Peloponnesian War.[28] In our twisted culture, power is based on big numbers, not on quality. The rise in consumption of food and energy (see Appendices A and B) is a measure of numbers – not of quality. The same is true of:

- birth rates,
- infant mortality,
- unemployment,
- sickness,
- persistence of poverty, and
- the growing list of social problems.

Social problems are multiple and perverse. They also tend to grow exponentially with the travails of the society of which they are part, while the ability of people to cope with them is rather limited. Beyond that, terrorism (including state terrorism) has become the order of the day. (As a friend observed, you can keep terrorists under lock and key, but not terrorism.)

The able handling of social issues affecting rapidly rising populations is an integral part of a successful turnaround. The basic sense underpinning such social issues is that they are the dual product of both *nature* and *culture*. *If* morality has an evolutionary component, argues Frans de Waal, *then* it must have its roots in pre-human species, in

27 Among the king-size mistakes of the George W. Bush administration was the starving of a commendable American effort to supply the developing countries with planned parenthood guidance and contraceptive devices.

28 Thucydides III, 83.

which the precursors of morality provided the raw material that natural selection acted on in the process of human origins.[29]

De Waal suggests using modern ethological methods of observation, combined with evolutionary theory focusing on the proximate causes of behaviour, rather than on its ultimate functions. In his opinion, this allows us to understand much more than previous generations of animal behaviourists. In the background lies the fact that:

- by limiting the scope of inquiry, researchers can attain greater certainty about the questions they try to answer, and
- the key to better certainty lies in explicit and precise definition of terms, so that investigators can make testable predictions.

This is not the policy followed today in most social studies. Instead, a common error is that of adding multiple layers of interpretations (some of them contradictory to one another) to everything one is watching or studying. Such a policy confuses the issue, blurs vision and makes back-testing impossible.

In the opinion of José Ortega y Gasset, the well-known scholar and clear thinker of sociology, the accession of the masses to complete power has altered the social equation as well as public life. The consequence, Ortega suggests, has been widespread because cause and effect has not been solely political. It is also, and even primarily,

- intellectual,
- moral,
- economic, and
- religious.

As a sociologist, José Ortega y Gasset has been visionary; he foresaw that a rapidly rising population will shift political power to the masses. His analysis particularly targets a characteristic of our times: the predominance of the mass which is frequently acting directly, outside the law; and imposing its aspirations and desires by means of material pressure.[30] That's exactly what has happened in mid-December 2009 during the Copenhagen circus (see Chapter 14).

When he spoke of the revolt of the masses, Ortega was way ahead of his time. Eight decades later this has become global, and the time has come for hard choices. Today, society's survival in an orderly form requires that our *priorities* are set correctly. It was Frederick the Great who said: 'He who would defend everything defends nothing.' To be successful in our defence of the earth's resources – including the human resources – we must first decide:

- what we wish to save, and
- at what price we are willing to do so.

29 Frans de Waal, *Good Natured: The Origins of Right and Wrong in Humans and Other Animals*, Harvard University Press, Cambridge, MA, 1996.

30 José Ortega y Gasset, *The Revolt of the Masses*, W.W, Norton, New York, 1932.

To find our way to sanity, we should learn from social research focused on other animals than *Homo sapiens*. As Konrad Lorenz, who studied the social behaviour of geese as proxy to that of mankind, emphasized,[31] today far too many people are alienated from nature because they seldom encounter anything but inanimate, man-made things in their daily life. As a result, they have lost the ability to:

- understand living creatures, and
- effectively interact with them.

The loss of real contact with nature helps explain why *Homo sapiens* exhibit such vandalism towards the living world: from other animals to the plants, the trees, the soil, the sea, the air to the water resources. It is unfortunately true that somehow the majority of people have lost the ability to understand that it is *nature* which makes our life possible.

Let me take an example from other species. Most animal populations are limited in size by some material constraint which they have the ability to perceive, and the will to adjust to it – for instance, a temporary scarcity of food. But humankind continues to reproduce well beyond the level it can sustain and then depends on charitable or aid-related assistance for rock-bottom survival.

Ecologists have also noticed that animals living in risky, unpredictable environments tend to have many offspring, but don't do much in the way of looking after them. This makes a sort of sense. If the world is hostile, many individuals will die of simple bad luck, so investing heavily in a particular offspring is risky. But *Homo sapiens* well-wishers don't allow Mother Nature to follow her course.

- They step-in to halt the process of death and recycling, and
- they do so without taking the elementary precaution of regulating the process of birth, as other animals do.

So-called religious leaders are among the more myopic, and are therefore the worst offenders. The Vatican was successful in correcting the practice of young priests, particularly in Latin America, who confused the word of Marx with the word of the Lord. But it has failed to take a much more important initiative: controlling the population explosion which threatens to destroy *all* creatures on earth. The same is true of all religions and their priests, mullahs and imams. This is self-defeating because:

- the problem of priorities is simultaneously religious, social and cultural, and
- not only politicians, parents and educators but also religious leaders must actively support birth-control practices.

Some years ago, in an Urban Institute study, economist Gregory Acs analysed the behaviour of a nationwide sample of women aged 14 to 23. He found that education, family, race and income have a much larger impact on child-bearing decisions than the generosity of welfare cheques.[32] But the latter should not be discounted.

31 Konrad Lorenz, *Das Jahr der Graugans* [*The Year of the Greylag Goose*], R. Piper & Co. Verlag, Munich, 1979.

32 Business Week, 31 January 1994.

Authoritarian politicos (elected or selected presidents, dictators, generals, archbishops, ayatollahs and monarchs included) have largely misled their folks as they press for higher birth rates without accounting for the fact that, when the youngsters grow up, they will lack:

- jobs, and
- hope for a better future.

This error in putting one's priorities right creates havoc for the coming generations, as well as a permanent state of distress for natural resources. It also increases the likelihood of future conflicts on a planet crowded with *Homo sapiens* where the freedom of action increasingly diminishes. 'Your freedom of action ends where the nose of your neighbor begins,' said a New York judge to a defendant who claimed that the harm he created to somebody else was part of his freedom guaranteed by the Constitution. Rapid reproduction habits have seen to it that the nose of 'thy neighbour' is coming closer and closer.

7. Back to Basics: The Earth's Population must be Commensurate to Resources, Well-educated and Self-sustainable

Women in Western countries were limiting the size of their families long before the contraceptive pill came along. Practically all over Europe, the fertility rate since World War II has been plummeting, and in some countries the population is indeed shrinking. Italy, Greece and Russia are examples.

This policy of self-control is far from being a universal trend. According to a Geneva Association article, today the median age for the world is 28 years. That simply means that half the world's population is at or below that age, and this has very important negative consequences on countries and their people.

- The state with the youngest population is Niger, with a median age of 15 years.
- The state with the oldest population is Japan, with a median age of 44 years.[33]

Thinking rationally, it should be the other way round – because Niger is one of the poorest countries in the world, while Japan is one of the richer. In Niger's case this statistic is a demonstration of no governance at all; the country is just drifting, having been hit by major problems, such as annual drought and food shortages. What a median age of 15 years (or of 18 years in other African countries) reveals is that:

- the state simply cannot manage itself, and
- the country's impoverished farmers find themselves in full-blown crises without let-up, while newborns die of malnutrition.

Indeed, in several parts of the world, poor and sporadic rainfall decimates the cereal crop and, according to anecdotal evidence, in some years families are resorting to eating

33 *The Four Pillars*, No. 46, The Geneva Association, Geneva, March 2010. Geneva-based, the Association is the research and education arm of the insurance industry.

wild leaves to survive. But they do nothing to control their rampant reproduction. Crisis follows crisis, and the situation obviously worsens with unprecedented birth rates which are unaffordable in:

- human,
- educational,
- social, and
- economic terms.

The hypocrisy surrounding the taboos around birth control is further documented by the statistics brought to the reader's attention at the very beginning of this chapter (section 1). In developing countries, not only are birth rates still soaring, but the distribution of the sexes is also extremely skewed because of widely practised infanticide. Asia is 'missing' 96 million women said the United Nations on 7 March 2010.

A feature article in *The Economist* put it this way: 'Baby girls are thus victims of a malign combination of ancient prejudice and modern preferences for small families.'[34] Countries that are committing genocide of the newborn and those allowing rampant reproduction, taking no account of requirements for food, education, health and jobs, have no right to raise a voice against birth control. In fact, they would have done better to:

- educate their people from an early age about contraceptive measures,
- provide, like Brazil (section 5), millions of free condoms, and
- make serious, enforceable family planning, commensurate to available natural resources, the law of the land.

Political will can correct runaway birth rates and sex preference. A positive example is that of South Korea which had as bad a sex discrimination ratio as China's. The country's culture started to change. Female education and other factors made preference for sons seem old-fashioned. South Korea is now heading towards 110 newborn males for 100 newborn girls – still an imbalance but not as bad as China's.

'The old mindset with its preference for making children has now combined with modern medical technology' that makes it easier to predict and abort unborn girls, wrote Anuradha Rajivan, the lead author of a United Nations Development Program (UNDP) report. She added that it is not just female infanticide but sex-selective abortion of unborn girls that cause the so-called 'missing' females.[35]

- It is this massive genocide, not birth control, which should be taboo.
- If this wrong culture does not change, the problems associated with sex selection will get worse as the boys born in large numbers reach maturity.[36]

Because of this wrong policy, Kenneth Boulding's reproductive licence should be endowed with a database which tracks birth rates not only for population control

34 'Gendercide', *The Economist*, 6 March 2010.

35 A Hindi proverb says: 'Raising a daughter is like watering your neighbour's garden.'

36 There is research on this issue by the World Bank and Chinese Academy of Social Sciences, among others.

purposes, but also for sex discrimination and genocide. Restoring nature's balances has great merits, but where are the political leaders willing to take the initiative?

As the December 2009 UN-sponsored climate change 'summit' (Chapter 14) demonstrated, priorities have been turned on their head. The best proof of throwing money at the problems and getting no results is provided by the fact that the Danish capital quickly went from being 'Hopenhagen' to 'Brokenhagen' and '*Fiascopenhague*'. Hillary Clinton, the US Secretary of State, described it as the worst meeting she had been to since eighth-grade student council.[37]

Worse still, instead of seeing a determined effort to attack the salient problems of *Homo sapiens* and environmental protection, one hears and reads about a new campaign (probably by speculators) which proposes to give every newborn child in the world an online bank account with $100 in it. The goal of so-called FinancialAccess@Birth (FAB) campaign is to do something about the fact that:

- deep-in-debt consumers are not spending as much as they used to, and
- some new initiative is necessary to get them rolling again, no matter what might be the consequences.

The original idea comes from Bhagwan Chowdhry, a finance professor at UCLA, who suggested that funds would come through charities,[38] a handout-style approach projected as a policy. Nor is there any evidence that thought has been given to the fact that turning people into money machines from the moment they are born provides plenty of fuel for the next financial crisis – and we are not yet out of the 2007–2011 debacle.

Quite to the contrary, rather than opening bank accounts for newborn babies, indirectly inciting them to become big spenders of money they don't own, later on in life, or encouraging them from Day 1 to go hat-in-hand to charities, what the current and next generations need to do to avoid extinction is:

- reorganize,
- downsize, and
- deleverage.

It is time to get serious about the earth's inhabitants and their future. That means not only *Homo sapiens*, but also those being exterminated by them: elephants, gorillas, whales and bluefin tuna, among others. The strategic risks of looking the other way or of ring-fencing crucial issues as untouchable outweigh the risks of intervening to correct the balance. To improve ordinary people's likelihood of survival, the human population in every country and worldwide must be:

- commensurate to resources,
- well-educated, and
- self-sustainable in the longer term.

---

37 *Financial Times*, 17 May 2010.

38 Chowdhry no doubt knows that roughly one-third of the population of India lives in shanty towns. Will his proposal change this for the better? Would it not be more rational to limit the population explosion in order to improve the condition of people?

Doing nothing will eventually amount to a massive and crippling shift in the quality of life for all people, touching everybody's life essentials. The ability to sustain an ever-increasing number of newborn and elderly is currently being severely tested, and it will surely become a dead weight on the shoulders of the next generation.

Longer-range planning is a 'must', and no such effort should forget that the population of older people is also rising steeply. According to the Geneva Association, the fastest-growing population of senior citizens is that of the oldest – aged 80 years or over. Their numbers are currently increasing by 4 per cent per year,[39] faster than those of the newborn – and the elderly, too, have to be subsidized.

All these issues are highly relevant because we belong to human communities of one sort of another and thus share common obligations arising from the interdependence they involve. This includes our obligations to future generations. Contrarians might respond that deep reasons within our existing culture make it very difficult to provide a moral basis for an *obligation to the future*. But no generation should leave planet Earth and the economy worse than it found it.

39 *The Four Pillars*, No. 46, The Geneva Association, March 2010.

CHAPTER 14 United Nations and its Conferences on Environmental Collapse

1. The UN's Entry into Environmental Issues: UNEP and the 1987 Montreal Protocol

The environment and its protection were not part of the United Nations[1] original charter. This is important inasmuch as the UN lacks statutory authority to deal with energy regulation and environmental issues. The most it can do is to exercise a sort of inverse delegation to its member nations. This is a process which requires unanimity to reach any meaningful decision; it is also a tough task:

- with some 188 member nations, ranging from tiny countries to superpowers, and
- largely diverse interests on practically all subjects, most particularly on energy, water and environmental care.

The UN's involvement on environmental issues was preceded by repeated calls by authors and scientists who warned of the destructive effects of acid rain and global warming well before the first Earth Day in 1970.[2] There were also individual battles, like those to save the African elephant and rhinoceros, but the results were rather limited.[3]

The concept of a transnational agreement, in which the UN might play a role, has its roots in the 1972 United Nations Conference on the Human Environment. Held in Stockholm, this conference set up the United Nations Environment Programme (UNEP), which declared that 'States have ... the responsibility to ensure that activities within their jurisdiction or control do not cause damage to the environment of other States or of areas beyond the limits of national jurisdiction'. This is also part of Principle 2 of the Rio Declaration of 1992.

It goes without saying that this is an oxymoron, because environmental pollution does not recognize borders and jurisdictions. The notion that those who pollute should at least clean up after it, or even bear the cost of preventing environmental damage, was adopted in that same year, 1972, by member countries of the Organization of Economic

1 'Ce machine' as Charles de Gaulle used to call it.

2 To the best of my knowledge, Rachel Carson's *The Silent Spring*, published in 1962, was the first book to arouse wider public interest in the overuse of chemicals and the effects of deforestation on animal life.

3 Although only South Africa, Namibia and Botswana could claim full protection of the species. Activists championed a plan, sponsored by the World Wildlife Fund, for the consolidation of dozens of game reserves in South Africa into what would become the world's largest conservation area.

Cooperation and Development (OECD). Although of the two – UNEP and the OECD – it was UNEP that developed ideas which led to environmental negotiations, this in no way gave the UN the authority to take action.

However, there has been a swarm of discussions. For instance, in negotiations for the protection of the Mediterranean, UNEP came up with the proposal of a framework convention in which countries set out a few broad lines of agreements. As an example, the Vienna Convention for the Protection of the Ozone Layer, signed in March 1985, incited those who think with their heads to think more about environmental issues.

- Talks on a protocol began in December 1986, and
- in March 1987 they resulted in the Montreal Protocol on Substances that Deplete the Ozone Layer, aiming to halve chlorofluorocarbon (CFC) production.[4]

Indeed, looking at it in pragmatically, the first consistent cross-border effort to do something for the environment was this 1987 Montreal Protocol, ratified by the United States and 30 other nations. Although it pledged a rather vague 50 per cent reduction of ozone-layer-destroying chlorofluorocarbons by 1999, it can be seen as a first step. Two years later, the European Union took something of a lead in dealing with the threat to the ozone layer when, in 1989, environmental ministers, meeting in Brussels, agreed that EU countries:

- would reduce CFC production by 85 per cent 'as soon as possible', and
- try to ban the chemicals altogether by end of the twentieth century.

For all practical purposes, the UN's first entry into the battle to save the planet (if it can thus be described) took place a year after Montreal, in 1988, when the UNEP and the World Meteorological Organization set up the Intergovernmental Panel on Climate Change (IPCC) which is a controversial outfit, to say the least (see section 6). Two years later, in 1990, the IPCC published its first report, pointing out that gases such as carbon dioxide increase the natural greenhouse effect.

By 1990, however, other forces were already in motion, aimed to inform and warn people everywhere that the clock was ticking and that it might already be too late to save the earth's environment. In the United States, Europe, Asia or anywhere else, initiated and promoted by common citizens, various movements focused on carbon dioxide (CO_2) emissions and the greenhouse effect. Their weakness has been that:

- because of the prevailing taboos, they said nothing about the need for birth control – which, as Chapter 13 explained, is the earth's salient problem, and
- voices linking the worries about CO_2 to those about the galloping pollution of the land by chemicals, as well as waste and water poisoning problems, were hardly audible.

For the large majority of UN member states, including the United States, birth control has been off the table. Far-sighted organizations like the Population Council of New York have tried to open people's eyes, but, concerned about their re-election, politicians would

4 In 1987 in Montreal, 31 governments agreed to cut the production and use of chlorofluorocarbons, responding to the evidence that CFCs were eating a hole in the ozone layer over the Antarctic (Chapter 11).

not touch that issue. Even among the few first-raters, no elected officials would make birth control his or her priority.

In addition, the pollution of land, rivers, lakes and the sea by all sorts of chemicals and human waste, as well as the after-effects of deforestation, have been left by the wayside, no doubt because they are less visible issues than air pollution. It suffices to visit sprawling urban areas like Mexico City or Cairo to appreciate that just breathing the air is life-threatening – but not everybody appreciates the after-effects of chemicals and of waste.

There are exceptions to this statement. An example is a 1989 report published by the (then) still in business Supreme Soviet which stated that one out of five people in what used to be the USSR was breathing air containing ten times the maximum permissible level of dangerous chemicals. And then the same report added that nearly 600 Soviet cities had advised people to drink bottled water. This, however, went rather unnoticed.

Wider support of environmental issues had to wait till the mid- to late 1980s – and even then the political trumpet's sound was uncertain. For instance, in July 1989 President George Bush came up with the idea of an international conference on global warming and related environmental issues connected to the scheduled economic summit in Paris.

- But his administration was divided on how to tackle the global warming challenge, and
- Bush himself was subsequently accused of dithering and inertia.[5]

This ambiguity was confirmed in the 1992 Earth Summit in Rio de Janeiro which produced the United Nations Framework Convention on Climate Change (UNFCCC). It is indeed most regrettable that the United States failed to take the lead in seeking international action to counter the build-up of carbon dioxide in the atmosphere, and even more regrettable that the United States totally abstained from promoting family planning and birth control on a global scale.

2. The Road from Rio 1992 to Kyoto 1997

In mid-1992 Brazil hosted a global meeting of political leaders to mark the twentieth anniversary of the Stockholm conference (section 1). The groundwork for this climate change convention was to a substantial extent prepared by UNEP. Preparation for it was done more thoroughly than for both previous and subsequent environmental get-togethers.

After a meeting in Geneva to consider reports of the UN Intergovernmental Panel on Climate Change (IPCC), UNEP officials hoped it would be possible to start negotiations on a convention, rather than limiting the goal to a declaration of good intention. Various UN members, however, had contrarian opinions on this issue. The United States and a couple of other developed countries were too suspicious of even general outlines about a convention to start considering detailed measures and quantitative objectives. For their part, less developed countries would not start talking at all until they received assurances that any convention would offer them *cash*:

5 *Financial Times*, 11 May 1989.

- theoretically, to help meet the costs of complying with environmental protection requirements,
- in practice, to cover their huge deficits created by spending money on weapons and in feeding the output of their population explosion.[6]

The second bullet is not just a hypothesis. Some interesting figures had come out of the 15–26 May 1989 meeting of the UNEP governing council in Nairobi. Experts had calculated that in 1989 the world – including less developed countries – would spend a huge amount on armaments and little or nothing on environmental protection.

It is to the credit of Rio's UN Framework Convention on Climate Change (UNFCCC) that it did achieve some results, even if these were largely centred on the wider recognition that there indeed was a problem confronting the planet. What the delegates adopted at that Earth Summit was less than the posted aim 'to achieve stabilization of greenhouse gas concentrations in the atmosphere at a level that would prevent dangerous anthropogenic interference with the climate system', but it was better than nothing.

Rio 1992 also opened the way to the Kyoto Protocol five years down the line. The Kyoto Protocol of 1997, named after the ancient Japanese capital where the convention took place, set about realizing Rio's aims.

- It was signed by several nations and theoretically came into force in 2005.
- But, in reality, it was a half-baked agreement because the Clinton administration and the Chinese leadership – the top brass of countries which are also by far the largest environmental polluters – would have nothing to do with it.[7]

The 1997 protocol to the UNFCCC treaty set out targets and deadlines by which developed countries must cut carbon emissions. Although most developed countries ratified it, the United States did not do so and moreover, four years later, in 2001, George W. Bush delivered a speech in which he rejected the Kyoto Protocol and cast doubt on the reasons behind climate change.

For their part, the Chinese kept mum. They had no need to say anything since the US president did their job for them. The Indians did the same, rejecting the Kyoto Protocol by proxy. By contrast, in 2004 Russia agreed to ratify the Kyoto Protocol, but, with the world's biggest polluters languishing on the sidelines, Kyoto became a paper tiger.

True enough, in 2007, just prior to leaving office, Bush changed his position, agreeing for the first time to enter into vaguely defined international negotiations on a successor to the Kyoto Protocol. Subsequently, in December 2007 at Bali, several governments decided to start two years of negotiations aimed at forging a successor to Kyoto – which was the main aim of the Copenhagen conference (see sections 4 and 5).

Nevertheless, the Kyoto Protocol was by no means perfect. Critics have pointed out that it targeted far too little. It not only completely ignored the extremely urgent need for birth control, but also paid very little attention to an array of man-made pollutants. Roughly half of global warming comes from CO_2. The origin of the remainder is a variety of sources, including:

6 Corruption, too, had something to do with it.

7 Some countries signed up to the Kyoto Protocol later on. Australia did so at the beginning 2008, and also established plans for an emissions trading scheme.

- nitrogen dioxide (NO_2),
- sulphur dioxide (SO_2),
- hydrofluorocarbons,
- black carbon (soot),
- methane and more.

Even the targeted CO_2 emissions were poorly studied. Not only were developing countries, which are big polluters nowadays, left off the hook, but also putting everything together under a CO_2 umbrella without details is a prescription for inertia. All countries must meet specific quantitative targets, and this should be monitored and controlled by an international authority. According to critics, the Kyoto Protocol has achieved a great deal less than the Montreal Protocol of 1987 which:

- was designed with a single objective in mind, and
- aimed to achieve a specific goal: prevent the use of ozone depleting CFCs.

Judging from obtained results, Montreal was a successful agreement. Implemented in 1987 and originally expected to cut half of CFC gases in 12 years, it got rid of them in ten years. More to the point, according to a 2007 study, the Montreal Protocol prevented the emission of 189 billion tons of CO_2 equivalent. By contrast, the Kyoto Protocol has abated around 10 billion tons.[8]

Examining them post-mortem, one of the basic reasons why Montreal worked better than Kyoto is because the problem was of manageable size, and it targeted control over gases similar in nature and origin. This led some experts to the proposal that, to get results, the greenhouse effect should be unpacked and its different destructive emissions dealt with under different agreements – for instance, nitrogen oxide and methane produced by agriculture are one thing, SO_2 is a different case, and the same is true of CO_2. This approach has its merits, although it is wrong to say that the methane overkill comes only from the guts of cattle, goats and sheep. All species of nature's animal kingdom produce methane.

- A large volume of methane comes from the guts of people – nearly 7 billion of them, and
- the millions of methane-producing cattle and sheep are raised to feed the appetite for meat of those 7 billion men, women and children.

All this means that if birth control is left out of the environmental protection equation, nothing will be done worth talking about. Think of that next time you eat a hamburger or a steak. As for the other greenhouse gases (GHGs – see Chapter 10) to which little attention has been paid in the various environmental conferences, these are industrial gases with 140,000 per cent the global warming potential of CO_2, according to some estimates.

Because, like CFCs, NOs, SO_2s and the other GHGs are produced by a small number of industrial processes, cutting their emissions should not pose inordinate problems (which is not true of methane and black carbon), but they are not specifically targeted. In fact,

8 Statistics stated during meetings attended by the author.

it is much more difficult to bend the rising curve of black carbon because it is produced by primitive stoves burning wood and cow dung.[9] People, particularly in less developed countries, create that problem – and they are in the billions.

Let's be realistic about 'who's who' in environmental pollution. Energy consumption by billions of people creates a concentration of suspended particles, like soot, in the air. In industry this can be controlled by switching from heavy fuel oil to natural gas, as well by government requirements for electrostatic precipitators to be installed on smokestacks. But in less developed countries,

- massive urbanization continues to exacerbate the problem, and
- black carbon emissions in the countryside are not easy to control.

Approaches appropriate for dealing with large-scale emissions from power plants and factories are bound to have little impact on peasants and even urban-dwellers cooking and heating using wood. These millions of people create the destructive effect of black coal emissions – and trying to control them through UN protocols without family planning targets, is like trying to control a pandemic of syphilis by planting hair on the heads of those who have lost it.

3. Hollywood Enters the Environmental Protection Game

'Hollywood never tires to exploit our fascination with what grammarians call the *subjective conditional*,' says Niall Ferguson, adding, 'Of course, Hollywood and science fiction are not academically respectable.'[10] Yet, Nobel prizes are starting to be awarded to Hollywood movies.

According to an article by Helga Zepp-LaRouche, published in *EIR*, Al Gore's film on the environmental catastrophe, *An Inconvenient Truth* – for which he got a Nobel Prize for Peace[11] – was nothing more than a *climate hoax* – an effort by investment banks and hedge funds to deflect attention from the emerging subprimes fiasco and its present and real risks to the global financial system.

- Gore himself had been a hedge-fund manager since 2001, the proxy chairman of Metropolitan West Financial LLL (Metwest), and
- in 2004 Al Gore and David Blood, previously the chairman of the board of Goldman Sachs Asset Management, founded the London-based Generation Investment Fund, an entity that is very active in the CO_2 emissions trade. [12]

All that fails to add up to the credentials necessary to warn about an environmental catastrophe. To the contrary, it creates some serious doubts. 'The real subject of what's going on is the booming emissions trade,' wrote the *Daily Telegraph* on 14 March 2007.

9 Also by diesel.

10 Niall Ferguson, *Virtual History*, Basic Books, New York, 1999.

11 Just like Barack Obama got a Nobel Prize for 'War and Peace' in 2009, having done even less than Al Gore to deserve it.

12 Helga Zepp-La Rouche, 'Climate Change is a Diversion: Real Danger is Financial Meltdown', *EIR*, 23 March 2007.

'Gore recognized a trend that CO_2 emissions certificates trading would be the hottest business.'[13] (The amusing story of cap-and-trade CO_2 has been discussed in Chapter 10.)

The biggest global investment banks are leaders in the CO_2 trading market, a Tony Blair brainchild. To better appreciate the extent of this intensively marketed misinformation, the reader should look up the documentary broadcast by Channel 4 in Britain: *The Great Global Warming Swindle*.[14] It took up many of the scientific counterarguments against the allegedly man-made climate catastrophe, tracking the beginnings of this campaign to the Thatcher era.

Stated in simple terms, there has been a suspicion not only that most governments knew that the man-made climate catastrophe was – to say the least – exaggerated, but that they also understood that the financial system associated with globalization was coming to an end. Confused about this strange convergence, they looked the other way as big banks called on Hollywood to do something to distract the people from the oncoming 2007 financial and economic catastrophe.[15]

Significant is the fact that on 10 October 2007, a few months after the film's appearance, a British judge ruled that *An Inconvenient Truth* could only be shown in British state schools *if* teachers informed the students that it might not necessarily be depicting the truth. The judge issued his decision on the basis of British education laws which prohibit the promotion of unbalanced political views in the classroom.[16] This legal case was initiated by a school governor who sued to stop the Department of Children, Schools and Families from sending copies of Gore's movie to schools across the country on the grounds that it contained:

- 'serious scientific inaccuracies,
- political propaganda, and
- sentimental mush'.

High Court judge Michael Burton cited nine specific scientific errors in the film, made in 'the context of alarmism and exaggeration,' including Gore's assertion that a sea-level rise of up to 20 feet would be caused by melting of either West Antarctica or Greenland in the near future. The judge called this 'distinctly alarmist', as it is 'common ground that if Greenland melted it would release this amount of water – but only after, and over, millennia'.

Likewise, the British judge stated there is 'insufficient evidence' that global warming produced Hurricane Katrina, as Gore asserted. As for the claims that polar bears have been drowning, searching for ice, Judge Burton wrote that 'the only scientific study that either side before me can find is one which indicates that four polar bears have recently been found drowned because of a storm'.[17]

---

13 Cited in ibid.

14 Produced by Martin Durkin; first broadcast on 8 March 2007.

15 D.N. Chorafas, *Financial Boom and Gloom. The Credit and Banking Crisis of 2007–2009 and Beyond*, Palgrave Macmillan, Basingstoke, 2009; and D.N. Chorafas, *Capitalism Without Capital*, Palgrave Macmillan, Basingstoke, 2009.

16 *EIR*, 19 October 2007.

17 'Al Gore's "Nine Inconvenient Truths"', *Daily Telegraph*, 11 October 2007.

Critics of Hollywood's foray into environmental issues also noted that Al Gore was the least qualified person to talk about the risks to the environment not only because of his aforementioned personal involvement in CO_2 trading but also for two other reasons:

- The United States accounted for 23 per cent of global CO_2 emissions, making it by far the largest polluter, and
- although during the 1990s the world looked to the United States for decisive leadership on this issue, the Clinton administration remained immobile – and Gore was the number two person in the government.

Moreover, critics pointed out that although more research was needed on the issue, waiting for research results cannot excuse inaction. Yet the American representative to the Kyoto conference had taken the stand that scientific evidence was not yet conclusive on the greenhouse effect – and this happened while Al Gore was vice president of the United States.

In 1994, during Gore's years as vice president of the United States, Maurice Strong, the Canadian businessman-diplomat who organized the Rio Earth Summit said: 'If you want to judge Rio's effect solely on the basis of government action, then you would have to give a poor report. Fundamental changes at the government level have not occurred.' James Gustave Speth, administrator of the United Nations Development Programme (UNDP), put it more straightforwardly: 'The momentum that was generated in Rio is faltering, in Washington and elsewhere governments dither.'[18]

Cynics viewed Gore's attention to global warming seven years after he was out of office as just another excuse to just do something even though during eight years in power he was not deeply troubled by the absence of US action during events like the 1992 Stockholm conference and the 1997 Kyoto Protocol (section 2). If he were, he would have insisted that the Clinton administration developed a coherent approach to the control of greenhouse effects not on one, but on a variety of fronts such as:

- family planning,
- energy efficiency,
- renewable energy,
- deforestation, and
- international negotiations to limit the use of fossil fuel.

None of these issues reached beyond the scope of protecting the earth's environment although they did involve international relations, globalization,[19] the future of technology and trade as well as issues connected to nuclear proliferation. It is ludicrous to suggest that, all of a sudden, an individual wakes up to environmental dangers – unless there is some other reason, like distracting attention from the financial bubble which exploded in mid-2007.

In addition, American environmentalists were not particularly thrilled with Gore's film because they remembered that their hopes and sense of purpose that surrounded Rio were tempered by disappointment and disillusionment as the Clinton administration's

18 *Time*, 7 November 1994.

19 D.N. Chorafas, *Globalization's Limits. Conflicting National Interests in Trade and Finance*, Gower, Farnham, 2009.

response ground along at a painfully slow pace. This had led to the interpretation that while environmental protection was popular with US voters, it receded as political priority and generated less press coverage than it warranted.

Right after Al Gore's Hollywood movie and prior to Copenhagen 2009 (see sections 4 and 5) came another event worth recording. Between 30 April and 3 May 2007 climate change experts met in Bangkok to discuss ways to combat global warming (the third such meeting in 2007 held by the UN's Intergovernmental Panel on Climate Change (IPCC) at taxpayers' expense). Those attending reportedly hoped to finalize a document on how the world can mitigate rising levels of greenhouse gases. But nothing came out of that tourist gathering.

Yet Bangkok 2007 was well attended. Some 400 scientists and other experts from about 120 countries took part in the meetings which ended with the foregone conclusion 'that the Earth was already warming considerably'. That was just an empty statement, falling way short of what was pre-announced as a turning point in environmental action.

'The science certainly provides a lot of compelling reasons for action,' Rajendra Pachauri, the IPCC chairman, who also chaired that Bangkok meeting announced beforehand. Who would argue with that? 'We look forward to very fruitful days ahead of us, so at the end of the day we can declare we are part of the solution, not part of the problem,' Ogunlade Davidson, co-chair of the meeting, told reporters.[20]

Yes, but the problem remained, and it came down comprehensively in December 2009 in Copenhagen. Most important, however, was the fact that the IPCC and the 400 scientists and experts were so happy with what they had achieved. 'The IPCC plays an incredibly important role in the political negotiations so people can point and say "Look, this is what is going to happen in 50 years, these are the options available for us to take actions",' a UN Environment Programme spokesman stated proudly.[21] What followed was the Copenhagen circus.

4. Copenhagen 1995 and Copenhagen 2009

The most hilarious of all UN-sponsored 'save the environment' events was the one which took place on 7–18 December 2009 in the beautiful Danish capital. It was preceded by an equally unsuccessful March 1995 'Earth Summit' which did not produce any deliverables – but as a tourist gathering it was well attended.

In December 2009 15,000 people showed up in Copenhagen, a quarter or so less than the estimated 20,000 people who had been in and out of the Bella Centre in Copenhagen, 14 years earlier. In both 1995 and 2009 well-documented optimism was hard to unearth. As always with mass rallies, the only issue on which opinions converged were vague pronouncements. For instance:

- poverty, unemployment and social breakdown are bad, and
- remedies have to be found even if they are hard to come by.

20 *BBC News*, 30 April 2007, at: http://news.bbc.co.uk/go/pr/fr/-2/hi/scienec/nature/6606465.stm.

21 Ibid.

Many of the less developed countries went in 1995 and 2009 to Copenhagen to ask for everything and offer nothing in exchange, so there were no takers. Not surprisingly, the poor wanted debt cancellation and lots more aid. The rich nations told them to govern themselves better and cut their CO_2 emissions (but stopped short of suggesting birth control). As these meetings went on, proposals were launched by one group of people and shot down by another.

At the time of the March 1995 Copenhagen event, critics said that most discussions involved plenty of nonsense, and one of the wilder ideas was that of a tax on speculative currency transfers. This was originally suggested at Copenhagen 1995 by James Tobin, an American economist who won a Nobel prize in 1981. Supposedly, such tax could raise $50 billion a year for the UN to put towards development projects:

- *if* one is able to define how to draw the lines between speculative and non-speculative financial movements, and
- after having done so, tax them, collect the proceeds and distribute them in an equitable way (whatever this might mean).

The *Tobin tax* has been an on-and–off idea for many years because it has its proponents and detractors, with each party having developed numerous arguments to suit its opinions and needs. One of the latest was Gordon Brown's November 2009 proposal to institute the Tobin tax, in a move supposed to lift the Labour Party's dismal fortunes and his own. But the kite did not fly.

According to its critics, the Tobin tax is impractical because it would be difficult to define who should pay and impossible to collect the money. Cynics add that what makes the suggestion hilarious is that it turns the UN into a tax collector. But who would refuse $50 billion a year if he could get his hands on it?[22]

Legal problems will be legion because the UN has never had the authority to tax, and there is nothing like that in its charter. No wonder that rich countries (and their bankers) hate the idea and so do the livelier poor ones, which fear a distortion in capital flows from which they benefit a great deal through foreign direct investments (FDIs).

Having tried to pull itself up by its shoestrings (the Tobin tax) Copenhagen 1995 was a non-event and, as misfortune never comes alone, Copenhagen 2009, too, provided zero results. In fact, its hiccups were more serious than those of 1995. Two weeks before the conference, Barack Obama, newly selected Nobel Prizewinner for War and Peace, announced that the United States would cut emissions to around 17 per cent below 2005 levels by 2020.[23] But, like Tobin, Obama, too, could not deliver.

Nor did Obama's verbal promises excite any enthusiasm. Other developed nations looked at them as derisory, as they represented something just 4 per cent below 1990 levels, significantly less than the figure offered by them (although it was roughly equal to a vague offer by China). The only positive comment came from Todd Stern, the US special envoy on climate change, who claimed: 'Several different countries have come up to me and said, "You've got a path that's pretty good, even if we don't like your 2020 number".'[24]

22 Madoff didn't. In fact, his jackpot was $65 billion.

23 'Obama to Go to Copenhagen with Emissions Target', *The New York Times*, 25 November 2009.

24 'Closing the Gaps', *The Economist*, 3 December 2009.

China, too, was under pressure to come up with a significant offer on pollution control. What it suggested was a vague 40–45 per cent cut in the 'carbon intensity' of its economy by 2020 – whatever that is supposed to represent. India did not consider it necessary to come forward even with the most vague and uncontrollable offer. While various nations were unable to decide on their commitments, this December 2009 UN conference confronted three big practical questions:

1. *By how much does the world need to cut emissions?*
   A factual answer is not forthcoming because of two handicaps. The first is conflicting scientific opinions about the warming or cooling of the earth's climate (discussed in Chapter 1). The second is that the word 'emissions' is at the same time impressive and confusing. Unless we are specific about which GHG we are talking about (Chapter 10) and negotiate on each of them quantitative targets applicable to every nation, we can achieve nothing in terms of emissions control. This cannot be done in the course of noisy conferences attended by thousands of people, who turn them into a circus. This leads to the second question.

2. *How much preparation should precede 'Earth Summits', who should organize them, and what should be the exact deliverables?*
   Whether we are talking of Montreal 1987, Rio 1992, Copenhagen 1995, Kyoto 1997, Bangkok 2007 or Copenhagen 2009, the least that can be said is that proper preparation was not on call. Under these conditions, the one who carries the day is the one who shouts loudest – or, alternatively, who is the most obstinate.

   In Copenhagen 2009 there was a great deal of crying and plenty of attempts at arm-twisting, including massive abstention from the conference particularly by African nations. Cheap politics also had its day – for instance, Venezuela's Hugo Chavez used the UN conference as a platform to speak against America (on 16 December) – and nobody called him to order.

   An integral part of appropriate preparation for an important conference like the 'Earth Summit' is to know who has the credentials to participate. Quite unwisely, because it is not part of its charter to enter into ethnic disputes within a country's borders, the UN has condemned Sudan because of Darfur and is looking to arrest its president, bringing him to an international court with the accusation of genocide.

   Led by China, the so-called Group of 77[25] chose as its representative the Sudanese ambassador to the UN. Di-Aping may be a most honourable person, but:
   - *if* Sudan is under UN sanctions for genocide,
   - *then* it cannot, at the same time, lead a group representing the majority of UN member states in a UN conference.

All this does not add up, and it leads to the conclusion that Copenhagen 2009 was a conference aimed to produce anything else than fun, inconsistency and anger. It can best be described as a circus – hence its nickname, *Copenhagen Circus*, although

25 A most diverse pressure group of developing countries with its richest members 50 times as wealthy, on a per capita basis, as the poorest ones. Contrary to what its logo seems to imply, Group 77 is an agglomeration of 130 developing and less developed countries.

the alternative, *Fiascopenhague*, coined by the French, describes even better what took place.

3. *How much give and take should characterize a global conference on environmental protection?*
The basic principle of any negotiation is that each of its participants gives something and takes something. Short of that, it is not a negotiation but a capitulation, which was never supposed to be the UN's goal. Although six and a half decades of UN experience document that little, if anything, comes out of UN conferences, these cost lots of money – and *if* the deliverables are zero, *then* there is no reason to assume all these costs.

It therefore comes as no surprise that in Copenhagen 2009 the developing nations, which were asked to pay the bill for the global control of emissions, wanted concrete promises from developing countries. Orally, and only orally, Hu Jintao, the president of China, which is the world's second biggest emitter of greenhouse gases,[26] pledged 'this' and 'that'. In reality, he committed himself and his country to nothing.

In direct opposition to what its vague statements might have suggested, China objected to any numbers being written in the final three-page communiqué of Copenhagen 2009. For its part, India said that it would accept only a limit on emissions *per person* that matches that of rich countries which, given India's present-day development, is no limit at all.

Despite having assumed no obligations of any importance, and having refused inspections of how well they are performing in reducing emissions, developing countries wanted the West to pay big money to the poorer ones – supposedly to adapt to climate change and shrink their use of carbon. The irony is that even filthy rich China (with over $2 trillion in foreign exchange) wanted money, and set the price-tag by insisting that industrialized countries should pay 1 per cent of their GDP a year. Using their abacus, the Chinese had evidently found that this would amount to $400 billion.[27]

The call was to spend billions without any plan, let alone rules regarding inspection and control of deliverables. One of the hilarious aspects of the Copenhagen circus was that no standards, objectives, timetables or programme were worked out and agreed by all UN members.

There was also another sticking point: completely out of context in an environmental conference, the less developed countries wanted debt relief. Who will bear the cost of the debt relief? It is highly unjust that Western taxpayers are brought to their knees in order to finance other people's unnecessary spending on armaments, filling of pockets, tourism and other pleasures.

26 And is on its way to becoming the biggest emitter. (See also the reference to China's production of greenhouse gases at the end of this chapter.)

27 To this, Gordon Brown, who is always forthcoming when it comes to spending other people's money, counter-offered $100 billion a year. 'Closing the Gaps', *The Economist*, 3 December 2009.

5. *Fiascopenhague*: The End Game

Originally, the 7–18 December 2009 UN environmental meeting in Copenhagen was projected as the event that would produce a treaty to replace the Kyoto Protocol, since the latter expires in 2012. Some 180 countries were expected to take part in these negotiations. But because they account among themselves for about 50 per cent of CO_2 emissions, China and the United States represented a sort of 'G2' that could determine the global post-Kyoto agenda.[28]

They did so, in the most negative way, assisted in these zero-point-zero deliverables by some 100 (or 119, according to some accounts) heads of state who flocked to Copenhagen for the tourist event. Even the sessions of *Fiascopenhague* shown on television and in newspaper photos have been hilarious. Practically all the big heads of the world participating in what was supposed to be an important negotiation:

- had their mobiles in front of them, and used them frequently rather than paying attention to the conference.[29]
- This happened as a matter of fact, without considering the incompatibility with what was supposed to be their main task.
- Not one of them consulted his neighbour or asked for his opinion on matters being discussed.

All this inaction on issues which matter has come at an exorbitant cost, and not only in money. An army of 15,000 UN conference participants from all four corners of the planet represented plenty of CO_2 emissions when 5–10 representatives from each country would have sufficed, turning the climate change conference into a comedy characteristic of present-day morals.

No wonder that the news from Copenhagen was grim, and negotiations for a new climate treaty stumbled toward stalemate. The talks between over 100 heads of state and their thousands of representatives were deeply divided over:

- broad emissions targets,
- specific measures each country should take to fulfil its quota, and
- how much developed countries should pay other countries to help them meet these targets.

There have also been differences over how to verify whether nations are living up to their obligations, or are just taking the money and running. Furthermore, the idea of developed nations having to pay underdeveloped countries not to destroy their forests is an aberration. It was first floated in the Kyoto talks in 1997 and nothing was done about it; then it resurfaced at Copenhagen 2009.

By contrast, the need to exercise birth control, the earth's salient problem, was not even discussed in Kyoto, Copenhagen or elsewhere. This was a colossal blunder for which

28 Indeed, to prepare the 'G2' ground, Todd Stern, the US special envoy for climate change, prior to Copenhagen went to Beijing where he said that although China may not be the 'alpha and omega' of the international conference, it was close to it. Details of these talks were, however, scanty. 'Heating Up or Cooling Down', *The Economist*, 11 June 2009.

29 In fact not only the heads of state but also their representatives to the earth's climate conference were glued to their mobile phones.

the world will pay a high price. But it was not the only blunder. The fact that in the Kyoto Protocol of 1997 forests were the big absentee and in *Fiascopenhague* they were only a side issue is both irrational and inadmissible.

The reason, I am told, is that sovereign-conscious countries are unwilling, at any price, to accept limits on their freedom to fell. Nobody from the UN or any other entity really bothered to explain to the great crowd of less developed countries that, as far as environmental pollution is concerned, even if they are tiny economies they no longer enjoy what Martin Wolf, the economist of the *Financial Times*, calls 'the liberty of insignificance'.

- Even the small economies are polluters.
- As for China, India and Brazil (the bigger economies), the scale and consequences of environmental disrespect are unprecedented.

Therefore, unless the preparations for a Save-the-Earth UN conference establish iron-clad guarantees for both transparency and deliverables – including political liberalization, because modernization is impossible without political liberalization – the money paid for environmental protection will simply drift and get lost. Self-serving states destroy the global environment, oppress their own people and don't give a dime about the earth.

The zero result of *Fiascopenhague* and its 15,000 attendees reflects most badly on these 100 or 119 heads of state – the biggest ever global collection of top brass to this day – and on the United Nations. Let's not forget that the UN bureaucracy took two years to organize, at great cost, that ill-fated conference. As an article in the *Le Canard Enchainé* aptly pointed out:

- *if* failure is the key to success,
- *then* a fiasco should lead to total victory,[30] or shame.

With all its haggles, whistles and twists, this was a hilarious summit. It was learned afterwards that, while some of the 100 or 119 heads of state were telling journalists that an agreement was around the corner, these same journalists were receiving text messages from the same heads' assistants saying exactly the opposite.

Fiascopenhague, however, did help in revealing what some of the heads of state were made of. Among the better known, the Nobel Prizewinner for War and Peace was trapped in a meeting 'of last chance' of which he was not informed in advance (more on this in section 7). For their part, most of the Europeans were distinguished by their self-effacing attitude. They did not come up with any firm propositions except to throw money at the problem.[31]

Diplomacy was at its lowest. Among other hit-and-miss initiatives, the Americans and Europeans asked Brazil's President Lula to help in bringing the African and other less developed countries close to their ill-defined position. The plan backfired, as the latter

30 *Le Canard Enchainé*, 22 December 2009.

31 Unwisely following a meeting in Brussels the European Union said it would contribute $10.5 billion over the next three years to help poorer countries deal with climate change. The United States stated that it would make a contribution but did not specify how big or how small it would be.

flocked to the anti-Western camp of the so-called Group of 77 – but still demanded big money from the West (which they did not get).

'If climate was a bank, they would have had already saved it,' said Venezuela's Hugo Chavez, pointing to the United States, Britain and the continental Europeans. Chavez's joke makes a point and misses another one, because with Tony Blair's and other EU leaders' cap-and-trade active in business (Chapter 10), the world's climate is indeed on its way to benefiting all big banks.

Another one of the ironies connected to the Copenhagen circus is that while emissions in many developed countries have peaked and are declining, just about all of the growth in greenhouse gases is coming from the developing world. This is expected to increase between now and 2030, half of it originating in China. These are precisely the countries currently selling air-pollution indulgences to the West.

- The industrialized nations must still reduce their emissions.
- But CO_2 and other emissions will not improve by giving the major developing countries a pass, as Todd Stern correctly pointed out.

Stern also scored a hit when he said that while Washington is prepared to help those who need it, given China's huge foreign currency reserves and go-go economy –which is projected to grow by 8.5 per cent in 2010 – he could not 'envision public funds, certainly from the United States, going to China.'[32] There were also some scandals. On 8 December 2009 a draft agreement, which had been discussed some weeks earlier, was leaked to the *Guardian*.[33] It caused a furore.[34]

Instead, the furore should have come from the depressing lack of transparency in the control of emissions by the large majority of the participating nations, particularly the big developing countries. At Copenhagen 2009 transparency was one of the more difficult issues that had to be resolved, but was not. There is no point in setting targets, or threatening penalties for non-compliance,

- unless countries are required to report emissions accurately, and
- emerging countries accept regular inspection and audits by an international body, because they themselves don't have sophisticated monitoring systems.

How much of an alien concept the issue of transparency was at the UN Copenhagen conference was attested by the fact that deliberations gave way to a flurry of ephemeral negotiating texts for a pact. As different alternative texts hit the buffers and were rejected because of transparency, secret alternatives to the alternatives saw the light, but the stumbling blocks remained intact.

---

32 'Copenhagen Summit: US "Will Not Subsidise China on Climate Change"', *The Times*, 9 December 2009. China has been demanding that rich nations contribute hundreds of millions of dollars a year to help poor countries, including itself.

33 'Copenhagen Climate Summit in Disarray after "Danish Text" Leak', *Guardian*, 8 December 2009.

34 Subsequently, it was suggested, but not proved, that China leaked it.

6. Money, Money, Money Going Down the Drain

French physicist Serge Galam is right when he says that by concentrating too much on CO_2, the UN and various world leaders ignore the geopolitical risk which is present and mounting. Although the scientific uncertainty on the origin of changes in the earth's environment is no excuse for doing nothing, carnivals *Fiascopenhague*-style are all about:

- enjoying the free ride at taxpayers' expense, and
- forgetting about tomorrow's salient problem and its challenges.

If carnival was the mood within the walls of the UN climate conference, as usual the loud and colourful chorus took place outside the conference building; and it got out of control. Thousands of so-called environmental demonstrators, their numbers swollen by the familiar anti-capitalist agitators, provided the open-air version of the Copenhagen circus.

- The stagehands did not view the climate meetings as a crucial juncture in the future of the planet.
- To them, it was nothing more than a nice backdrop for troublemaking and for expressing their anger which is vibrant for unspecified reasons.[35]

Part of the blame for *Fiascopenhague* goes to the 15,000 delegates and over 100 heads of state who attended it. Inevitably, responsibility also falls on the shoulders of the UN, which organized it, and its managers, Yvo de Boer,[36] the United Nations climate chief, and Dr Rajendra K. Pachauri, head of the UN Intergovernmental Panel on Climate Change[37] (IPCC).

It is indeed interesting that, to justify the big money spent by the IPCC to produce virtually nothing, Pachauri named 'four achievements' brought about at Copenhagen. First, he said that among the participants were over 100 heads of state and heads of government and they went there 'essentially because they thought that this was an important issue'. This does not add up, because:

- *if* they really thought so,
- *then* by going back home empty-handed they confirmed their impotence.

Second, Pachauri mentioned as a 'plus' the vague acceptance by at least some countries 'of limiting temperature increase to 2 degree [*sic*] Celsius'. This is plainly superficial.

35 Obscure groups of 'protesters', decrying what they saw as corporate influence over the UN climate talks, had many skirmishes with the police. The groups fanned out across the city in an attempt to disrupt operations at several companies, including Maersk, the Danish shipping firm and (as usual) McDonald's.

36 Who in the meantime left to join KPMG.

37 And at the same time director-general of TERI, an Indian Research Institute. A notable(?) contribution to environmental protection by Dr Pachauri is a novel he recently published. As *Le Figaro* (the well-known French newspaper) relayed it on 11 March 2010: 'Between the war against deforestation and spiritual reflections, there are a lot of sexual exploits.' More on Dr Pachauri and his travails in section 7.

Anyone can see that 'vague acceptance' by 'some of the participants' is worth absolutely nothing.

Then Pachauri stated as a success that '[t]hey have also referred to the science and the fourth assessment report of the IPPC on the basis of which they have arrived at this figure of 2 degree increase in temperature'. This is a sort of self-congratulation, unfit even to be mentioned in a scientific context. *If* they were scientists, *then* the folks at IPCC should have known that:

- science is never based on majority opinion, and
- scientific principles are never established through one-man-one-vote, or by means of consensus.

The great power of science is to challenge established notions and principles. Very often, one new and nasty fact has demolished long-held scientific theories. Trying to prove 'the truth behind its findings' – which are massively undocumented in a scientific sense – by invoking the majority opinion of the 15,000 Copenhagen circus delegates, the IPCC opens itself to the accusation of a plot which has misfired in unforeseen ways.

The fourth fake reason presented by Pachauri to beef up the IPCC's image is also the most beautiful, because it hits the vital point: 'The developed countries are talking about providing around USD 30 billion as assistance in the period 2010 to 2012.'[38] This is, most obviously, an overstatement, but at least it contains a grain of truth since the Western countries have been silly enough to throw down the drain in this way. All told, Copenhagen 2009 was nothing more than a big-hearted charitable love affair.

All this casts a very dark shadow on the IPCC and other outfits which pose as scientific laboratories, but in reality are pressure groups, costing plenty of money and contributing practically nothing. Moreover, their output is dangerous because they try to justify their existence:

- by giving one-sided advice to various heads of state and the UN top brass, and
- by attracting headlines for pronouncements which, to say the least, are not ientifically validated.

How wrong this advice can be is documented by none else than Pachauri himself. As an article in *The Economist* puts it:

> *His intemperate defence of mistakes about Himalayan glaciers in the most recent IPCC report had to be followed by a public statement of regret as it became clear that IPCC had indeed been wrong – and that its source has been a magazine article rather than a piece of scientific literature.*[39]

Pachauri mentioned that the mistaken claim about the melting glaciers never came to his attention. But doubts on the degree to which the IPCC chief is in charge persisted. The glacier story and bias in giving one-sided advice to heads of state about global warming led the IPCC's critics to demand:

38 All quotes from an interview with Dr Pachauri on *CNBC*.

39 'A Time for Introspection', *The Economist*, 6 February 2010. For more on the Himalayan glaciers see section 8.

- greater transparency, and
- higher scientific standards.

At the eye of the storm raised by such incidents is also the so-called *grey literature*, a term used for reports by governments and other entities that circulate internally and are neither published commercially nor passed through scientific channels. Hence, they are never subject to peer review.

All that misinformation did not come free of cost. Plenty of money has been thrown out of the window. As Serge Galam aptly suggests,[40] here we are talking about scarcity of resources on earth, but money also is a scarce resource. Financing is limited and when capital is spent for nothing in one direction, it will be missing in another where results are more critical.

The cost of the UN environmental congresses and other global gatherings is indeed staggering. I am not privy to how many hundreds of millions IPCC spreads around for its budget and its unwarranted global gatherings. However, Britain's Taxpayers' Alliance (TPA) has published Research Note 51[41] with some very interesting figures regarding the huge expenses to be paid by the UN, the Danish government and other governments of countries from which came the 15,000 delegates who descended like locusts on Copenhagen on 7–18 December 2009.

The TPA's very conservative estimate of expenses associated to that event is at the level of $214.7 million – of which $62 million was put up by the Danish government and $90.3 million represents the salaries paid to these thousands of 'delegates' who roved around Denmark's capital city. Another $13.4 million was the cost for conference facilities,[42] and $49 million for hotel and food expenses.

Nearly 60 years ago I had a professor at UCLA who taught his students that it is not enough to make a budget; you must also carefully ensure that it is religiously observed. Otherwise, chances are that the real expense will be double the financial plan. With such a diverse and heterogeneous gathering there is no evidence that somebody has been watching over such extravagant expenses.

- Even *if* costs eventually were kept at the $214.7 million level, this is a wholesome amount to spend on a UN conference which led nowhere.
- Moreover, it would be wrong to believe that the Copenhagen circus was an exception in spending money for zero results. Tourist UN affairs masquerading as 'important conferences' have become an industry.

The same Research Note 51 by the Taxpayers' Alliance names some other big-spending 2009 events which went over the top in expenses, with the bill sent to the taxpayers. Two examples are the G8 Summit at Gleneagles, and G8 Summit in Japan. Nothing has come out of them except their costs (and a hell of a lot of CO_2) – at $132 million and $464 million respectively.[43]

40 During a personal discussion with the author.

41 Taxpayers' Alliance, 'The Cost of Copenhagen' Research Note 51, at: http://www.taxpayersalliance.com/costofcopenhagen.pdf.

42 Including a curious expense item of 'hostess 6pm to 7am', ibid., p.7.

43 Ibid.

Somebody among the participating heads of state and other delegates must have the decency to state that enough is enough. It is high time that this intensive and highly costly tourism of 'summits' and UN global 'conferences' came to an end. As Mathew Sinclair, TPA's research director, wrote: 'Politicians need to stop this expensive jamboree and instead focus domestically on bringing down the ruinous cost to ordinary families of green taxes and regulations.'[44]

7. Imperial China: Reborn at the Copenhagen Circus

By default, our generation is deciding on what sort of planet our descendants will be living on and on the amount of debt they will be burdened with. Whether we talk of birth control, energy conservation, deforestation or environmental protection, Copenhagen 2009 made no change to the status quo – while the various G8, G20 and other summits did nothing to discourage increases in public debt.

As the conference came to a close Europe offered a 20 per cent reduction of CO_2 in 2020 on the basis of 1990 emissions. To show bigger numbers, both the United States and China calculated their proposed CO_2 on 2005 rather than 1990 emissions, which meant a great deal less in terms of cutting down *their* pollution. To better appreciate what is meant by these numbers, which have not been confirmed in writing, the reader should know that today worldwide:

- Europe accounts for only 13 per cent of CO_2 emissions,
- the United States and China share among themselves up to 50 per cent, and
- all the others together account for the remaining 37 per cent, which is 300 per cent more than Europe's emissions.

In terms of CO_2 reductions all that adds up to a triviality. Even *if* the stated numbers were honoured by those who pledged them orally but did not sign up to them before the end of January 2010 deadline, 20 per cent of 13 per cent plus 4 per cent of 50 per cent equals a mere 4.6 per cent of total emissions. Way too little to crow about.

This is just part of the story. While the total theoretical emission cuts are less than 5 per cent – hardly worth the Copenhagen circus – there have been some other results much more significant in terms of geopolitics. Between 7 and 18 December 2009, Imperial China was reborn. Reaction to it has been swift. In an article, 'Chinese New Year', Paul Krugman, the economist, stated: 'I predict that 2010 will be the year of China. And not in a good way.'[45]

Krugman is not alone in that assessment. In his book, *When China Rules the World*,[46] Martin Jacques, columnist of the *Guardian* and expert on China, argues that China will not only displace the United States as the major superpower. It will also:

- marginalize the West in history, and

44 Ibid., p. 2.

45 Paul Krugman, 'Chinese New Year', *The New York Times*, 31 December 2009.

46 Martin Jacques, *When China Rules the World. The End of Western World and Birth of a New Global Order*, Penguin Press, New York, 2009.

- alter our core notions of what it means to be 'modern'.

In fact, for some years China has been manipulating the United States and the European Union to make them serve its plans and desires. But there has been nothing more profound so far than the humiliation which it visited upon the politicians who, by misfortune, happened to represent the West's interests during the Copenhagen circus. One of the politicians who got short shrift was none other than President Obama. On the final day of the UN Copenhagen conference, he was obliged to conduct negotiations with comparatively junior Chinese delegates. 'At one point', said an article in *The Economist*, 'Mr Obama expected to meet his Chinese opposite number one-on-one but instead found himself with the leaders of South Africa, Brazil and India as well.'[47]

Instead of politely greeting and saying goodbye when he saw that he had been deceived, Barack Obama sat down with his flanks exposed. That trick explains how it happened that in that fateful final meeting on the night of 18 December 2009, none of the European Union's big shots was present. Being inexperienced in international relations, if not in politics at large, while

- taken by surprise, and
- outnumbered three-to-one by the BRICs (minus Russia plus South Africa),

the new Nobel Prizewinner for War and Peace stayed there and kept talking while the BRICs took him to the cleaners. If the Chinese can get what they want to hear out of him by ambush, they can also get anything else they want. The man has no cunning, and maybe he is afraid to take a stand.

The argument heard in some quarters in Washington that the United States cannot displease China because the Chinese would wreak havoc with the US economy by dumping their hoard of dollars on the market. Krugman aptly states that:

> *This is all wrong, and not just because in so doing the Chinese would inflict large losses on themselves. The larger point is that the same forces that make Chinese mercantilism so damaging right now also mean that China has little or no financial leverage.*[48]

Martin Jacques is even more forceful in his thesis regarding what lies behind Chinese policy. Starting with the assumption that nothing will derail the political stability and economic dynamism that China enjoys today, he follows with the bold proposition that, having largely set aside its socialist ideology, the ruling Communist Party has become a modern version of *an imperial dynasty*. China's communist leaders:

- have flirted with Chinese unity, the state's traditional role, and
- have given their country a reputation for low-cost, high-volume, flexibility and speed.

This has allowed China to sweep through world markets, aided by the WTO agreements which it negotiated as 'a less developed country'. Moreover, Jacques suggests, the large

47 'China's Thing about Numbers', *The Economist*, 2 January 2010.

48 Paul Krugman, 'Chinese New Year', *The New York Times*, 31 December 2009.

majority of Chinese will back their leaders, with or without democratic reforms, as long as the country keeps getting stronger. This may well lead to a Pax Sinica if nothing happens to derail current plans.[49]

8. Pax Sinica?

Not unexpectedly, in Copenhagen China's viewpoint was shared by most developing countries. This states that the developed world bears the main responsibility for dealing with CO_2 and other pollution problems. Nobody dared to voice the fact that about 70 per cent of China's electricity supply comes from coal-fired power stations and that, being the rich country she is, she cannot demand that developed countries:

- foot the bill, and
- also provide the technology.

Admittedly, after that double whammy aimed at the West misfired, China said that she would not ask for money. But she also made it plain that she wants no quantitative objectives or CO_2 reductions in the Copenhagen communiqué. When drafts of a nebulous 'compromise agreement' began to circulate on 18 December 2009, the 'original 80 per cent by 2050' formula was still readable. Hours later, it vanished at China's insistence.

'Why?' journalists asked Lars-Erik Liljelund, the Swedish government's spokesman on climate. Why would a pledge that applied only to rich nations, and to which all those nations seemed to agree, have vanished from the final document? The way it has been reported in the press, after a few seconds of silence, came the reply: 'China doesn't like numbers.' This was one of the first acts of Pax Sinica.

The Chinese government probably has good reasons for taking that position. Since 2000 China's surge of investment in heavy industry and power production has resulted in retreating energy efficiency levels and soaring pollution measurements. In one year alone, 2006, China added power capacity equal to the entire grids of Britain and Portugal combined – and 90 per cent of it is coal-fired. The surge was necessary to feed its growing stack of:

- steel,
- aluminium, and
- cement plants.

By 2009 China had overtaken the United States as the largest emitter of energy-related greenhouse gases, which explains the reason why climate change is an area in which the two countries are locked in an alliance of inaction. This is also the reason why the Chinese denied Obama's request for a *tête-à-tête* with their president, and trapped him in an all-night meeting with the BRICs (section 7).

The Copenhagen circus has been the first tournament between reborn imperial China and superpower America; and the game ended 1-0 in the former's favour. The weak hand

49 Martin Jacques, *When China Rules the World. The End of Western World and Birth of a New Global Order*, Penguin Press, New York, 2009.

played by President Obama against a shrewd and calculating opponent ended badly. Pax Sinica exercises too much pressure.

Confronted by a weak opponent, pressure changes almost everything, but it does not change the fact that both China and India suffer from acute air and water pollution. Recent meetings brought up interesting figures. In more than 80 Indian cities for which air quality monitoring data is available, 85 per cent of the population is forced to inhale poor, bad or dangerous air. Only 3 per cent seems to have access to air that was rated good.

For its part, China is home to 16 of the world's 20 most polluted cities, with dirty air causing an estimated premature death of 400,000 people a year. As for water supplies, a rumoured 350 million people, roughly one-quarter of the population, do not have access to clean water – with all that means to health and agriculture. Clean water may prove the Achilles heel of Pax Sinica.

The fact that *their* environment is going to the dogs should have made both China and India eager to see global standards on clean air and clean water established and policed – with every UN member's obligations clearly spelled out.

- But they also understand that the cost of bending their ever-rising curve of pollution is not money but a much slower pace of development.
- Therefore, they turned the tables during the Copenhagen circus and consequently hurt themselves more than they hurt others.

All this should have been drummed up by Obama and the European Union's heads of state abundantly present in this United Nations' 2009 environmental conference. But each probably had second thoughts and a fear. A mercantilism fear, partly in recognition of the fact that today's China is above all:

- a giant mercantile nation, and
- a vast depository of foreign currencies.

China does not get an 'A' grade as a mercantile nation, because it has been imprudent about the wares it manufactures and exports. Illegal copying of software, music and other entertainment copyrights is current currency; these, however, are not the only missteps.

In February 2010 a police raid on a sprawling complex of warehouses in Rome turned up about 500,000 tons of Chinese goods, including fakes of some of the world's most famous designer labels. This operation was carried out over several days, and its results were announced by the Italian police on 13 February. The merchandise included fakes of big brands, boxes with designer logos, as well as toys that did not meet European safety standards.[50]

Theoretically at least, thanks to its intensive manufacturing and merchandising activities, China is now a vast depository of foreign currencies (mostly in dollar-denominated US treasuries). The country's leaders view that as a success story, but this is not entirely true. China has imitated Japan's policy in the 1970s and 1980s, which led to the collapse of the Japanese economy. Numbers don't tell the real story:

50 *International Herald Tribune*, 15 February 2010.

- In 2001 China's share of international foreign exchange reserves was 18 per cent, and that of emerging Asia 8 per cent.
- In 2009 China's share had more than doubled to 37 per cent, while that of emerging Asia increased only slightly to 9 per cent.[51]

Qualitatively, however, the results have not been that brilliant – and not only because in the process the country polluted its land, water and air. This surge in collecting truckloads of foreign exchange looks too similar to Mao's Great Leap Forward, which ended in total disaster. In addition, by going wholesale on the exports bandwagon, through an asymmetric strategy, China neglected its internal consumer market which has remained underdeveloped. In essence, what the Chinese government did was to ship abroad real goods, receiving in exchange pieces of printed paper and electronic credits. The flaw in China's strategy can be summed up in two bullets:

- Keep the currency cheap.
- Push exports over the internal market.

Consequently, by 2010 internal consumption was a meagre 20–25 per cent of GDP. This is hardly a base on which the Chinese government can underwrite Pax Sinica. To sustain growth and increase the country's global weight, internal consumption should stand at 50 per cent of GDP – double the current level. I see a China bubble bursting, with real estate the trigger.

To make matters worse, all that vast amount of money is kept in vaults, rather than being invested in a massive import of clean technology and pollution-control skills. Both are urgently needed to significantly improve the country's environment. By keeping the lavish foreign reserves in vaults, the reborn Imperial China risks fulfilling Dr Rajendra Pachauri's (so far) wrong prophecies.

As section 7 brought to the reader's attention, what Pachauri said in late 2009 lacks any scientific foundation. But that is true today. As pollution of the land, water and air increases by leaps and bounds in developing countries, and continues to remain unchecked, what is not the case today can turn into a nightmare 20, 30 or 40 years later.

- True enough, Pachauri's warning is not based on scientific evidence.
- But this can also be seen as a worst-case scenario, which is good to keep in mind, not in the closet.

Here is in a nutshell what is contemplated. The Tibetan plateau is the origin of the great river system in China, south-east and south Asia: the Yangtze and Yellow Rivers, the Brahmaputra, the Indus, the Mekong and the Salween. As the Himalayan glaciers melt away, these rivers – already hit by unprecedented pollution – are going to see some of their sources dry up. (Also highly polluted is the Ganges, which rises on the Indian side of the plateau's Himalayan rim.)

51 International Monetary Fund and Merrill Lynch Wealth Management. The 2009 figures are estimates.

- Any clear-minded person would appreciate that there is no way of turning back the clock.
- However, there is still time for damage limitation if China and India give a greater priority to survival than to 'development'.

The problem is that Chinese and Indian politicians don't seem to appreciate that these rivers, fed by thousands of glaciers and supporting 1.3 billion people, are at a risk of running dry. Yet, even Chinese experts predict that by 2050 the icy area on their side of the Himalayas will have shrunk rather alarmingly,[52] while projections for the Indian side are gloomier still.

Nothing like that disturbed the fun at Copenhagen circus. At the twelfth hour Cuba, Bolivia, Nicaragua, Venezuela and a few other UN members insisted that, because it had not been drafted by any official procedure, a 'pseudo accord' struck by hand-picked heads of state in a midnight session was just a miscellaneous document of no practical consequence. Well done. At least this was a truthful statement which put *Fiascopenhague* to rest, though not yet Pax Sinica.

---

52 A guesstimated level of shrinkage is more than a quarter since 1950, according to some sources.

CHAPTER 15

Can Science and Technology Solve Mankind's Practical Environmental Problems?

1. The Helping Hand of Science

Ancient history teaches that sooner or later *Homo sapiens* adapt to the environment and to the changes it undergoes. With adaptation they find ways to suit their needs without a imposing a great deal of damage on the world in which they live. However, in modern times this has changed for two principal reasons:

- Development became too fast to adapt to, and change took on its own momentum.
- As the number of people increased exponentially, the masses became a brake on the management of change.

With adaptation taking leave of absence, practical solutions are not forthcoming. Examples are a legion. We have already spoken of the torrent of CO_2 from massive transportation and energy production; air pollution from all sorts of sources, including methane from human excrement; disappearing clean water supplies; sewage drains; sprawling household and industrial landfills; and garbage spoiling once majestic seashores and rivers.

The careful reader will also recall that alternative methods of energy production do not provide a whole solution, and the mastering of other sources of man-made poison is neither easy nor forthcoming. Ideas on how to escape from this vicious cycle –, such as using common garbage for energy production – come and go. But the disposal of waste has some questionable side-effects.

- When waste is burned, it spews dangerous gases into the air.
- Dumped industrial waste turns lethal when corrosive acids, long-lived organic materials and discarded metals leach out of landfill sites into groundwater supplies.

Myriad factors contribute to the hole in the ozone layer, the greenhouse effect, the contamination of drinking water and the destruction of farmland's topsoil. Taken together, these factors raise the question as to why science does not come up with an innovative approach able to correct these problems. There are three answers to this question:

1. *Science is not sanctioning bad policies and acts. This is the remit of ethics, laws and, in a way, politics* (see also section 7).
   The lack of environmental ethics characterizing the mass of *Homo sapiens* has promoted very bad habits from multiplying existing populations in a completely unaffordable way, to massively misusing energy resources and disregarding the fact that the earth is not given a chance to regenerate its resources. These are not issues in the domain of science; rather, they are squarely the responsibility of political leadership.

   Once it has been informed by scientists that nature does not have time to recycle what we are unwarrantedly throwing away, at the rate and volume we are doing this, the political leadership should take the initiative to do whatever it takes to reverse the trend. And, as Chapter 13 documented, this task includes taking steps to reduce the population explosion.

   Instead, in several countries the political leadership is doing exactly the opposite. For instance, the United States has contributed nothing to the United Nations Population Fund (UNFPA), the organization whose objective is the distribution of birth-control information and devices as well as assistance in planned parenthood.

   The stated reason for this is that UNFPA conducts programmes in countries like China, where abortions are encouraged – a situation (supposedly) politically unacceptable to the Reagan, Bush, Clinton, George W., and Obama administrations.[1] The United States also refuses to give money to the International Planned Parenthood Federation because it actively supports abortions.[2]

   Science has provided the means for radically reducing abortion as last resort, but the political and religious will to apply them is absent.

   In matters connected to limiting waste, engineering has provided commendable support, because engineers understand that garbage is not an inevitable result of a production process. In factory production, for example, it is the outcome of:
   - poor quality standards,
   - imperfect materials, and
   - bad workmanship.

   If we analyse the history of any technology, we see that what has been considered in the past to be garbage often becomes material of value in new, more sophisticated and better-tuned production processes. We have to underline and accelerate such processes, but science and engineering alone are not going to solve the waste problem we are talking about.

   Several projects in scientific research have focused on the preservation of energy and resources, including the effective use of materials, recycling of water and treatment of waste products. This is indeed welcome, but it is not the 'big picture' which inevitably involves a higher-level political will.

   The destruction of the soil provides an example of the lack of this higher-level political will. Its absence has been at the origin of the elimination of 2 billion hectares of productive land. As mathematicians, we know how to calculate the exploitable part

1 In fact, Obama did much worse than his predecessors. In exchange for a senator's vote for his universal health plan, he sold the freedom of American women to buy insurance against unplanned parenthood. The case of Kane and Abel repeated.

2 Abortions should be only a case of last resort, and they would be vastly reduced if birth-control information and contraceptive devices were widely distributed – as well as training for boys and girls on how to avoid unplanned parenthood.

of the planet worldwide. This currently stands at less than 0.2 hectares per inhabitant, and it is diminishing every year through the dual impact of the population explosion and further destruction of fertile areas. But mathematicians have no power to stop this descent into the abyss; politicians do.

In conclusion, scientists can study and suggest environmental protection criteria and standards, as well as ways and means, but it is up to politicians to apply them and ensure that they become compulsory. A scientific contribution is important inasmuch as environmental protection standards must be arrived at after thorough consideration of the widest possible range of factors and probabilities (not through street riots). Hence, the scientific contribution is important, but it is not the whole story.

2. *While research into the very large and the very small,[3] which are the more recent domains of scientific investigation, is proceeding, a political decision should add to these fields of endeavour a policy for effective environmental preservation.*
Scientists may act as advisors to political leaders, but the tough decisions as well as the courage to put new draconian rules into action belong to the latter not to the former. Politicians know very well that the massive elimination of forests has led to the diminution of oxygen in the atmosphere to the tune of 10 billion tons per year, but they have done nothing to reverse the scales.

It is a talking example of ineffectiveness that of all the UN earth protection tournaments – Montreal 1987, Rio 1992, Copenhagen 1995, Kyoto 1997, Bangkok 2007, Copenhagen 2009 – only the latter involved something resembling a feeble (zero results) effort to save the forests. Environmental protection agreements are completely useless if they don't lead to comprehensive global laws against environmental degradation.

In addition, because people and lobbies find always ways to circumvent the laws, it is not far-fetched to suggest that criminal trials should take place if they are violated. In this connection, the contribution of scientists is vital in establishing a monitoring and control system. Environmental protection laws will be workable *only* when a system is in place to monitor their implementation and abuses.

For their part, to measure up to the task, legislators and judges must be able to access professional consultancy services for technology assessments and project evaluation. Certified environmental protection laboratories should be available to undertake immediate technical and ecological studies of possible disasters – including the task of ascertaining their cause and associated responsibilities.

In conclusion, scientists should be on tap, not on top. It is the job of governments to criminalize environmental and ecological offences. Environmental courts should be endowed with wide powers to help maintain ecological checks and balances for the conservation and preservation of natural resources – with authority ranging from birth control to soil, forests, water, air and generally all important environmental factors.

3 The research on the 'very large' targets the universe. Particle physics concentrates on the 'very small'. These two endeavours seem to be quite separate from one another, but open-minded scientists appreciate that the very large and the very small are a continuum.

3. *Scientific research, particularly applied research and development necessary for effective environment protection, must be given specific goals – not abstract notions.*
   These specific goals have to be set by governments and financed through taxpayers' money and private organizations. The latter will have profit motives which, for instance, can be found behind the spreading concept of *cleantech* (clean technology industry – see section 2).

   It should be noted, however, that, for the time being, cleantech is more label than substance. Its stated aims range from reducing the global carbon footprint and water recycling to drawing up a longer-term plan and finding ways to co-finance demonstration projects designed to commercialize potentially important technologies for the preservation of the environment.

   Because industry will not move into new and (sometimes) treacherous uncharted waters, it is up to governments to draw up a long-term clean technology policy – as well as to depoliticize research decisions. They should engage voters on the need to support research initiatives able to tackle non-polluting energy production, clean water resources, waste recycling and the other vital environmental issues addressed in this book.

   It is also the obligation of governments to ensure that research in universities[4] is channelled into the revamping of current products, devices, processes, policies and practices towards goals of environmental sustainability. This would be no different to the policies followed in connection to weapons systems during the Cold War years. Some of it is probably part of President Obama's late 2009 package which contains $4.5 billion for smart grids (section 3) and related projects.

   The obligation of businesspeople is to cope with disruptive innovation which threatens their product lines and be transparent about the risks and benefits of new technologies. Correspondingly, it is the responsibility of scientists to be investigative and efficient in their work, fully appreciating opportunities, consequences and risks of new technology developments.

   Scientists should also spend time communicating with the public and explaining what new environmental technologies can and cannot do. Examples of urgently needed projects include intelligent power grids, large-scale energy storage, non-polluting but effective recycling processes and much more. But the drive to see that environmental protection is at the top of the agenda lies with politicians.

2. Disruptive and Sustaining Cleantech

Cleantech is an umbrella concept under which come both existing and relatively newer environmental protection or recycling products, devices and processes. These are largely promoted by existing firms and also by start-ups. In this sense, cleantech is not a new industry, as some of its proponents claim, although it might become one at a future date. For these reasons the definition of clean technology is bound to be tentative.

Down to its most basic functions, cleantech can be sustaining or disrupting. *Disrupting technology*, also known as *killer technology* is newer (and often less expensive), entering a market where more traditional products and processes do not provide the

4 D.N. Chorafas, *Education and Employment in the European Union: The Social Cost of Business* , Gower, Farnham, 2011.

needed solutions. By contrast, *sustaining technology* provides product performance improvements in small steps, replacing established products and processes. Whether they belong to the one or the other class, cleantech products, devices and processes basically concern:

- environmental sustainability, and
- greater efficiency in the use of natural resources.

Under these two bullets are included energy production through solar, wind, geothermal and other sources; energy storage, smart grids, green building materials and conservation; water purification; household and industry waste recycling; and much more. Regrettably, laboratory activity is not included in this list in the way it is usually conceived.

Decisions made by politicians at national or international level affect the business prospects of cleantech in a big way. It goes without saying that clean technology executives want to see a robust, global agreement to cut emissions of greenhouse gases. But, as Chapter14 documented, this did not emerge from the December 2009 climate summit in Copenhagen, and the chances of it emerging in the near future are far from good.

The fact that the 119 heads of state who flocked (at taxpayers' expense) to the Danish capital failed to agree on a global mechanism to put a cap on emissions was bad news for cleantech firms. It also made it harder for energy firms to justify big investments in unproven green technologies, such as advanced biofuels or carbon capture and storage. In the aftermath, almost all areas of clean technology will attract less interest from investors because there is no mandate.

Ultimately, the sustainability of the earth's resources is less about technology and more about attitude – and *Fiascopenhague* has proved that a change in public attitude cannot be counted on as the main driver for cleantech investments.

Nevertheless, some companies, like Siemens, recently stated that they will press ahead with their plans to invest in green technology, despite the disappointing outcome at Copenhagen. On the other hand, utilities have been more ambiguous about how big their market for cleantech will be. This is not unexpected because, particularly in the current economic climate:

- private industry will not invest in finding out if greener technologies will work unless governments share the risk, and
- this extends cleantech's adoption timetable, as new products and processes need demonstration projects to prove that proposed solutions will really work.

This creates a vicious circle. The lack of international agreements on emissions and other issues leading to environmental damages leaves it to governments and parliaments to look after their countries' competitive interests. Environmental technologies are more expensive and will not be implemented fast enough without support from public policies.

Real-life applications-based testing is necessary because the implementation of any new technology involves surprises. Take waste recycling as an example. Researchers at the Tema Institute in Linköping University, Sweden, have been developing a technique that employs ultrasound rather than pumps to break up the sludge. According to available information, this increases methane yields by 13 per cent but, at the moment, the process generating the ultrasound consumes more energy than it yields and also seems to be more costly.[5]

It should be appreciated that there is no silver bullet in cleantech. Unwanted after-effects are inseparable from the benefits we may derive from it. Therefore, real-life tests are necessary to assess the health effects of cleantech's wares. For instance, products and processes using nanoparticles are already available on the market, but some of them are suspected of causing health hazards. An example is carbon nanoparticles already known from sources such as photocopier toner, diesel soot and welding fumes. Unwanted effects have to be clearly defined and isolated, because, generally, products of *nanotechnology* are welcome. They have brought:

- improvements in medicine,
- cheaper materials,
- manufacturing efficiencies, and
- agricultural benefits that might assist in reducing the use of chemicals.

A well-known controversial case is that of genetically-modified (GM) foods, which encounters public disapproval because its further-out after-effects are unknown. Particularly in need of intense testing is disruptive technology which often involves *unknown unknowns*, but also carries the greater benefits.

Sustaining technology, by contrast, presents improvements which add efficiency and/or control functions using new advances. Because its risks involve mainly *known unknowns*, neither its exposure nor its benefits are outsized. Examples include smart grids (section 3) alternative current/direct current (DC) converters, intelligent measurement systems and more.

To repeat, sustaining technology provides numerous benefits over more traditional approaches, but no dramatic breakthroughs. In-between sustaining and killer technologies fall solutions such as arrays of solar panels and farms of wind-generated electric power (Chapter 5). When their introduction made the news, both attracted a great amount of interest and investment, but by all indications their thrust in the market may have been broken. Returns have been particularly disappointing in solar energy, but other sectors, too, have not been performing as expected. As a result, analysts have taken a neutral view of cleantech.

Governments are still mostly pro-renewables, but they are short of funds. Companies are no longer sure that renewable are an area in which they want to be.

Shell's strategies changed in late 2009 when its then chief executive, Jeroen van der Veer, stated of wind, solar and hydrogen, 'I don't expect them to grow much at Shell from here' and further commented that further investments in renewable energy would focus on biofuels.[6] By contrast, BP insisted that the role of renewable energy in its strategy has not been altered, but admitted that investment in it fell from $1.4 billion in 2008 to roughly half that in 2009. The company:

5 'The Seat of Power', *The Economist*, 30 December 2009.

6 'The Green Slump', *The Economist*, 5 December 2009.

- has been selling some of its renewable energy assets, including three wind farms in India, and
- has cut its solar-cell manufacturing capacity in Spain and America.[7]

Like Shell, the one renewable energy source in which BP still seems to be a serious investor is biofuels. In the European Union private individuals who went ahead with solar electricity production (particularly through photovoltaic cells) got burned when state utility monopolies significantly changed the prices for which they buy electricity. The problem was that these prices stood way above cost of energy produced through coal-fired plants or nuclear reactors, but short of what was necessary to break even.

3. Intelligent Grids for Power Distribution

Power grids in the form we know them today may well be something of the past. Traditional concepts characterizing power transmission are being turned upside down to make way for a new and more efficient technology: the *intelligent grid*.

In October 2009 Barack Obama rated its first incarnation, known as the *smart grid*, as one of the main themes of his stimulus package, earmarking $3.9 billion for it. Countries currently wanting smart grids range (in alphabetical order) from Australia, Brazil, Britain, China, Germany and Italy to Japan. Companies getting ready for what promises to be a lucrative market include diverse entities like Cisco, Computer Associates, General Electric, Google, IBM, Microsoft, Oracle, SAP and Siemens.[8]

Let's first examine what the term 'intelligent grid' means. The classical electricity grid is designed for specific maximum charges in a system-wide sense. Even if some limited portions of the network fail, it is still able to transfer and deliver the required load by using engineering tolerances. Accidents, however, do happen – the great blackout in eastern Canada in 1998 as a result of ice storms is an example.

The premise of embedding intelligence into the grid is that such accidents might be avoided if the system knows its status. Sensors can monitor the state of nodes and links and inform, in real time, a control computer endowed with *knowledge engineering* constructs. This serves all the way:

- from the design optimization of the electricity grid to,
- its day-to-day more efficient operation, preventive maintenance and medium- to longer-term restructuring.

Another application in which network intelligence can prove its value is the avoidance of gridlocks. As the use of electricity spreads widely and loads fluctuate, peak demand can rise 25 per cent while capacity rises just 6 per cent. With traditional power distribution networks, shortages cause brownouts[9] and also result in huge price spikes. This has happened time and again in America, both in California and in the Midwest.

7 Ibid.

8 'Clever, but Unprincipled', *The Economist*, 8 October 2009.

9 A reduction in the voltage of commercially supplied power instigated by the utility when supply exceeds demand.

- With the growing appetite for power the margin for error is sinking.
- This serves to wake up politicians, regulators and consumer bodies to the need to find better solutions for power distribution.

Lessons can be learned from the California debacle. The state had chopped up its electrical system too finely, breaking it into hundreds of groups of buyers and sellers, each of which had to maintain a constant balance of supply and demand among its members. It was easier and cheaper for them to trade power with each other, hoping that the grid as a whole would keep in balance (outages cost the American economy $150 billion a year). The facts did not validate that hypothesis. When supplies were adequate, they were not always going where they were needed. When they were inadequate, the balance fell out of bed. Hence the need for both a wider electricity grid interconnection and for intelligence in its nodes to make optimal load balancing feasible.

In Britain there is an ambitious plan for an electricity supergrid in the North Sea. It was launched in London in March 2010 by a group of ten leading European companies and is supposed to connect England, Germany and Norway. The expected cost is €34 billion, which will probably prove to be an underestimate.

A pressure group known as 'Friends of the Supergrid' (which includes Siemens and Hochtief of Germany and Areva of France) says that the project would make it possible to develop high volumes of offshore wind power in the North Sea and reduce electricity prices.[10] Only someone who believes in miracles would buy that argument. Another initiative, along the lines of using high tech, is smart buildings. Utilizing sensing devices and data analysis software, their promoters aim to:

- make buildings more sustainable, and
- reduce operational costs, while assuring occupants' comfort.

All these are bets on technology which hopefully will prove fruitful, but currently they are unsupported by solid evidence. Speaking from a background of 60 years' experience as an engineer, particularly in large information technology systems, I believe that there are reasons for not getting too optimistic. These can be best expressed as a poem:

> *The technology you have is not the one you want,*
> *The technology you want is not the one you need,*
> *The technology you need is not the one you can obtain,*
> *The technology you can obtain is not the one you can afford.*

This is written to inform, not to discourage. The theoretical principles are fine, but real-life practice does not necessarily conform to them. The way it is currently used, the term 'smart grid' encompasses almost anything that would make power transmission more reliable, flexible and economical. It's a vague concept whose underpinning notions range:

10 'Power Transmission Equipment Makers Launch Friends of the Supergrid', *Power-Gen Worldwide*, 8 March 2010, at: http://www.powergenworldwide.com/index/display/articledisplay.articles.powergenworldwide.t-and-d.t-and-d-infrastructure.2010.03.power-transmission.QP129867.dcmp=rss.page=1.html.

- *from* metering that sends in readings automatically to a computer
- *to* software that detects damaged nodes or cables, and
- reroutes power supplies around them.

While in the longer term the whole power transmission system is up for rethinking and revamping, putting upfront all these goals serves little purpose because it confuses the issue rather than providing a roadmap. *If* grid intelligence has any future, *then* its planning and implementation must start with:

- a few concrete goals,
- a couple of pilot projects, and
- targeted deliverables that can be measured against, as well as contrasted to, plans.

An example is short-term results based on experimentation made feasible through smart-grid facilities, which lead to the reduction in power system components – with positive effects on cost and space savings. Another example is the ability to configure, simulate and debug the software (section 4) to reduce development time. Real-time telemetry of voltage, current and temperature for each device:

- facilitates the programming of fault-management scenarios, and
- provides great system visibility as well as reliability improvements.[11]

Design-wise, a complete power system can be configured by using programmable components such as inventoried parts, subassemblies and subsystems. Properly projected intelligent nodes support combinations of turn-on, ramp-up, sequencing and shutdown processes that have been too cumbersome to implement using old technologies. Energy savings can be obtained from:

- improved electrical efficiencies, and
- power turn-on/off sequences, optimized to system-level functional requirements.

Advances in intelligent grids are not yet sufficient to permit meeting all these goals at once. However, the fact that these results can be independently obtained makes it possible to plan for their aggregation. Reducing losses in power transmission is another of the many important reasons why accurate real-time monitoring, measuring, simulating (section 4) and controlling is vital. Significant amounts of power are currently lost to technical problems (and theft) connected to electricity grids.

- Such losses are estimated to be up to 10 per cent in the United States and Europe, and
- they allegedly rise to 50 per cent or more in some of the developing countries' big cities where technology is substandard and the law enforcement industry turns a blind eye.[12]

---

11 D.N. Chorafas, *Statistical Processes and Reliability Engineering*, D. Van Nostrand Co., Princeton, NJ, 1960.

12 This 50 per cent figure and the reasons for it come from anecdotal evidence provided in one of the meetings I have had.

Underpinning the improvements outlined in the preceding paragraphs is the addition of plenty of information technology, such as sensors, digital meters, a communications network (like the Internet, but dedicated), databases, knowledge banks and expert systems, which radically changes what has been so far dumb wires.[13] Embedded intelligence will help to avoid outages, save energy and distributed generation of electricity in a more efficient way.

If intelligent grids are seen as the upper level of sophistication under present-day technology, and smart grids are seen as the lower level, *then* the currently discussed smart-grid project will not accomplish all of the aforementioned benefits. It will, however, provide a testbed on which to experiment with further advances.

A higher level of sophistication will be characterized by a fundamental integrative representation which can effectively serve in command-and-control. Qualitative models (not necessarily part of smart grids) will supplement quantitative algorithms representing primitive elements and subsystems used as network components made of hardware and software (section 4). Knowledge engineering artefacts will:

- assist in explaining operational characteristics, and
- exhibit expert behaviour in problem-solving situations.

Through model representation, experimenters can study grid behaviour under assumed rare events and extreme conditions, develop stress scenarios and evaluate the effects of assumptions about forces that are currently not well understood, but impact on the problem area.

Key to grid intelligence is *profiling*. Information technology has made strides in this direction, an example being IBM's profiling of its 50,000 consultants in order to improve the management of the firm's intellectual resources. Since the middle of the first decade of this century, data miners, statisticians and knowledge engineers working at the company's research centre have been developing methods for scrutinizing personal profiles of thousands of IBMers.[14] The process is known as *stochastic analysis*, mapping the random behaviour of humans into mathematical models. Using these mathematical profiles, expert systems (section 4) can:

- pick the best team for every assignment,
- track each consultant's contribution to the project's progress, and
- hour-by-hour rate each person's quality of work and deliverables.

The potential that intelligent grids represent in improved efficiency and reliability of power distribution networks is nearly an order of magnitude over current practices. This has attracted venture capitalists who have put more than $1 billion into smart-grid start-ups (although 2009 investments did not match the heights of 2008).[15]

13 In technical jargon a dumb wire is one which does not benefit from expert systems.

14 *Business Week*, 23 January 2006.

15 Two of these start-ups, GridPoint and Silver Spring Networks, raised $220 million and $170 million respectively. Among the better-known companies, Siemens hopes to win orders worth €6 billion ($8.5 billion) in the 2010 timeframe. 'Wiser Wires', *The Economist*, 10 October 2009.

Cisco projects that the intelligent grid's underlying communications network will be 100 or 1,000 times larger than the Internet.[16] This is largely based on the fact that such a power network will be characterized by communications intensity.

In the background of the need to use advanced technology in power distribution also lies the fact that systems transmitting electricity are vulnerable. The wholesale trading of power has put heavy strains on the capacity of high-voltage transmission lines, and this is going to get worse in the coming years, not better.

In conclusion, major improvements to the quality, efficiency and dependability of power transmission networks are the alter ego of improvements in energy supplies. People and businesses are using more power than ever. They are also demanding clean power for their computers and sophisticated industrial equipment. Spikes and dips in the flow of electrical current are anathema to high technology.

4. Software and its Impact on Modern Technology

Sophisticated knowledge-based routines are one of the more advanced forms of computer programming, typically referred to under the umbrella name of *software*. The term includes not only new computer programs, but also plenty of others written in obsolete languages (like Cobol), as well as 30-year or more old programs that are difficult to maintain and are error-prone.

As a term, software came into existence during the late 1958 convention of the Association of Computing Machinery (ACM) in Philadelphia. It was coined to stand in contrast to the (then) widely used reference to *hardware* – the electromechanical machine – and it incorporates two broad classes of computer programs:

- *Basic software* such as operating systems (OS), programming languages, a variety of utilities,[17] database management systems (DBMS), teleprocessing routines (TPR), and more – for instance, shells, platforms, assemblers, compilers and generators. Basic software is the alter ego of hardware and serves either of two purposes: helping to run the machine and supporting the applications running on it.
- *Applications software*. These are primarily designed to serve end-user needs – for instance, accounting, customer handling, supply chain management, engineering design, inventory control and so on.

The number of computer applications has been soaring, even if the paperless office remains a distant dream. Practically all computer user organizations have rapidly growing libraries of application programs from bread-and-butter procedural accounting routines to sophisticated models.[18] (A *model* is a simplified representation of the real world. This is done by abstraction, retaining only the most important variables whose interaction and values are mapped into the computer by means of algorithms and heuristics.)

---

16 Ibid.

17 Such as input/output and housekeeping.

18 Mobile computing, for example, is one of applications domains where software developers have been particularly busy. By the end of 2009 they had created more than 12,000 games and other applications that run on Google's Android OS.

The examples discussed in section 3 in connection to intelligent and smart grids have been essentially based on modelling software. This has typically included high-level constraints for defining and manipulating objects (voltage is an object), node reliability and line losses.

- Mathematical constructs describe objects using modelling coordinates.
- Such modelled objects are mapped into world coordinates using a transformation applied to all primitives.

The feed-forward function of an intelligent grid, to which reference was made in section 3, is supported through simulation. *Simulation* is a working analogy. By means of analogical thinking we are able to emulate real-life systems to derive values which, so to speak, permit us to read tomorrow's newspaper today. Simulation can be done either:

- through hardware (analogue simulation) which has been successfully used since World War II to study power distribution networks,[19] or
- by means of mathematical models which – depending on the area of study – may represent power grids, communication networks, business enterprises and more.

It is essential to keep in mind that when analogous systems are found to exist, experimentation on one of them may reveal valuable data on the performance of the other. The mathematical construct simulates the power network and, by means of analogical thinking, makes feasible planning, experimentation, real-time monitoring and instantaneous control – activities which turn a dumb network into a smart network. Intelligence will be provided by knowledge engineering constructs.

A critical element is the way in which expert knowledge is represented. The representational framework plays an important role in the way in which obtained data is interpreted and assimilated into a computer-based decision framework. Expert knowledge can be encoded as rules (the so-called production rules) which describe condition and action in a simple forward form.

IF condition A

THEN do action B

ELSE do action C

A production rule sends a signal (fires) if its activation responds to condition A, which is determined by input signals. More advanced expert systems are able to learn on the job, and they are robust against noisy data. Assisted by neural networks they can also exhibit some other advantages, such as non-linear processing.

Models and expert systems for knowledge-enriched software are best developed in conjunction with the designers' work of *functional decomposition*. This consists of splitting up designs into a component hierarchy corresponding to functional modules. Knowledge

19 The first job I did in 1951 in Paris with Electricité de France (EDF) as a trainee engineer.

engineering artefacts are restating design problems in a manner that allows certain critical issues to be considered:

- early on, and
- across the full breadth of the grid or other system for which they are intended.

If a power transmission network has, and can use, a data structure which can be said to represent something (an object, a procedure and so on), *then* the system itself can also be said to have intelligence applicable to its daily operations. This significantly improves the software's utility.

The knowledge embodied in that representation about the physical or logical supports makes the system tick. The foregoing brief definitions and examples explain advanced software's role in engineering and other projects – for instance, tracking consumptions and emissions while targeting energy optimization. Tesco recently did such an application in connection to its 4,300 stores, to:

- gather data about each store's energy consumption,
- distinguish between sources of usage – light, air conditioning, refrigeration and so on,
- compute emissions of greenhouse gases from existing stores and distribution centres,
- test each energy usage, and associated emissions, against pre-established targets.

Efficient energy administration is an important domain within the new *carbon-management software* market. While in the past several firms have tracked energy consumption in an effort to save money, the new drive is to identify and monitor types of emission in order to comply with regulations on pollution. Leading-edge companies are now driving for holistic system solutions using knowledge artefacts.

Unlike Cobol programs and other procedural routines, intelligence-enriched software does not need to be large. In the late 1980s, when expert systems came to the foreground, many of them consisted of thousands of rules – like DEC's XCON (expert configurer). Today, agents (mobile expert systems) are tiny, and a myriad of them serves a wide variety of purposes.

Pharmacology provides a good example of future trends. Many firms have not come up with enough treatments for their drugs to enable them to replace the original patents when they run out. To diversify and stabilize their revenue, some drug manufacturers are now selling ancillary services tied to their pills – for instance, enabling these pills to relay data about a patient back to the doctor after they have been swallowed.

- Stomach fluids activate the edible communications device contained in the pill, and
- this sends wireless signals through the body to another chip worn as a skin patch or embedded just under the skin.

Coupling smart pills with wireless networks and mobile phones allows the information the pills capture to be shared with whoever needs this data to monitor treatment, such as another expert system designed to extend the reach of a doctor's surgery, which monitors, in real-time, patients under intensive treatment for adverse reactions or non-compliance.

Kalorama, a research group, estimates that sales of such services will leap from a guesstimated $4.3 billion in 2009 to $9.6 billion by 2012.[20] Patients who clamour for more data about their health, as well as governments providing medical care, are beginning to demand that drugs firms prove the effectiveness of expensive new medicine. Power networks are not alone in need for intelligence-enriched solutions.

5. The Disruptive Impact of Advanced Software

The revolution in energy management will be most felt in those domains where only minor technical advances have taken place during the last decades. Power distribution systems are a good example, because they have changed remarkably little over the post World War II years with centralized utilities delivering electricity to passive consumers. By contrast, a grid endowed with intelligence:

- will use technology to collect, communicate and react to data, and
- will be proactive in making power distribution more efficient, reliable and easier to manage.

Other benefits are for the time being guesstimates, but some of them may well materialize. General Electric believes that installing smart meters in just 25 per cent of American homes will obtain environmental gains equivalent to removing 1.7 million cars from the roads.[21] But there are also conflicts of interest. For utilities, reducing consumption means reducing revenues, while consumers are not wholly convinced that smart technology will cut down their bill.

Engineers and information technology experts are forthcoming with suggestions on significant improvements to today's status quo. They point out that advanced software (models, simulators, expert systems) can be used in a vast domain ranging from searching for new energy resources to energy production and distribution to energy preservation. Already advanced software solutions assist in providing a clearer picture of oilfields through:

- three-dimensional seismic imaging,
- advances in deepwater exploration,
- improved ways of extraction from oil reservoirs, and
- multi-directional drilling.

Lessons on the effective usage of technology in energy conservation can indeed be learned from the transformation taking place in seismic technology. Geophysicists are developing systems enabling them to remotely sense in real-time the motion of fluids, using both acoustic and electromagnetic mapping, a combination which improves production from the reservoir.

20 'Potential Encapsulated', *The Economist*, 16 January 2010.

21 'Smart Move', *The Economist*, 21 March 2009.

Developments in technology which provide food for thought also come from *cloud computing.*[22] According to a number of opinions, which I detail in my book, *Cloud Computing Strategies*,[23] the best applications will not be those from the conversion of old IT, which in a way resembles sustainable cleantech (section 2), but from domains still in their infancy: Smartphones, iPhones, personal assistants, mini-laptops.

This is the disruptive generation of engines demolishing the legacy of personal computers – just as PCs demolished the mainframe computer in the 1980s. They are devices that can be connected almost anytime, anywhere. In cloud-computing terms, it is wrong to look at the network infrastructure just as a vast array of interconnected machines managing the data and software running on all sorts of IT equipment. The better way is to perceive the aggregate in novel terms – as an expanding combination of mobile and fixed technologies which:

- will shape up advances in the computing universe in the next decade, and
- already pose a long list of challenges characterized (and to be characterized), by fairly stochastic technical requirements.

Both intelligent (not simply smart) grids and cloud computing are examples of software's *disruptive impact*. To better appreciate the reasons for this statement, as well as the meaning and reach of the cloud's evolution, it is necessary to turn back to the early 1970s and the developments associated with cellphones.

Much of the credit for downsizing the volume and cost of cellular phones goes to Marty Cooper who, confronted with what used to be unwieldy two-way radio devices built into dashboards, conceived the idea that they needed to be significantly redesigned and eventually made portable. Cooper's concept developed into a tangible instrument after Motorola put him in charge of its carphone division in 1972.

- It took 90 days from idea to prototype, as Cooper sponsored a design contest among Motorola engineers.[24]
- By contrast, the wider spread of mobile telephony required more than two decades and a great deal of hardware and software development.

To understand the magnitude of the breakthrough, which must now be repeated with intelligent grids, one should compare Cooper's engine to present-day smart phones and the advent of mobile Internet. Smartphones have evolved into versatile handheld computers, an example being Apple's 3G iPhone and Google's response with Android software. Note, however, that:

- as the competition among different vendors intensified, hardware features became secondary to software, and

22 This reference is no wholesale endorsement of cloud computing – which presents great interest in terms of on-demand software and platforms, but significant risks with remote, very large databases.

23 D.N. Chorafas, *Cloud Computing Strategies*, Auerbach/CRC, New York, 2010.

24 Called DynaTAC, the handset had 35 minutes of talk time and weighed one kilo. Cooper reduced DynaTAC's weight by half, and it was finally launched in 1983 with a price tag of $4,000. This process, incidentally, is still in evolution, and great steps in technology are still ahead of us.

- to succeed, their makers had to inspire third-party developers to write clever applications that attracted an army of end users.

There are almost endless new breeds of mobile devices and functions. In late 2009 Google acquired AdMob whose technology and network allow ads to be placed on mobile websites and within iPhone and Android applications. Google needs to enhance its advertising power because, despite its success, it still has a long way to go to challenge the iPhone's dominance of the market.

Some critics say that the array of cellphone manufacturers producing Android phones could result in a less consistent product than competitors like the iPhone and the Palm Pre, in which the hardware and software are tightly integrated.[25] Google answers that the variety of Android systems is no different than the situation in the personal computer market, where multiple versions of Windows exist. It is way too early to have a well-informed opinion about which strategy will pay the greater dividend.

It should also be appreciated that sometimes 'more' means 'less' because there are problems associated with software, as well as software/hardware failures. On 19 November 2009, a computer glitch caused widespread flight delays and cancellations in the United States. The system was practically down for four hours, and the failure was attributed to software connected to the Federal Aviation Administration (FAA) infrastructure. A news item on Bloomberg television called it 'the 20 billion dollar glitch'.

Along with software, a crucial factor in information technology's renovation has been bandwidth. Cooper's Law says that *spectral efficiency* has doubled every 30 months since Guglielmo Marconi patented the wireless telegraph in 1896.[26] Present-day devices have a spectral efficiency more than 1 trillion times greater than Marconi's original late nineteenth-century invention.

- Smart antennas will, most likely, assure that Cooper's Law continues to apply, as Moore's Law did for four decades, and
- this will permit cloud computing to encompass a vast array of applications from sensors to computers and social networking.[27]

Sensors embedded in a wide range of machinery currently in use can directly gather data about the reliability of component parts as well as noise level, temperature, humidity, ice formatting and more. Adding digital sensors and remote controls to the transmission and distribution of a power network, for example, and combining their input with data-mining and expert systems, will significantly increase its utility by making it better able to:

- cope with new sources of renewable power,
- assure coordinated use of different energy sources,
- provide information to consumers about their energy usage,[28] and

25 *International Herald Tribune*, 23 November 2009.

26 Which broadcast in Morse code over a wide frequency range.

27 Including areas of vital importance to an aging society, such as health care.

28 Studies suggest that when people are made aware of how much power they are using, they reduce their consumption by roughly 7 per cent.

- permit utilities to monitor and control their networks more effectively.

In a fairly similar way, sensors and digital relays installed on other types of transmission and distribution network will make possible a higher level of supervisory control, as well as permit the optimization of the deployment of resources. They will also, eventually, provide a new array of services which find precedence in the introduction of supply chain management, a decade and a half ago.

6. Environmental and Other Risks Associated with Technology

It would be unwise to state that technology's contributions are free of environmental risks. Pollution due to technology is part of the context of environmental challenges posed by all sorts of engines. One of society's most potent weapons for self-destruction is the car – or, more precisely, the intensive use of motor vehicles by people (see also Chapter 10).

It is not difficult to find plenty of other examples. If Ralph Nader, champion of consumer protection, is to be believed, hospital patients run no small risk on the operating tables not only from incompetent surgery, but also from electrocution and bacteria. Poorly administered hospital procedures are another type of exposure that cause the death of patients. But people still do go into hospitals. Quite similarly, aircraft accidents, even on the ground, have taken toll of many lives in a few seconds. But we still fly.

When we know how to use it wisely, technology can be helpful, but this does not mean that there are no consequences. According to some estimates, as a result of recent research on toxic waste disposal, solvent wastes generated at semiconductor facilities could be reduced by up to 70 per cent and arsenic compounds could be cut by up to 90 per cent. This, however, significantly increases the cost of production and of clean-up.

Pollution due to technology is directly connected to its hardware: from manufacturing to disposal of waste after the devices have been discarded. Almost all the major electronics companies in Silicon Valley, and elsewhere in the world, had problems with contaminated sites.

As semiconductor technology progresses, its production of toxic chemicals also increases. The quantities of hydrofluoric acid required to make 4-megabit dynamic random access memory (DRAM) chips are greater than that of 1-megabit DRAMs by a factor of about five. Some years ago, IBM was rumoured to have spent more than $42 million pumping large quantities of contaminated water through cleansing systems at its plant in San José.

Several years ago, based on research findings,[29] *Fortune Magazine* published a list of how different US firms pollute their environment. A group of companies were rated 1 to 5, with 1 being the best rating. For instance, Apple Computer and H.B. Fuller each got 1.5; Louisville Gas and Electric, Amoco Nucor and WalMart got 2. The worst scores of 4 and 5 were awarded to the following industries and firms:

---

29 The research was done by Franklin Research & Development of Boston, an investment firm which based its ratings on several factors. Each company in its list started with a score of 3, which then rose or fell based on corporate actions that harm or help the environment.

- *Chemicals*: Monsanto, **4**, pesticides, toxic dumps, offset by clean-air efforts; W.R. Grace, **5**, toxic dumps, several environmental lawsuits
- *Computers*: IBM, **3.5**, high CFC emissions
- *Electrical utilities*: Southern, **5**, high SO_2 emissions contributing to acid rain
- *Environmental services*: Browning-Ferris, **5**, numerous landfill violations; Waste Management, **5**, numerous landfill violations
- *Forest products*: Louisiana-Pacific: **4**, air and water pollution problems
- *Oil*: Exxon, **5**, poor response to *Exxon Valdez* oil spill
- *Photo equipment*: Eastman Kodak, **4**, substantial leaks on Rochester sites
- *Steel*: Bethlehem Steel, **4**, old mills with numerous environmental problems
- *Other*: Borden, **4**, toxic dumps, air and water complaints; General Electric, **4**, major PCB clean-up problems; General Motors, **5**, toxic dumps, air and water problems

True enough, at the time this study was made cleantech was not the darling of industry, but it would be wrong to think that it can produce miracles. As defined in section 2, cleantech aims to reduce emissions of CO_2 and other GHGs, recycling waste and so on. Ironically, this does not mean that cleantech itself is clean in an environmental sense.

Science is far from having mastered all problems connected to environmental pollution. Generally, the extraction of ore, production of materials and their disposal poses severe challenges to the environment – and this is as valid of the new technologies as of the old.

It is not only the manufacture of semiconductors but also the discharge of advanced manufacturing processes – for instance, for ceramics and organic resins – that contributes to pollution. Such refuse is no less poisonous to the environment than the refuse of a chemical plant, foundry or factory for the treatment of leather products. Man has not yet succeeded in effectively protecting his environment from high technology. If anything, pollution is becoming worse because:

- the more the population multiplies, the greater is the rate of production, and
- with a large part of manufacturing now being done in less developed countries, pollution problems are skyrocketing.

It is also quite proper to emphasize that in reality materials are not final products per se. They are intermediate stages critical to the fabrication of further, more sophisticated new products. The history of goods and processes made by man teaches that the amount of pollution increases over time, no matter what public relations officers may say.

The risk of the greenhouse effect – by no means the exclusive domain of COs – is emblazoned in newspaper headlines and preached about in UN environmental conferences. It is an omnipresent operational risk of our society, closely associated with well-being, travel, comfort, employment and technology – including computers, databases and communications.

Until the early years of this century nobody really bothered to look at high power consumption and the environmental impact of computer centres, PCs, networks and data warehouses. But in February 2007 a study released by Lawrence Berkeley National Laboratory found that:

- power consumption of data centres alone doubled between 2000 and 2005, and

- by 2006 it accounted for over 1.2 per cent of all US electricity consumption, and was growing.[30]

Power consumption at US data centres presently stands at over 2 per cent of total US power usage. This amount has doubled in less than five years and there is no clear view of an upper limit. In fact, if all other information technology gear, including the Internet, is added, this brings information technology power consumption to more than 5 per cent of US power usage. Although such figures are lower on a worldwide scale because many countries do not have the technological sophistication of the United States, they are rapidly moving upwards, and their background causes are next to impossible to weed out as all countries try to imitate the pattern of American statistics.

7. Power Consumption Increases with the Spread of Technology

The reader would be absolutely right to conclude that what the previous sections have highlighted is a wholesale reversal of what was supposed to be the contribution of technology (particularly of IT) to the conservation of the environment. This reversal is the direct result of three factors:

- a major global increase in the population of *Homo sapiens*,
- the relative rise of standard of living, and
- technology's insatiable thrust for power which gives no sign of abating.

Power consumption by information technology has increased not only because of the rise of the Internet and of nomadic computing, but also because computers have been historically designed to maximize performance. Measures taken by some computer companies, like Dell, the PC-maker, prompting their customers to plant trees to offset the carbon emissions generated by their computers are nice but totally ineffectual when compared to the magnitude of the problem.

Installing more efficient power supplies and switching to modern cooling systems is advisable, but not enough to counterbalance the huge annual increase in computer usage. The inescapable conclusion is that better performance, high-density servers, more customers, more portables and an increasing level of usage all combine to produce more CO_2 and other pollutants, which greatly affect the environment despite what:

- euphoric heads of state may say, and
- ineffectual, but highly polluting, UN conferences may target.

According to published statistics, from 2000 to 2005 the total electricity use for *servers* in the world increased from less than 60 billion kWh per year to nearly 130. For 2010 this can be updated to 200 billion kWh, through a conservative linear progression (which is an underestimate). Following a similar algorithm from, 2000 to 2010:

30 'Going Green', *The Economist*, 1 March 2007.

- The number of common servers increased by 500 per cent.
- The number of high-end servers soared by almost 800 per cent.
- Power needed for cooling and auxiliary equipment quadrupled.

To bend the power consumption curve, AMD, Dell, HP, IBM, Intel, Microsoft and Sun launched the *Green Grid* consortium with technological fixes such as multi-core chips (2.4 cores) by AMD and Intel (with criterion *performance per watt*), using DC rather than AC. (DC is more 'efficient', but there are no standards for a multi-supplier policy.)

'On-demand' use of cooling systems may be a way of slowing the almost exponential progression of power consumption, but it will not do miracles. Sensors, actuators and expert systems can help. However, despite all these efforts, US companies now spend 10 per cent of their IT budget on electricity consumption, and this is expected to rise to a level equal to hardware expenditures. The same trend exists in the European Union, Japan and other parts of the world.

Private individuals do no better. Statistics concerning the global mobile industry, and its billions of subscribers, indicate that it is now responsible for about 1 per cent of the world's CO_2 emissions. Just consider all the 15,000 'delegates' who participated in Copenhagen 2009 on 'environmental protection' and spent all their time on their mobiles (Chapter 14).

To get a better picture of the contribution of mobile technology to global pollution, one must also consider the impact of waste associated with such equipment, as huge numbers of mobile phones, PCs, printers and other IT devices are discarded each year due to 'upgrading'. It should be remembered that these are not biodegradable, yet in most countries they are thrown into general garbage. (Switzerland and Germany are among the very few countries to collect technological waste and handle it carefully.)

As all this suggests, there is plenty of scope to redefine, through properly focused research projects, the choice of materials used for technological devices all the way to waste disposal and treatment.

Superficially, paper cups seem to be far less harmful to the environment than plastic foam ones. Paper is biodegradable and made from wood, a renewable resource; polystyrene takes centuries to decompose and is derived from non-renewable oil and gas. These are the well-known 'facts'. However, a comparison of the environmental impact of the manufacturing process for one polystyrene coffee cup versus one paper cup reveals some surprises, according to *Science* magazine.

To start with, six times as much wood pulp as polystyrene is required to produce a cup. The paper cup consumes 12 times as much steam, 36 times as much electricity, twice as much cooling water – and produces 580 times as much waste water. The Greens who actively promote biodegradables don't seem to have done their homework.

Moreover, when the paper cup does actually biodegrade – a process which may take decades after it has been discarded and is buried in a landfill – it will produce methane, a far more potent greenhouse gas than the pentane used to make plastic foam. And the paper cup contains waterproofing resins that make it impossible to recycle, while polystyrenecups can be turned into lawn furniture and flowerpots.

In conclusion, nothing can be taken for granted in regard to waste produced by technology, as well as in other domains of man-made products. 'Biodegradable' may sound like a nice term, but on its own it says nothing. The whole concept of design has to be turned inside out, all the way from using less energy and reducing throwaway materials to significantly improving waste disposal.

8. Conclusion: The Badly Misunderstood Role of Scientists and Technologists

'If you ask a man "what is science",' said Konrad Lorenz, 'the usual answer will be physics and chemistry based on analytical mathematics. It is really a mental illness of humanity to believe that something that cannot be defined or described in terms of the exact nature of science, or cannot be verified by analytical mathematics, has no real existence. [But] our values are emotional ... To cut off the subjective, emotional side of humans is a dirty lie.'[31]

Fundamentally, science has to do with the whole range of issues impacting on man and his future. Therefore scientists bear *responsibility* for their work and deliverables. We should keep stressing this to ourselves and to fellow scientists. Part of this responsibility is keeping the communications channels open, so that everybody appreciates the risk:

- of man-made destruction, which is integral part of civilization, and
- of what is done to improve quality of life and protect the environment.

Niels Bohr, the physicist, was passionately opposed to secrecy in science believing that information should be shared. With no secrecy, he thought, there would be no point in an arms race. The same premises are valid today about birth control and issues relating to the preservation of forests and water resources; avoidance of pollution of land, seas, and oceans; the safeguarding of the ozone layer and much more.

For roughly 130 years, since the institution of organized research through laboratories, we have successfully used basic and applied science to better our standard of living;[32] in fact, we have made great strides in this direction. Now we must apply science in protecting and safeguarding our environment, being both forceful and realistic in the goals we set out to reach.

A study of the direction taken by scientific research effort is necessary because research acts both as catalyst and incubator of new ideas. Research requires goals, findings and regular reviews of obtained results. Without deliverables, it is daydreaming; but deliverables, once obtained, must be put into action. Research provides the ground for, but on its own does not create, change.

The goals of applied science have to be realistic. Pollution control does not mean that the pristine waters and clean air that used to prevail will return on planet Earth. They are gone forever *even if* all further pollution stops at this very moment. And we all know that it is unrealistic to think that such a radical change in attitude will happen so quickly.

Apart the fact that nobody knows how to recreate the earth's pristine environment, there are too many vested interests – too much profit – in continuing the earth's

31 Konrad Lorenz, *Das Jahr der Graugans* [*The Year of the Greylag Goose*], R. Piper & Co. Verlag, Munich, 1979.

32 D.N. Chorafas, *Education and Employment in the European Union: The Social Cost of Business*, Gower, Farnham, 2011.

destruction. There are also too many taboos to overcome, like birth control, and too much general apathy to permit an instantaneous reversal of our current path.

But scientists should show the way. Change, and most particularly the management of change, is part of the scientists' responsibilities. But they cannot do that alone, and they cannot achieve it without active support from politicians and an appreciation of what is at stake by large segments of the population.

- The problem is that politicians, as well as the general public, have a short-term horizon.
- By contrast, science works in the longer term, not day-to-day, even if it is based on consistent and concentrated daily effort.

Furthermore, it is a fundamental but not well appreciated fact that science is *amoral* – not immoral. It can eliminate friend or foe with equal ease. While basic science is primarily interested in theoretical truths and principles, applied science focuses on processes and products which often get their own momentum from outside the area of what has been projected.

In doing their work, scientists are expected to be both *moral* and *virtuous*. It should be remembered, however, that ancient wisdom based *virtue* on a foundation of knowledge without outlining which one of the external sources of knowledge is the most moral. Morality is a man-made concept which varies:

- from land to land, and
- from time to time, through history.

If *virtue*, as Socrates said to Protagoras is *knowledge which cannot be taught*, then there are neither teachers nor students of virtue. But there are teachers and students of knowledge, which is not the same thing. (The Socratic method of analysis through a torrent of challenging questions is the best available process to help people to define by themselves the virtue they espouse.)

A logical sequel is that there are *virtuous people*, but not all people are virtuous. Those who are virtuous build themselves up in their social environment in response to their beliefs and societal goals. But there exists a mismatch, which has largely to do with the absence of limits in one's behaviour, as:

- laws are largely structural, and
- they impose limits but do not promote deep ethical values.

Since man became a self-conscious entity on earth, he has mostly laboured with the geophysical environment. This is true from housing and agriculture, to research on astrophysics as well as molecular and genetic substructures. In contrast, man's achievements are small in regard to logical and ethical principles, which are not built on physical premises.

Physical premises are still important, but in today's society they have been overtaken by logical (more precisely, virtual) premises. This is true all the way from the concept of money to the notion of preservation of vital resources (land, air, water).

Taken together, virtual and physical premises create a megasystem, which does not fit into any one of the current UN or governmental programmes and initiatives – no matter how much these are enlarged, stressed or twisted. Big, holistic projects require *macro-engineering*,[33] in which both financial and technical considerations play a key role. Macro-engineering is the science of big systems, and big systems are not small systems which have outgrown their past boundaries.

A fundamental characteristic of all big systems is that for their conception, management and sustenance, they require high skill and lifelong learning. This is true of all activities emphasizing the role of a virtuoso. On his way to Carnegie Hall, Arthur Rubinstein was once stopped by a young man who asked him how he can become a virtuoso, and he answered: 'Practise, practise, practise.'

As for himself, Rubinstein used to say: 'If I don't practise for a month, the public will notice it; if I don't practise for a week, my friends will notice it; if I don't practise every day, I myself will notice it.' Scientists and technologists must practise every day, training themselves and their minds to move forward. A scientist or technologist falling behind in his art is definitely misusing science's own principles.

33 The intelligent grids themselves, incidentally, are a macro-engineering project.

PART V Appendices

Appendix A: The Best Way to Judge Progress towards Sustainable Fertility is by Agricultural Self-sufficiency

It is not enough to downsize the population to a level sustainable by natural resources. These people have also to be educated, provided with opportunities for work, as well as taken care of in terms of clean water, healthy food and energy supplies. Agriculture is key to the last two requirements and therefore sufficiency in agriculture is a good gauge by which to judge self-sufficiency.

Back in 1798, in his seminal work *An Essay on the Principles of Population*, Thomas Malthus predicted today's plight of less developed nations. It should surprise nobody who has read Malthus's work that exponential population growth is exerting a tremendous strain on the world's food supply and energy sources. As a United Nations study conducted by Wassily Leontief, the Nobel prizewinning economist, pointed out several years ago, to meet even a moderate rate of global growth over the years ahead, it will be necessary to increase:

- food production fourfold, and
- the production of common minerals fivefold.

The years that past proved Leontief right. With his algorithm in mind, think about the exponential growth in newborns now experienced by the poorer countries – whether because of omission or commission. These are precisely the countries which cannot feed their folk and depend almost wholly on Western aid to continue limping on and avoid an outright famine.

By large majority these are African countries, formerly parts of the British and French empires. Theoretically, they are independent, but in reality they have done nothing to take their future into their own hands. As an old friend suggested, 'They wanted their independence, they had it. Now they say that they starve. But their independence starts at home, with their ability to sustain themselves, which they have not done so far.'

While reproduction rates in these countries reach for the stars, malnutrition is widespread, the citizens are living in violence-saturated big-city ghettos, many are dying for lack of health care, and they are polluting the earth through their misuse of natural resources. In both Africa and Asia the gap between rich and poor increases by the day. Some sociologists suggest that, in Pakistan, 5 per cent of the population, the upper-crust, is richer than the large majority of Europeans or Americans will ever be. Yet the country is in chronic depression.

It is surprising that people parading as heads of state fail to understand that independence starts at home, and that agriculture is a focal point of real independence because it represents the source of both food and energy (Chapters 6, 7 and 8). Obviously, we are speaking of sustainable agriculture without poisoning the soil, water and air. One does not expect developing countries to have a first-class agricultural model. That would simply be daydreaming. What should be expected is that they:

- feed their population beyond mere subsistence,
- avoid destroying the earth through rampant deforestation and other disasters, and
- provide themselves with environmentally-friendly and sustainable energy sources.

Apart from hard work, which is the prerequisite to any successful effort, a meaningful solution requires that the politicians and bureaucrats of these independent nations don't fill their pockets with the money they are getting as foreign assistance, but put it to productive use. Corruption is poison to society. Every system has to be tested and, in this case, the test is: are these countries able to rely on themselves?

Using already acquired experience in the most intensive way is another 'must'. In November 2005, in Zimbabwe, a minister in Robert Mugabe's government admitted that the forcible seizure of mostly white-owned farms had contributed to the successive massive crop failures in the country. Farms had been given to 'people without the faintest idea of farming', the minister added.[1] Cheap politics is the best way to destroy a country's physical economy.

A solid agricultural model requires that those governing, the law-makers and, in fact, all citizens are trained to understand the *physical economy* and how it works. They must also appreciate the need for emergency measures to stop the current agricultural collapse which characterizes large stretches of the less developed world.

- Solutions aimed at saving local agriculture must be proactive and oriented towards a clearly established national goal.
- Policies which are defensive, reactive and largely dependent on interminable subsidies are the best prescription for failure.

Note that this advice is also valid for the European Union's and any other agriculture. A proactive solution would look at alternatives correlating agriculture and energy supplies. *Biofuels* is an example.[2]

Based on these premises, developing countries would be well advised to learn not only from the successes but also from the failures of developed nations. For example, the bad news in the European Union is that in some countries' conversion to alternative fuels like biodiesel and ethanol is facing political obstacles in the form of lobbyists. This is:

- delaying the implementation of EU targets, and

1 'Zimbabwe Farm Seizures Led to Crop Failures',*Guardian*, 5 November 2005.

2 See also Chapter 6 on biofuels and Chapter 2 on how Sweden is becoming independent of imported oil.

- perpetuating a reliance on exporters of oil and gas, the energy sources which Europe is generally lacking.

(Aiming to reduce Europe's reliance on fossil fuels, Brussels wanted biofuels to make up 5.75 per cent of fuel used in transport by 2010, EU-wide. This goal has not been reached because several EU countries were not interested; only Germany and France rose to the challenge, followed by Spain which concentrated on ethanol.)

Not everybody agrees with using agriculture's produce for energy production. They also raise other objections. Talking of his studies and experience regarding the state of US agriculture, Lyndon Larouche suggests that two main features characterize the lowering of US food and agriculture output capacity over the past 35 years. Both involved the imposition of trade practices intended to serve the interests of:

- a select few commodities multinationals, and
- the financial circles associated to them.[3]

Larouche also singles out as negatives to agriculture the 1980s Uruguay Round of the GATT, the 1989 Canada–US Free Trade Agreement (CFTA), the 1994 North American Free Trade Agreement (NAFTA) and the creation of the World Trade Organization (WTO) in January 1995. The good news has been the failure of the Doha Round.[4]

3 *EIR*, 8 July 2005.

4 D.N. Chorafas, *Globalization's Limits. Conflicting National Interests in Trade and Finance*, Gower, Farnham, 2009.

Appendix B: Independence in Energy Supplies is as Important as Independence in Food

In parallel to satisfying the requirements for healthy food of a downsized earth population, there is significant scope, and great potential, in directing agricultural production towards renewable energy. A growing part of the current huge energy expenses of a modern nation, paid in foreign currencies, should be raised at home. There is no point in paying between $60 and $145 for a barrel of oil and not knowing whether this expensive supply pipeline will continue to function.

- Independence in energy supplies is as important as independence in food.
- This is also true of clean water,[1] because water is the common ground of food and energy.

As an article in *Science Magazine* put it:

> *As population grows and developing countries require more resources to sustain themselves, the agricultural, energy, and municipal sectors must increasingly compete for finite or dwindling water availability. At the same time, many countries have insufficient water infrastructure, forcing people to use the same water sources for both cooking and sanitation purposes.*[2]

Poor sanitation, water shortage included, is having a wide range of consequences.

- from declining national productivity,
- to increasing child mortality and malnutrition.

Energy and water correlate in other important ways. For either and both of them, investments needed for uninterrupted supply are too large, new technologies too important, but returns too uncertain for the private sector to stump up. The bottom line is that the state must intervene, and the prerequisites are:

- capital,
- technology, and
- sound policies.

---

1 See Chapter 12.

2 *Science Magazine*, Winter 2010.

Some countries complain about being bypassed by the market, or even of market backlash, but the entrepreneurial market is no charitable enterprise. Guaranteed returns also almost certainly summons forth higher private investment, but they also translate into higher bills – and this runs contrary to the policies followed by almost all developing countries, as well as by populist governments worldwide.

Governments find themselves in a bind because of competing and even contradictory ends, as well as because of lack of skills. Therefore, developing countries can gain more by asking for technology transfer rather than for outright handouts – improving return on investment by making the right choice as to where they put their money:

- with agricultural produce, and
- energy supplies being at the top of the list.

Brazil provides an excellent example. Besides the fact that it is already a leader in ethanol production, it does the best it can to lure direct foreign investment precisely in the aforementioned areas. In early February 2010 Royal Dutch Shell announced the biggest foreign investment to date in Brazil's ethanol industry when it set up a joint venture with Cosan, a producer of biofuels.

- The venture, worth $12 billion, will control around 4,500 fuel stations in Brazil, and
- this wholesome decision focuses on a country where almost all new cars sold can run on any mixture of ethanol and oil.

As discussed in Chapter 6, while non-OPEC oil supply is pressured by accelerating decline rates and lower capital spending, biofuels are a consistent source of growth. Estimates by Bank of America Merrill Lynch analysts suggest that total non-OPEC biofuel production will increase to 2.7 b/d by 2015, compared to 1.6 million b/d in 2009. And it will account for 5 per cent of the total non-OPEC supply.[3]

If Brazil can do it, so can other developing countries, particularly those with similar climates. The chief executives of developing countries, and their assistants, must also account for the fact that if they don't solve the 'energy for transportation' problem facing their economies, the result will be a return to underdevelopment – even if other plans do well.

Developed countries provide very interesting statistics. An estimated 63 per cent of oil consumed in the United States goes to transportation, a great deal more than the country itself produces. Of course, transportation is also the largest environmental polluter. The burning of fossil fuels to move people and cargo around the world releases in excess of 1 billion tons of carbon into the atmosphere, greatly contributing to the greenhouse effect.

Pollution due to transportation being what it is, car efficiency should definitely be a goal, but developing countries don't seem to care about that – India being an example. Its car manufacturers, particularly those with new entries to the markets, have probably not

3 Bank of America Merrill Lynch, 'Global Energy Paper', 16 February, 2010.

read the statistics that between 1978 and 1988 the average American car pumped its own weight into the atmosphere every year – and the same is true today.[4]

This leads our discussion to the hypocrisy of current save-the-planet campaigns, shown by the fact that while everyone knows that railways and ships use only 30 per cent of the fuel required to move an equal volume of freight by truck, the world still heavily relies on the trucking business. The consequences of this wrong policy are known, but nothing is done about it.

The clock is ticking, and half-baked measures will fail to achieve results. In both food and energy, developing countries should establish a policy of accumulating resources and building reserves rather than living day-to-day. One of the destructive notions that is doing the rounds these days with the World Trade Organization (and with Doha) is the so-called *just-in-time* (JIT) food supply. JIT made a great contribution by keeping costs down in manufacturing, but a just-in-time delivery system for food means:

- dependence on long lines of communication for perishable foodstuffs,
- an unwarranted increase in energy usage, due to long lines of transportation, and
- high vulnerability of food supplies all the way to shortages, while the Amazon and other forests are systematically destroyed to cut a few cents off the cost of hamburgers.

Note that in manufacturing, too, JIT has hit a wall because anything that disrupts the supply chain system – including strikes, accidents and terrorist attacks – works against the availability of wares and has also (in many cases) adverse effects on quality.[5] For staple foods, disruption will come pretty rapidly, and the whole aggregate of 'free trade' agreements in food projected and (partly) implemented by theoreticians will crash far and wide.

---

4 Christopher Flavin and Alan Durning, 'End of the Profligate Age?', *International Wildlife*, Vienna, VA, May–June 1988.

5 Look, as an example, at the very severe quality problems which hit Toyota – formerly known for the quality of its products. Beyond a certain point, lowest-cost supplies and solutions are counterproductive.

Appendix C: Wastage has become Second Nature

The first basic reason why energy conservation is not on the cards is because energy means power and greater energy consumption has become a status symbol. But at the same time there are limits to power, and it is simply not possible to continue with current policies and practices. *If* we let wastage run wild, before too long there will be tough measures on energy consumption and, along with Orwell's Thought Police, we are going to have an Energy Police.

- Reining in energy consumption to safeguard the earth's non-renewable energy resources can only be done through an international agreement.
- This should have been the order of the day for the various UN conferences on the environment, but it was not.

Figure C.1 shows that, according to several estimates, while over the next two decades the gross domestic product (GDP) in OECD countries is projected to increase, the GDP in the developing world would more than double. By necessity, this rapid growth will be propelled by huge energy consumption – hence pollution of land, waterways, the seas and the atmosphere.

Preoccupied by feeding a population boom, which they should have controlled but did not, the leaders of developing countries will not have the technology, management skills, mind or money for a disciplined investment in clean energy – or even in the basic infrastructure needed to support efficient energy production and distribution. (Even today what they ask most often in the various environmental conferences is for 'more money' – money widely believed to be subsequently wasted.)

The second fundamental reason for failing to control energy wastage, and with it pollution, is that in the go-go years which followed World War II the concept of wastage itself became second nature. A dramatic example is the wastage of food, as a late 2009 research project, carried out by the prestigious National Institute of Diabetes and Digestive and Kidney Diseases in Bethesda, Maryland,[1] revealed about the United States.

The average American wastes a wholesome 1,400 calories a day, which adds up to 150 trillion calories a year for the country as a whole. This represents 40 per cent of the US food supply, up by 45 per cent from four decades earlier (a level which was anyway too high). Worse still, these thrown-away calories account for:

- over 25 per cent of America's consumption of freshwater,
- an estimated 300 million barrels of oil a year, and

1 Reported in 'A Hill of Beans', *The Economist*, 26 November 2009.

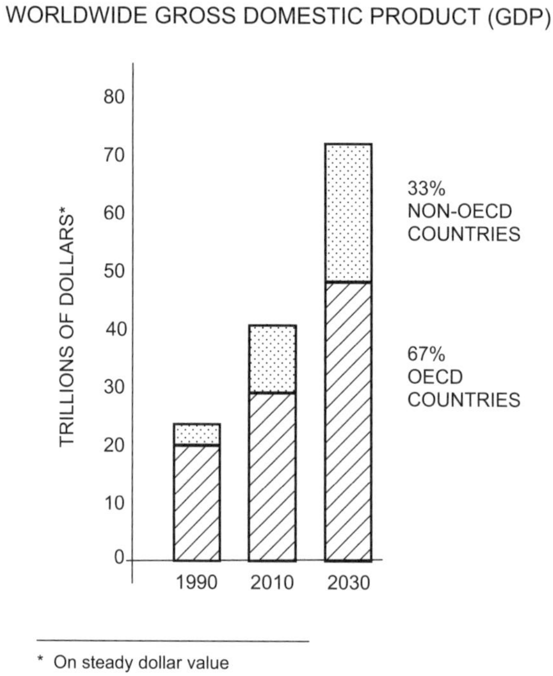

Figure C.1 Projected growth of worldwide gross domestic product (GDP)

Source: Adapted from OECD statistics.

- a huge volume of methane released in the atmosphere, as the wasted food rots – methane being a far more potent greenhouse gas than CO_2.

The researchers also linked food wastage to obesity, which is rampant, and suggested that increased food availability and its intensive marketing is having a 'push effect'. The outcome of more food in the stomach and in the waste-bin is an unmitigated disaster.

- Furthermore, the cheaper food is, the more likely it is to be thrown away even before it is sold to someone who might actually eat it.
- This habit of the consumer society, which is copied the world over, creates a supply chain of waste because throwing away rotten food is seen as better business than running out of stock.

What is happening with food is also representative of man's habits with other forms of energy. Even countries that have a policy and a record of relatively good conservation are confronted with the fact that the demands of the consumer society are nowadays rising faster than at any other time in the past – and this is not only true of energy and food but also of other consumer goods like clothing, housing, vacations and travel (Chapter 10).

Nor can all the blame be heaped on developing countries for resenting CO_2 constraints and the push towards alternative energies. Developing economies (particularly China and India) generally continue to prefer coal because:

- coal provides a certain amount of supply security and offers,
- the lowest-cost option for huge and increasing populations with their surging demand for energy.

Western countries, by contrast, are constrained by their Greens and environmentalists. Take Switzerland as an example. Because energy consumption is increasing by more than 3 per cent per year, the country needs a new power factory every three years. It is impossible to meet this requirement and limit CO_2 emissions because of vigorous objections to nuclear power by the Greens. Hence the only solution is to buy energy across the border – for instance, from France where 80 per cent of power production (ironically) comes from atomic energy. Germany and Austria face exactly the same problem.

- It would have been a much better solution if the Greens and other environmental activists demonstrated *for* saving energy rather than *against* nuclear plants.
- But energy saving is not (yet) a political asset, and the Greens are awfully political in what they say and in what they do. Their true business is politics, not environmental protection.[2]

Theoretically, but only theoretically, new technology can reduce energy wastage, but such applications (Chapter 15) are neither widespread nor can they single-handedly reverse past trends. Nevertheless, the US Electric Power Research Institute estimates that simply adding variable speed controls to many electric devices would be enough to offset most of the added energy that will be used by the new electrical industrial technologies projected to be introduced in the next few years.

Using the same type of reasoning, some experts guesstimate that major energy savings could be made in the production of chemicals, cement and steel. Statistics, however, tell a different story. Traditional open-hearth steel furnaces have largely been replaced by more efficient oxygen furnaces and electric arc furnaces that recycle scrap steel. Theoretically, these have the potential to cut energy needs by half, but, in practice, such a reduction has not been achieved because:

- investments have not been made in a big way,
- increased production cancelled out the savings, and
- technology alone is changing nothing in terms of the habits of our throwaway society.

It takes leadership to change this worldwide *déclinisme* in man's respect for nature and his ethical standards. During the years following the second oil crisis of the 1970s, the United States as a whole became 25 per cent more energy-efficient and 31 per cent more oil-efficient than prior to the crisis. But since the early 1990s energy efficiency has been put on the backburner.

2 The same is true of all other political parties and their leadership.

Appendix D: Recycling is Wanting Because Progress in Materials Science is Lop-sided

The fact that energy means power is by no means a recent finding, and the way in which man has shaped his mastery over materials is proof of it. For economic reasons, civilizations began to transform materials back in remote antiquity. Initially, these were of natural origin: argil, stones, vegetable fibres, wood and animal skins. The fusion of metals and glass manufacturing extended man's mastery of physical processes, but in reality:

- the search for materials is more profound than the use to which we put them, and
- as a result, civilizations are closely linked to finding and exploiting materials and energy sources.

Theoretically, the earth has a wealth of energy and materials sources; in practice, these are not inexhaustible and nowadays consumption has reached unprecedented levels. Without even counting products of a petroleum origin, each man, woman and child on earth consumes, on average, an estimated ten tons of irreplaceable raw materials per year. One of the ironies is that:

- the greater the strides made by the science of materials,
- the greater, not less, becomes the demand for materials and the more polluting is their production, usage and subsequent disposal.

About 10,000 years ago, man discovered that clay made of argillaceous soil became hard when treated by fire. This first purposeful transformation of a raw inorganic substance brought a new beginning. A different material was created with properties resulting from human handling, and this began a process which, over centuries, became extremely sophisticated. (Let's not forget that terracotta has not only been used for making storage containers for goods, but has also been put to many other uses – such as a medium for scripts and for invoices.)

Today, the science of new materials is the cradle of a colossal field of activity, and it is instrumental in determining tomorrow's economy. Countries which lag behind in materials research are out of luck as far as future development is concerned. Countries which depend for their living mainly on old raw materials are feeling the pinch of the market's wild gyrations because the demand for them is volatile.

The baseline of this transformation is our *scientific know-how*, which permits us to fabricate both widely used and custom-made materials for each of our needs. A high level of customization is one of the facilities human civilizations never had before. There has been a great deal of evolution from the general-purpose clay of early cultures to today's special-purpose alloys, with intelligence, research and energy at the fulcrum.

- Science is the enabler of this far-reaching transformation.
- But where science has made no significant strides so far is in the domain of efficient recycling and non-polluting waste disposal.

To better appreciate some of the basic notions underpinning this issue let's return to the fundamentals, taking *ceramics* as an example. Throughout thousands of years argil, the first clay used by man to make items ranging from pottery to bricks, served as raw material. Hence, for many people ceramics evokes rustic items. Not so for the scientist and the engineer.

For the materials specialist the term 'ceramic' signifies a solid which is neither metal nor polymer, but can contain metallic or polymeric elements in the form of components or additives. As technology progresses, ceramics are elaborated from very different raw materials than those characterizing earlier processes – and some of these new materials do not exist in a natural state.[1]

- We now have basic knowledge of molecular structure, which was not the case in the past.
- What we do not have is the ability to dispose of thousands of tons of structures that we no longer use, without creating a permanent environmental hazard.

The disposal of waste did not matter in the remote past, or even some time prior to the Industrial Revolution, because the quantities of refuse did not fill the land and the oceans. Today, the pieces of clay archaeologists find from past civilizations adorn museums. But recycling and waste technology (Chapter 11) greatly matter today because of the thousands of tons of new waste which we do not know how to handle.

This suggests that our research is lop-sided with too much emphasis on development and too little on disposal. Because materials are vital to our civilization (and to energy production) all effort is oriented towards new inventions and product design, with precious little attention paid to recovery, recycling and disposal. The latter three activities lack the elements of focus and attention which brought materials research to new heights. Among the missing ingredients are:

- a solid scientific base which permits new leaps forward, and
- movers and shakers of government-sponsored projects like space research which requires sophisticated heat-resistant materials.

This is another basic issue that the circus of UN and other environmental conferences has missed. They did not develop the awareness that the energy crisis is here to stay: metals, ceramics, plastics and other waste are not easily recyclable, and new solutions have to be found – which means intensive research on materials disposal. With singular

1 Some 2,500 years ago, ancient Greek philosophers were the first to formulate the hypothesis that matter is constituted of minuscule invisible particles, all of the same nature. They could not prove what they said, but this tentative statement helped man, over the centuries to, investigate – then understand – the structure and science of materials.

lack of political, social and scientific foresight, no real attention has been paid to how to dispose of mass-produced materials which no longer serve their original purpose.

On the one hand, better materials allow us to build more efficient engines that are lighter and work at higher temperatures – which means greater efficiency. For instance, the weight of the cars we drive has been significantly reduced if we compare like with like.[2] But, on the other hand, we know precious little about how to dispose of the 50–60 million of annually produced cars in an efficient recyclable manner when they become junk full of poisonous oils, plastics and electronics. (China alone boasts an annual manufacturing capacity of 10 million cars and this number is increasing as new factories are built.)

The argument that, as materials science progresses and more intensive use is made of new alloys and ceramics, the weight of cars and other engines will decrease, is very short-sighted. Also ignored is the fact that not only is the number of cars, TVs, PCs, mobiles and other modern gadgets continuing to soar, but the population explosion also further increases the need for such goods and for greater energy production no matter what may be their lasting effects on pollution and quality of life.

In conclusion, the problem of effective environmental protection is difficult to solve for many reasons:

- There is disagreement over how real and dangerous the threat is.
- Intensive research on massive recycling and waste disposal is nowhere to be seen.
- Current solutions conflict with other goals, like employment and living standards.
- It is unclear who will supply the money and effort required to cope with the problem.

Everybody is looking for relatively 'inexpensive' strategies which so far have proven to be ineffective, and the issue of materials disposal has no prominent position on the radar screen of scientific research.

2 By this I mean that if we constructed a vintage sedan in today's materials, it would be lighter in weight. However, because of added safety features and the current trend for bigger cars, modern cars would probably be heavier.

Index